A Modern Contagion

Iran during the Qajar era

A Modern Contagion

Imperialism and Public Health in Iran's Age of Cholera

AMIR A. AFKHAMI

Johns Hopkins University Press

Baltimore

© 2019 Johns Hopkins University Press
All rights reserved. Published 2019
Printed in the United States of America on acid-free paper
9 8 7 6 5 4 3 2

Johns Hopkins University Press
2715 North Charles Street
Baltimore, Maryland 21218-4363
www.press.jhu.edu

Library of Congress Cataloging-in-Publication Data

Names: Afkhami, Amir Arsalan, author.
Title: A modern contagion : imperialism and public health in Iran's age of cholera /
    Amir A. Afkhami.
Description: Baltimore : Johns Hopkins University Press, 2019. | Includes
    bibliographical references and index.
Identifiers: LCCN 2018018037 | ISBN 9781421427218 (hardcover : alk. paper) |
    ISBN 9781421427225 (electronic) | ISBN 1421427214 (hardcover : alk. paper) |
    ISBN 1421427222 (electronic)
Subjects: | MESH: Cholera—history | Public Health—history | Cholera—
    epidemiology | Epidemics—history | History, 19th Century | History,
    20th Century | Iran
Classification: LCC RA644.C3 | NLM WC 264 | DDC 614.5/14—dc23
LC record available at https://lccn.loc.gov/2018018037

A catalog record for this book is available from the British Library.

*Special discounts are available for bulk purchases of this book. For more information,
please contact Special Sales at 410-516-6936 or specialsales@press.jhu.edu.*

Johns Hopkins University Press uses environmentally friendly book materials,
including recycled text paper that is composed of at least 30 percent post-consumer
waste, whenever possible.

*To Hastie and my father, who longed to see his beloved Iran
prosperous and free*

Oh, how lovely is my homeland!
It is far from the reach of city folk.
There is no pretension there, no adornment.
No fetters, no cheating or treachery.

Nima Yushij, "The Tale of Pallid Color / Cold Blood"

CONTENTS

Any pioneering multidisciplinary historical scholarship on the late modern era is a daunting undertaking, even more so when the subject is Iran. The current Iranian government's mistrust of American scholars and the country's fledgling archival institutions are just a few of the complexities of conducting historical research on this part of the world. Overcoming these hurdles requires patience, tenacity, and the willingness to explore an array of primary sources, in addition to what is available inside Iran. This is why it took me almost two decades to assemble and study the vast records used in this book, including documents from dozens of international archives. The methodological rigor and investigative stamina required in this project was inspired by Abbas Amanat. Over the years, he has conveyed his commitment to meticulous archival research and balanced analytic approach to a generation of modern Iranian historians. I hope that this work will validate his decision to take on an extremely young, long-haired, cowboy-boot-wearing, and utterly unserious-looking doctoral student as his mentee. I also owe a debt of gratitude to John Harley Warner, Frank Snowden, and Paul Kennedy. My discussions with them prepared me to address the complex sociopolitical determinants of health and the wider intellectual and diplomatic currents that are covered in this book. I am also enormously grateful to Farzaneh Milani and Duane Osheim for helping sow the seeds for this undertaking during my undergraduate years by respectively introducing me to literature from the Qajar period and the history of epidemics.

I would be remiss if I did not acknowledge the unwavering support of a number of friends and colleagues over the years of working on this project, including Ahmad Ashraf, Houchang Chahabi, Ali Gheissari, John Gurney, George J. Makari, Rudi Mathee, Abbas Milani, Howard Markel, Mohammad Tavakoli-Taraghi, Ehsan Yarshater, Magnus T. Bernhardsson, Ranin Kazemi, Arash Khazeni, Ramin Ahmadi, Michael Rubin, Michael Soussan, Daniel Tsadik, Farzin Vejdani, and Heidi A. Walcher. As a clinician, educator, public health practitioner, and historian, I have often struggled to balance the rigid demands of academic medicine with my scholarly interests. The ability to keep my head above water during the course of this project is in no small part due to the support of my mother and sister and to the good counsel I received from Jeffrey S. Akman, Jack D. Barchas, Marie E. Michnich, Julia B. Frank, Stephen J. Trachtenberg, and Kenneth B. Wells.

I am grateful to the American Association for the History of Medicine for recognizing the significance of this work with the Jack D. Pressman–Burroughs Wellcome Fund Career Development Award and the Shryock Medal. Fellowships from the Henry Hart Rice Family Foundation, the Smith Richardson Foundation, and the Andrew W. Mellon Foundation allowed me to travel abroad and conduct the initial archival investigations for this book. I am indebted to Brian Rohlik, David Nickles, and the anonymous readers for their invaluable feedback and to my research assistants, Azad Amanat and Cristine Oh, for checking the citations and transliterations and helping with the overall formatting of the manuscript. I am also thankful to Johns Hopkins University Press, particularly the editorial and faculty boards for their unwavering support of this project and the exceptional editorial staff for their determined effort in bringing this work to publication.

Finally, this book would have never seen the light of day without the love and support of my wife, Hastie. She not only did her utmost to provide me with the time and space to write and revise, but her exceptional editorial advice vastly improved the pace and clarity of the manuscript. This book is dedicated to her.

1796   Coronation of Aqa Muhammad Khan, founder of the Qajar dynasty

1797   Fath 'Ali Shah Qajar crowned

1801   British East India Company cements trade and political ties with Fath 'Ali Shah

1804   Start of the First Russo-Iranian War

1813   Treaty of Gulistan ends the First Russo-Iranian War

1816   Movable type printing press introduced into Iran

1821   Asiatic cholera reaches Iran for the first time

1826   Start of the Second Russo-Iranian War

1829   Treaty of Turkmenchkay ends the Second Russo-Iranian War

1834   Muhammad Shah Qajar crowned

1835   The first lithographic newspaper printed in Iran

1838   The British thwart Muhammad Shah's claim on Herat

1847   Second Treaty of Erzurum settles Ottoman-Iranian boundary disputes

1848   Nasir al-Din Shah crowned

1851   Iranian Polytechnic College inaugurated; Mirza Taqi Khan Farahani Amir Kabir murdered

1856   Start of the Anglo-Iranian War over Herat

1857   The Anglo-Iranian War ends with the Treaty of Paris

1865   Completion of the Indo-European telegraph network

1866   Iran takes part in the international sanitary conference in Istanbul

1867   The Iranian Sanitary Council meets for the first time

1869   Nationwide famine begins

1871   Husayn Khan Mushir al-Dawla becomes Iran's premier

1872   Julius de Reuter receives concession to develop and control Iran's public works

1873   Reuter concession repealed; Mushir al-Dawla dismissed

1888   Nasir al-Din Shah opens the Karun River to foreign commercial fleets

1891   Tobacco Protest

1896   Assassination of Nasir al-Din Shah; accession of Muzaffar al-Din Shah

1904   Haffkine's cholera vaccine administered in Iran for the first time

1905   Demonstrations against Belgian customs officials and the government

1906   The First Iranian National Assembly gathers in Tehran

1907   Anglo-Russian treaty divides Iran into spheres of influence; Muhammad 'Ali Shah crowned

1908   Coup d'état by Muhammad 'Ali Shah; bombardment of the National Assembly

1909   Muhammad 'Ali Shah deposed in favor of his twelve-year-old son, Sultan Ahmad Shah; the Second Iranian National Assembly opens

1910   Hygiene and Smallpox Vaccination Act ratified

1911   Medical Practice Act ratified; Iran joins the International Office of Public Health; coup d'état by the regent with support from Russia; Britain ends the Second National Assembly

1914   The Third Iranian National Assembly opens; declaration of Iran's neutrality in the First World War; the Ottoman Empire invades Iran's Northwest

1915   British and Russian forces invade and occupy Iran's eastern provinces and force the Third Iranian National Assembly to disband

1916   Amir Khan Amir A'lam presides over the Iranian Sanitary Council

1917   Nationwide famine begins

1918   Spanish flu overruns the country

1920   Establishment of a Soviet Socialist Republic in Gilan

1921   Coup d'état by Riza Khan; the Pasteur Institute of Iran inaugurated; Iranian Ministry of Health established

1925   Abolition of the Qajar dynasty

1926   Coronation of Riza Shah, founder of the Pahlavi dynasty

1931   Razi State Serum and Vaccine Institute inaugurated

1941   Riza Shah Pahlavi abdicates in favor of his son, Muhammad Riza Shah

1955   Underground piped water system inaugurated in Tehran

All non-English terms are in italics. The transliteration system adopted for Persian and Arabic is a modified form of the *International Journal of Middle East Studies* transliteration guide without the use of diacritical marks. The silent *h* in Persian and Arabic is represented by *a* and the silent *waw* in Persian with *w*. The letter *'ayn* is represented by ' and *hamza* by '. With few exceptions, the Arabic definite article *al-* is not assimilated to the noun. When possible, I have used the familiar English forms of Persian and Arabic such as bazaar, ayatollah, and Ramadan.

Well-known place names are written in their common form (as in Google Maps) and less familiar place names are transliterated. I have also used the modern rendition of city names (e.g., Gorgan instead of Astarabad), although, when possible, I cite earlier names from the period in parentheses. To reflect the common Western usage of the name for the country after 1935, Iran is used throughout the text instead of Persia, although Persia is retained in quotations and citations. Persian is also preferred over Iranian for culture and society. In translations for *sadr-i a'zam* prime minister and premier are preferred to grand vizier. Dates in the Islamic lunar calendar are indicated with AH (for *hijra* era) and dates in the Iranian solar calendar are indicated with AS (*shamsi*); in all instances, equivalents are given in the Christian (Gregorian) calendar.

A Modern Contagion

Introduction

Everything has a remedy except death.—Persian proverb

In the summer of 2008, the Iranian government initiated emergency measures to stop a growing Asiatic cholera epidemic in its cities. The culprit, a gram-negative, comma-shaped bacterium called *Vibrio cholerae*, caused profuse watery diarrhea, vomiting, and body cramps. Without treatment, its victims succumbed to rapid dehydration, electrolyte imbalance, systemic shock, and, in the most severe cases, death. The bacterium spreads through water and food contaminated by feces and other secretions from infected individuals, though in Iran raw fruits and vegetables washed in untreated sewage were the most common culprits. A similar outbreak three years earlier had cost the Iranian government millions of dollars, an expense it could now ill afford after two years of heightened Western financial sanctions over its nuclear program.

Like the previous epidemic, this cholera outbreak reached the country from neighboring Pakistan and Afghanistan, across conflict-ridden borders. Despite their decades-long expenditures in treasure and lives, Iran's military and police forces had failed to secure this lawless frontier against narcotics traffickers who fueled a growing scourge of addiction in the country. Similarly, the much less equipped Ministry of Health could not halt the reintroduction of cholera, despite draconian interventions such as quarantines, which Iran's ruling Shi'ite Muslim clerics found objectionable because of their reluctance to restrict pilgrimage into the country; this was alluded to in the interview of an Iranian parliamentarian:

> "Cholera has spread from neighboring countries which makes dealing with the illness difficult. It is not possible to contain an illness like cholera independently," said Amir Hossein Qazizadeh, a member of [the Iranian] parliament's health committee. "Enforcing quarantine is very difficult. Travelers can't be held at border points for a week," he said . . . Qom, 183km south of Tehran, is Iran's second largest holy city. Millions of Iranians and Shiites from other countries, including Afghanistan and Pakistan, flock to the city every year to pay their

respects to the shrine of Lady Fatemeh Masoumeh, the daughter of the seventh Shiite Imam. The first cases of cholera were reported in Qom a month ago.[1]

In spite of these initial setbacks, the government eventually managed to slow and then stop the epidemic by banning the sale of raw fruits and vegetables in the country and increasing civic sanitary awareness through educational campaigns on state television and radio.[2] Officials also attributed their success to Iran's early detection and monitoring systems to control diarrheal diseases.[3] What many did not know was that cholera itself had played a singular role in Iran in the institutional and intellectual development of public health and disease prevention, paving the way for the successful intervention against this latest outbreak. Even more broadly, the disease had a formative role in the development of various aspects of modernity in Iran by steering the country's social, economic, and political currents in the nineteenth and early decades of the twentieth century.

I first came across cholera's profound impact on Iranian society rather unexpectedly, in a course on feminism and literature as an undergraduate student more than twenty-five years ago. Class readings included the autobiography of Taj al-Saltana, a progressive Iranian princess who lived through a terrifying cholera epidemic in 1904. Her dramatic description of mass casualties, widespread hysteria, and the breakdown of governance in Tehran evoked Boccaccio's account of the Black Death in fourteenth-century Florence. She also depicted a twentieth-century capital largely bereft of the sanitary advances of the Victorian era: "Though this epidemic was a sign of divine wrath and chastisement, we can still say that it was engendered by an inattention to hygiene and the contamination of water. Every Government's first duty is to see to the cleanliness of the streets and water, as well as the tranquility of the people. There was a municipality in name, but, like other arms of the Government, none in actual fact—and yet the employees felt entitled to their undeserved salaries."[4] My doctoral research, conducted several years later, revealed that Taj al-Saltana's *cri de coeur* was not unique among her contemporaries. First-person chronicles of Iranian society in the nineteenth and early twentieth centuries often described the devastating impact of cholera and the country's sanitary shortcomings—making the absence of cholera epidemics from Iran's historiography particularly remarkable. In fact, contemporary historians of Iran have generally shied away from studying diseases and other social determinants of health in favor of topics with obvious and immediate implications for the region's tumultuous current affairs. Pervasive ideological and methodological rigidity among Iranian historians of this period has also obstructed the study of epidemics, which inherently depends on a multidisciplinary approach to the subject matter.[5] This focus has resulted in the omission of both a key chapter in the history of pandemic cholera and, more broadly, an account of the development of sanitary modernity at the geographical and cultural intersection of Europe and Asia. Likewise restricted is our understanding of the important historical forces that continue to shape Iran's current domestic and international policies on public health and disease prevention.

My goal in writing this book has been to rectify this historiographical deficit by describing the Iranian experience with pandemic cholera, also known as Asiatic cholera, from its advent in the early nineteenth century to its attenuation shortly after the First World War. The time span covered roughly falls within the confines of the Qajar dynasty (1796–1925), an era of globalization that established the intellectual and institutional foundations of modernity in Iran. Rather than give a simple tally and overview of recurrent epidemic outbreaks, I examine cholera's complex and transformative impact on medicine, the science of sanitation and hygiene, and broader sociopolitical developments in the country. As the ensuing chapters illustrate, cholera not only shaped the adoption of new paradigms in medicine and health; it also changed Iranian perspectives on governance, influenced European imperial policy, unmasked social and political vulnerabilities, and caused enduring institutional changes during this critical time in the country's development. While the subject of this book is a relatively new one, the preceding decade has seen the emergence of an embryonic scholarship on the modern history of Iranian medicine.[6] The present work adds a new facet to this literature by showing the multiple, and often contradictory, social and political determinants of medicine and public health in Iran through the prism of the great cholera outbreaks of the period. These local, national, and international forces not only affected Iran's susceptibility to the epidemics but also shaped the country's emerging public health consciousness and disease-prevention philosophy.

The story of cholera in Iran begins with the Qajar dynasty's ascent of the Peacock throne on the cusp of the nineteenth century, ending more than six decades of intermittent civil war, fractured rule, and isolation that followed the fall of the Safavid dynasty in 1722. Chapter 1 illustrates how Iran's new rulers unintentionally created the right conditions for the introduction of an unfamiliar epidemic disease from the East and novel approaches to sanitation and medical practice from the West by centralizing governance, increasing international diplomatic engagements, and reestablishing transnational trade. Pandemic cholera reached Iran for the first of many times in 1821, assisted by Britain's territorial expansion and growing commercial pursuits in the Persian Gulf and the Indian subcontinent, where the disease was endemic. For thousands of years, the cholera bacterium had survived and evolved in the Ganges Delta, located in India's Bengal region, because of its tropical climate, its human ecology, and the unique chemical and organic composition of its brackish estuaries.[7] In the past, any outbreak of cholera that made it out of India would have likely died out on Iran's coast, but the revival of Iran's commercial arteries in the first two decades of the nineteenth century allowed the epidemic to spread inland and assume national proportions. While British expansion disseminated cholera to Iran and the rest of the world, Russia's successful invasion and annexation of Iran's northern territories in 1813 and 1828 initiated the new dynasty's pursuit of post-Enlightenment scientific and technological advances from the West, including new approaches to disease prevention and therapeutics. Iran's first European-styled Polytechnic College (Dar al-funun), established in 1851 to modernize its defeated military, accelerated this intellectual exchange through the translation and dissemination of European scientific literature, including medical works. Other innovations such as the printing press, the telegraph, and newspapers familiarized Iran's literate population with the sanitary revolution in the West during the second half of the nineteenth century. The country also began participating in international conferences on pandemic cholera, exposing its leaders to the scientific debates and rivalries of the period. This sanitary globalization gradually transformed the Qajar regime's notions of its obligations to disease prevention, both domestically and internationally, especially in light of the reported successes of European governments in eradicating cholera. The ruling elites increasingly viewed the maintenance and improvement of their nation's well-being as a proactive enterprise and a matter of governmental concern. This culminated in the creation of the Iranian Sanitary Council in 1867, establishing public health as a social institution in Iran almost half a century after Asiatic cholera first swept across the country.

Iran's emerging public health consciousness did not result in any substantial interventions to address its sanitary vulnerabilities to cholera, even though the disease was generally accepted to be water-borne by the 1890s. The Sanitary Council, lacking an independent budget and executive power, depended on the central government to implement its recommendations; but its proposals were largely ignored because of the regime's administrative crises and financial deficits, which had worsened with the country's growing economic globalization in the second half of the nineteenth century.

Chapter 2 demonstrates how Iran's integration into the steam-driven international marketplace accelerated the reach of Asiatic cholera to its borders during the 1889 and 1892 outbreaks, while the country's unaddressed social and infrastructural susceptibilities amplified the disease's domestic impact. Iran's urban features, particularly its vulnerable potable water systems, and its religious-cultural traditions, including Shi'ite pilgrimage and burial practices, shaped the dissemination and lethality of the epidemics. The almost complete absence of civic leadership during the outbreaks permitted local forces to subvert the central government's authority, revealing the country's deteriorating state of governance. The epidemics also unmasked simmering anger at the West's growing stranglehold on the country's commerce. This manifested itself in episodes of violence against European financial interests and religious minorities under the diplomatic protection of Western powers. Shi'ite clerics were particularly effective in mobilizing popular discontent in the wake of the epidemics and cementing their role as leaders against Western encroachment and government exploitation at that time.

Domestic administrative shortcomings were not the only obstacles to sanitary progress in the Qajar state. European imperial rivalries, particularly the growing economic and political efforts of Russia and Britain to dominate the region, also hindered the development of public health in Iran at the cusp of the twentieth century. Chapter 3 explains how the country's worsening financial condition after the 1889–1892 cholera epidemics obliged Tehran to increasingly subordinate the sanitary regime on its frontiers to imperial interests. Controlling Iran's overland and maritime quarantines along its borders became a new front in the Anglo-Russian competition for influence in Asia known as "The Great Game." By the closing years of the nineteenth century, each power jealously guarded its sanitary prerogative, Russia in Iran's Northeast and Britain in the Persian Gulf, and both used their respective quarantine authority to undermine one another and advance their political and economic goals in the region. Their sanitary struggle extended to the international stage, where the two powers pursued strategic regulatory concessions at the international sanitary conferences that were meant to prevent the extension of pandemic diseases from India into Iran and onward to Europe. They also resisted any attempts by a resurgent Iranian administration to gain control over its quarantines in the first years of the twentieth century. Iran's failure to shape an independent sanitary course prolonged its susceptibility to pandemic outbreaks that continued to decimate its population and its economy, despite Tehran's growing sanitary knowhow and medical manpower.

Meanwhile, the Iranian public's reception of the germ theory of disease grew in step with the overall progress in the fields of bacteriology and vaccinology during the first years of the twentieth century. A new cadre of native medical experts promoted this empirically informed view of contagions, eroding traditional nonscientific interpretations of epidemics as acts of God that had contributed to the country's fatalistic mind-set on disease prevention. Chapter 4 shows how this intellectual transformation changed Iran's response to a recurrence of Asiatic cholera in 1904, giving rise to unprecedented efforts to halt the epidemic. Increasingly accepted microbiological models of cholera's etiology and clinical course moved local administrations, provincial leaders, and civic groups to engage in sanitary and hygienic interventions to stop the epidemic and minister to its victims rather than avoiding their responsibilities, as they had done during previous outbreaks. But the religious leadership's stubborn opposition to quarantine measures, the country's political instability, and unaddressed sanitary vulnerabilities largely undid their work, allowing the epidemic to exact a substantial toll in lives and treasure during its nationwide assault.

While the popularity of microbialism did not have a significant impact on the 1904 cholera outbreak, it did contribute to the outbreak of the Constitutional Revolution a year later. The sanitary and hygienic pragmatism that grew out of Iran's shifting infectious disease paradigm increased the Iranians' views that their central government, led by the shah, was responsible for their wellbeing. The country's revolutionaries saw the latest occurrence of Asiatic cholera as evidence of the Qajar administration's broken contract with the governed and demanded sanitary reform and government accountability in matters of public health. The reigning shah's own fears of another cholera epidemic precipitated immediate institutional changes, which included convening the Sanitary Council on a regular basis and publishing its proceedings. But these changes were not enough to mitigate the growing dissatisfaction across Iran's social spectrum with the Qajar regime's mismanagement of the economy, made worse by the epidemic's ruinous impact on the country's commerce. The mercantile class, which bore the brunt of the downward spiraling economy, led the revolt against the government, less than a year after the latest outbreak subsided. This culminated in the establishment of a parliamentary system of governance in Iran, bringing a short pause in the country's long history of absolutism.

The efforts to improve Iran's public health by the nascent constitutional government rapidly ran up against a growing imperial hegemony over the country's economic and political life. Britain and Russia resolved their decades-old cold war in Iran with the 1907 Anglo-Russian Convention that divided the country into agreed-upon zones of influence between the two powers. This essentially transformed Iran into a proto-colony that left little room for the Iranians to enact the far-reaching structural reforms they needed to halt the ingress of pandemic cholera and other diseases. The bombardment of the First Iranian National Assembly by Russian-backed royalist troops in 1908 embodied the willingness of imperial powers to use violence to bend Iran's government to their will. The last chapter of this book shows how these types of external interventions caused the country's sanitary progress to falter and helped a series of deadly epidemics, including cholera, to cross its borders in the second decade of the twentieth century.

A yearlong civil war following the bombardment of the National Assembly ended with the defeat of the royalists and the establishment of a second, reform-oriented parliament. Western-educated sanitarians in the new government spearheaded initiatives to incrementally improve the country's public health. They regulated the medical workforce, mandated smallpox immunizations, and appropriated regular funding streams for the Iranian Sanitary Council. The Second Iranian National Assembly's willingness to back these interventions, despite the country's persistent economic morass, showed the extent to which public health had become ensconced in Iran's governance at this time. The Sanitary Council used the new funds to expand its reach and scope of interventions against outbreaks. Its increasingly vocal and nationalistic native members also called for new quarantine stations and upgrades to existing facilities in the Russian and British zones of influence. These and other policies to stabilize the country ran against imperial interests, leading Russia to actively support a coup that dissolved the Second National Assembly in 1911 and, three years later, turned Iran into a battlefield for belligerent European armies.

The First World War obliterated Iran's security and autonomy. Occupying military forces requisitioned foodstuffs and disseminated epidemics as they freely traversed and fought on Iranian territory. The resulting displacement and famine worsened the occurrence and lethality of both endemic and imported infectious diseases, including the almost yearly cholera epidemics between 1914 and 1918. Ironically Iran, a neutral country in the war, lost as many citizens to war-related diseases as belligerent countries lost in the trenches.

The final nail in this coffin came in the form of the "Spanish" influenza pandemic that killed up to twenty percent of the country's population in the last year of the conflict.

Iran's calamities in this period did not stop the public health momentum that followed the Constitutional Revolution's reforming undercurrents. The departure of European physicians to join their respective armies allowed a new generation of patriotic, French-trained, native doctors to take the helm of the Sanitary Council and expand the government's vaccination services during the war. But the Iranians were incapable of comprehensively reforming the country's public health under worsening economic and political circumstances. The Pahlavi regime, which rose from the ashes of the Qajar dynasty after the armistice, saved Iran from the brink of disintegration by rapidly stabilizing the country's administration and security. Building on earlier organizational and intellectual advances, Iran's new rulers centralized the country's essential public health operations and took over the contested British-controlled quarantines in the Persian Gulf. The Pahlavi regime's patronage of the newly inaugurated Pasteur Institute of Iran allowed for large-scale domestic vaccine production, expanding Tehran's ability to immunize vulnerable populations during outbreaks of Asiatic cholera. These reforms shaped the foundations of Iran's contemporary culture of public health and slowly brought an end to the era of major epidemics in the country.

Because the development of modern notions of disease, medicine, and prevention in Iran are a central theme of this book, I want to clarify my point of view on the contentious concept of Iranian modernity. Biomedical modernity in Iran followed the country's growing acceptance of a scientific explanation for the disease process during the nineteenth and twentieth centuries. The germ theory was the pinnacle of this intellectual transformation, which progressively sidelined ancient philosophical and religious notions of sickness in favor of empirically tested models of illness and therapeutics. Because much of this perspective grew out of the post-Enlightenment scientific method, it is difficult to pry apart Westernization from medical modernity in Iran. Disciplines like epidemiology, which matured in England between the late seventeenth and nineteenth centuries, were instrumental in decoding the incidence, distribution, and etiology of cholera and other contagious diseases. I passionately disagree with the postcolonial polemicists who might argue that this description of modernity perpetuates a Eurocentric biomedical discourse. Such a distorted and often politically motivated characterization of modern science as a construct of Western dominion has been used to justify

various regressive policies in the developing world, including the growing an-
tivaccination movement in some Muslim communities in Asia and Africa.[8]
This does not mean that I lionize the West's role in the development of public health in Iran; rather, my work argues that European powers exercised
multiple and often contradictory influences on the intellectual and practical
evolution of public health in Iran. On the one hand, European powers became a vessel through which Western professional and scientific ideologies
of health, disease, and disease prevention were transmitted into the country;
at the same time, their imperial interests led them to create barriers against
the effective implementation of these ideas. Similarly, overstating the role of
globalization in advancing public health in Iran ignores the powerful domestic forces that shaped the country's progress in this area. While Iran's integration into the international order and its intellectual exchanges with the
West helped introduce novel approaches to disease prevention, the ebbs and
flows of its evolving notions of public health were critically influenced by the
forces of tradition, modernity, and nationalism within the country itself. Iran's
encounter with Asiatic cholera, from its advent in the early decades of the
nineteenth century to its twilight after the 1920s, effectively unmasked the
forces, both domestic and foreign, that eventually shaped the country's modern culture of public health.

This book's examination of cholera as the Qajar period's prototypical and
transformative disease is also a matter on which I want to elaborate, as it can
be argued that Iranians encountered deadlier and more prevalent contagions,
ranging from endemic outbreaks of smallpox and malaria to pandemic visi-
tations of influenza. Cholera eclipsed other diseases in the Iranian psyche of
the period because of its terrifying symptoms, which Charles Rosenberg ap-
propriately described as "spectacular."[9] Severe abdominal cramps forced its
victims to remain reclined with their knees drawn up in a fetal position for
extended periods. Unremitting watery diarrhea, vomiting, thirst, and rigors
worsened their agony. The extreme loss of body fluids eventually made
them appear cadaveric, with sunken eyes, grayish-blue complexions, and dry
mucous membranes. Delirious and bedridden, patients became a shadow of
their former selves, sometimes within a matter of hours, searing terrifying
memories of cholera on both survivors and caretakers alike.[10]

Iran's religiocultural norms, particularly beliefs rooted in its Shi'ite sectarian identity, also magnified the impact of Asiatic cholera relative to other contagious diseases in the Qajar era. Nowhere was this more obvious than in the
country's popular standards of water purity that prompted one contemporary

Fig. I.1. Caravan transporting cadavers for burial in Karbala and Najaf. *London Illustrated News,* June 14, 1873.

British physician to comment that "Persians 'religiously' hold to the idea that running water cannot be defiled."[11] This belief, based on prophetic traditions, maintained that the flow of water made it suitable for consumption unless its color, taste, or smell indicated the presence of impurities. Even stagnant sources, such as ponds and wells, were considered clean and drinkable as long as they contained at least 350 liters (*yik kur*) of water. Because of their conviction that running water could not be polluted, Iranians often performed religiously mandated washing of cadavers in the same rivers and streams that they used to obtain drinking water and then buried their dead near the same potable sources.[12] These behaviors, which endured even after the cholera microbe was isolated by Robert Koch in 1883, were extraordinary multipliers of a fecal-oral disease like cholera, which spread through contaminated water and food supplies. Devotional visits to shrine cities and the practice of seeking to be buried at the holy sites by the predominantly Shiʻite Iranian Muslims also increased cholera's dissemination and virulence.[13] Iranians often hired caravans to carry both the living and the dead to these venerated locations, which included Najaf, Karbala, Baghdad, and Samarra in the Ottoman Empire and Mashhad in Iran (fig. I.1).[14] A contemporary observer noted that cadavers on caravans were "wrapped in sackcloth, and carried, slung across the backs of mules, to their distant resting place, sowing not improbably the seeds of a fresh outbreak."[15] Essentially, victims continued to disseminate cholera even after death, prolonging pandemic waves of the disease in the country.

Iranian rituals of cleanliness, especially the virtues of regular bathing, proved just as lethal as the drinking and burial norms in this period. Public bathhouses (*hammam*) in Iranian cities were customarily built below the street level to facilitate the flow of water into these structures. Their location below ground also allowed sewage, garbage, and the accompanying cholera microbe to more easily reach the baths from the surface above. Making matters worse, not only were bathhouses usually filled with recycled water from previous, potentially infected bathers, but the sick, who were advised to immerse themselves in water, often bathed beside the healthy.[16] Iran's urban bathhouses essentially became one of several important nodes for the oversized propagation of cholera during epidemic outbreaks in the Qajar era.[17] Despite the fact that an overwhelming segment of Iranian society led a rural life in this period, people in the countryside remained susceptible to a typically urban disease like cholera because of the subsistence networks that connected cities and towns with villages, rural settlements, and even nomadic tribes.[18] Farmers and herders sold their harvest and cattle at urban bazaars, which became focal points for the spread of cholera to the surrounding villages and campgrounds.[19] The corrosive demographic and economic impact of the recurrent outbreaks of Asiatic cholera, in turn, worsened the country's fiscal and administrative condition, hampering the central government's ability to take any significant steps to stop future outbreaks. The resulting unremitting cycle of pandemics contributed to the overall decline in governance, which was another hallmark of the Qajar era.

Iran's distinctive encounter with Asiatic cholera reveals the varied social and political forces, both foreign and domestic, that shaped the country's institutional and intellectual paradigms of public health. It also sheds light on cholera's seminal role in Iran's modern political history. The biosocial complexities that triggered the recurrence of Asiatic cholera in Iran during the

Qajar period unmasked unique Iranian sociocultural vulnerabilities to a global phenomenon. It is my hope that this work gives the reader an opportunity to understand Iran's encounter with a disease that played a seminal role in the emergence of global modernity and the enduring and persistent political and social determinants of health that continue to perpetuate Iran's susceptibility to both communicable and noncommunicable illnesses alike.

CHAPTER ONE

Cholera and the Globalization of Health in Iran, 1821–1889

The de facto fall of the Safavid dynasty in 1722 began a period of decline in Iran for much of the eighteenth century. War, disease, and deprivation reduced its population by a third and commerce by eighty percent.[1] Tribal warlordism and regional rule replaced the imperial administration of the Safavid shahs, undermining professional institutions and social services centered in urban areas. Iran's major cities, depopulated and battered by warfare, decayed to a shadow of their prior grandeur. Sustained national governance and dynastic stability eventually returned with the rise of the Qajars as Iran's preeminent rulers at the cusp of the eighteenth century. The newfound political stability allowed the country to reenter the global marketplace of ideas and commerce at a time of significant scientific and technological innovations in the West.

Improvements in shipping, manufacturing, and warfare were transforming European nations into imperial powers with expanding commercial and colonial interests in Asia. This placed Iran and the West on a collision course, beginning with the annexation of its territories in the southern Caucasus by Russia during the first decades of the nineteenth century. Battlefield setbacks against a superior Russian army drove the Qajars to seek the assistance of friendly European nations, opening the country to post-Enlightenment scientific and technical paradigms from the West. European expansion in the region also helped introduce Asiatic cholera into Iran during this period. Britain's growing colonial presence in India and its wide-reaching navy allowed the ancient disease to break out of its endemic home in the southern Ganges Delta region and assume pandemic proportions. Iran's rulers, like their Western counterparts, were powerless against the outbreak that killed indiscriminately and terrorized a population unfamiliar with the alien disease.

The twin traumas of cholera and military reversals against Russia planted the seeds of public health modernity in Iran. The country's political elites, impressed by advances in military medicine and vaccination in Europe, began gravitating toward Western perspectives on illness and therapeutics. They sponsored translations of manuscripts, sought out Europeans as personal phy-sicians, and began sending Iranians to study medicine in the West. These early exchanges established the basis for the gradual globalization of public health and the emergence of a proactive culture of disease prevention in Iran.

Dynastic Consolidation and the Rise of Cholera in Iran

The 1796 coronation of the first Qajar monarch, Aqa Muhammad Khan (1742–1797; r. 1796–1797), ended decades of fractured rule and civil war in Iran. Much of the country's northern territories had disintegrated into independent Khan-ates, and competing local notables dominated its major southern cities for most of the era that followed the de facto fall of the Safavid dynasty in 1722. The ascent of Aqa Muhammad Khan, who began as a powerful Turkic tribal chieftain from Iran's Northeast, was built on years of extraordinarily brutal wars of subjugation. In the North, his troops punished the defiant vassal king of Georgia by sacking Tbilisi, killing and enslaving thousands in the process; and in the South, Aqa Muhammad Khan's soldiers gouged the eyes of every surviving adult male in the rebellious city of Kerman. Fath 'Ali Shah (1772–1834; r. 1797–1834) succeeded his childless uncle a year after his coronation and continued his predecessor's centralizing policies. Traffic on Iran's caravan routes slowly recovered from decades of banditry and extortion as the new dynasty pacified regional warlords and enforced its writ on the country's provinces. Fath 'Ali Shah also increased Iran's transnational commercial en-gagements, beginning with an 1801 trade treaty with the British East India Company. Iran's growing commerce with the Indian subcontinent opened the avenues for cholera to spread westward from its endemic repository in the Ganges Delta, but not before Aqa Muhammad Khan's legacy of violence in the North caught up with his successor.[2]

A year into Fath 'Ali Shah's reign, the Russians, who had been erstwhile al-lies of the Georgian king, garrisoned Tbilisi in response to the late shah's atrocities, effectively annexing territory claimed by Iran's Qajar rulers. This triggered the First Russo-Iranian War of the nineteenth century five years later. The more advanced Russian army outmatched the Iranians in the open field, forcing Tehran to seek closer political and technical relationships with aligned European nations to improve its military standing. In 1807 the shah signed a treaty of alliance with the French, who were also at war with Russia. As part of the agreement, Paris sent a military and scientific mission under General Claude Mathieu de Gardane to help modernize the Iranian army. The delega-tion included a military surgeon who became the first in a long line of French

military physicians to treat courtiers in Iran.[3] The British replaced the French
military advisers several months later, after the mission folded as a result of
Napoleon's nonaggression pact with Russia. Despite London's help, Iran continued to face mounting setbacks on the battlefield, forcing Fath 'Ali Shah to
sue for peace. He reluctantly signed the Treaty of Gulistan in 1813, yielding
Iranian territories in Georgia, Dagestan, and parts of Azerbaijan and Armenia
to Russia.

The British continued to advise the Iranians on military matters after the
armistice, introducing new ideas and technologies from the West in the process. Andrew Jukes, a British surgeon and political agent, performed the first
Jennerian smallpox vaccination procedure in Iran several months after the
truce.[4] 'Abbas Mirza, Iran's crown prince and commander-in-chief, embraced
the technique in the hopes of reducing mortality in the ranks of his modernizing "new army" (*nizam-i jadid*). He commissioned John Cormick, a British
subject who served as his military's foremost surgeon, to write a treatise on
vaccination and had the work translated into Persian to increase the practice
among Iranian physicians. The subsequent *Discourse on Learning Vaccination*
(*Risala-yi ta'lim dar 'amal-i abila zadan*) was one of the earliest manuscripts
published using the movable-type press in Iran.[5] This printing technology, also
introduced by 'Abbas Mirza in 1816, was pivotal in disseminating post-
Enlightenment intellectual trends from the West in the ensuing years.[6]

'Abbas Mirza had good reason to popularize new medical procedures and
theories. The physicians and surgeons in the Iranian medical workforce, cor-
roded by decades of civil war and chaos, lacked the number and caliber of
professionals who practiced at the height of the Safavid era, and their knowledge base had not changed since the late seventeenth century.[7] Iranian ver-
sions of humoralism (*tibb-i yunani*), or Galenism, principally derived from
Avicenna [Ibn Sina]'s eleventh-century *Canon of Medicine* (*Kitab al-qanun
fi al-tibb*), dominated mainstream medical thought, and its adherents were con-
centrated in major metropolitan areas where only a minority of the country's
population lived (fig. 1.1). These Galenic physicians conceptualized the etiology of illness as an imbalance in the body's four elemental humors (*chahar
akhlat*) caused by harmful influences from the environment, diet, and daily
activity (fig. 1.2).[8] They remedied the disease-producing internal disequilib-
rium by purging and bleeding excess humors or by administering hot, cold,
dry, and moist treatments that affected humoral qualities.[9] The sick who could
not access mainstream physicians sought out dervishes or clerics who dispensed prophetic medicine (*tibb al-nabi*). These practitioners used the *sunna*

Fig. 1.1. Galenic practitioner Hakim Nur Mahmud (*seated center with book*), surrounded by his household and patients (detail). Myron Bement Smith Collection: Antoin Sevruguin Photographs. Freer Gallery of Art and Arthur M. Sackler Gallery Archives. Smithsonian Institution, Washington, DC. Gift of Katherine Dennis Smith, 1973–1985, FSA A.42.12.Up.58.

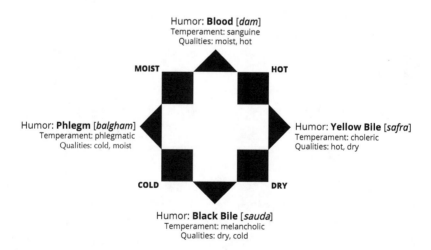

Fig. 1.2. Schematic representation of the four humors, associated qualities, and temperaments.

(a record of Prophet Muhammad's sayings and deeds) and the Quran to explain the cause, prevention, and treatment of disease.[10] Prophetic medicine viewed epidemic outbreaks as divine retribution, against which there was little defense or cure beyond God's intercession in contrast to mainstream humoralism, which framed plagues as the outcome of imbalances in the natural environment. The few orthodox practitioners in the country were unregulated and informally trained and were religiously prohibited from practicing on cadavers. Unsurprisingly, British military surgeons often outperformed their Iranian counterparts in both general therapeutics and surgery, prompting 'Abbas Mirza to sponsor the first Iranian to study medicine at Oxford in this period.[11]

While scientific knowledge trickled into Iran from the West, Asiatic cholera worked its way into the country from the east. British troops from Bombay introduced the disease into the Persian Gulf for the first time in 1821. The lethal epidemic made landfall in Iran's southern port city of Bushehr, where a British East India Company's factory was located, and followed caravan routes inland. Cholera reversed the country's fortunes in its months-old war against the Ottoman Empire when it reached Iranian troops encircling Baghdad, decimating the ranks of its most modernized battalions.[12] Mainstream physicians were powerless against the new disease, which did not respond to such humoral treatments as immersions in cold water and enemas with tea and hot spices. Even Fath 'Ali Shah's decree to inspect caravans for illness before allowing them to enter Tehran could not stop the epidemic from reaching the capital and spreading to the rest of the country's North. The virulence of the new plague forced the monarch to escape his seat for the perceived safety of the countryside, fueling rumors of his demise.[13] Those who could not leave the diseased cities sought safety in prophetic medicine. Healers prescribed prayers, invocations, and sanctified figures containing combinations of letters and numbers(*jafr*) associated with Quranic verses (fig. 1.3). People recited the prayers, wore the pictograms as talismans, or drank the ink from the pictograms dissolved in water, which they hoped would prevent and cure cholera.[14] The epidemic killed tens of thousands by the time it ended in the fall of 1822 and made no distinction between highborn and commoner in its lethality.

The loss of well-trained troops and able leaders to cholera contributed to Iran's battlefield setbacks against Russia four years later when war erupted once again between the two antagonists.[15] The shah capitulated to the Russians after two years of fighting, giving up more territories in the Caucasus and agreeing

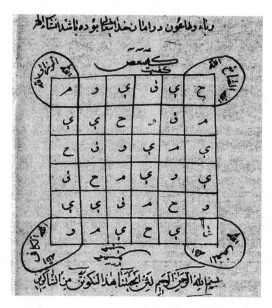

Fig. 1.3. A pictogram of letters referencing the hidden name of God used in the treatment of cholera and bubonic plague. The four revealed names of God: al-Mani' (The Withholder, The Preventer), al-Razzaq (The Provider, The Sustainer), al-Mughni (The Enricher), and al-Kati (The All-Sufficient) surround the 6 × 6 cell. In the manuscript in which this pictogram is included, the author discusses the impact of the 1853 cholera outbreak on the inhabitants of Tehran, its humoral etiology, and effective talismans against the disease. Hajji Mirza Musa Fakhr al-Hukama, *Dastur al-attiba' fi daf' al-ta'un va ilaj al vaba*, Melli Library, Tehran, MS no. 675F, 1269 AH [1852–1853].

to staggering reparations in the 1828 Treaty of Turkmenchay. Asiatic cholera returned a year later, reaching northern Iran from India via caravan routes through Herat. The epidemic, which was particularly severe in Tehran, lasted until 1830, when it gave way to one of the country's worst spates of bubonic plague.[16] These uninterrupted outbreaks likely contributed to mounting sectarian intolerance in Iran's North, resulting in the departure of the millennia-old Jewish population of Tabriz between 1826 and 1831.[17] They also marked the start of greater political instability, beginning with the premature death of 'Abbas Mirza in 1833 and followed by the demise of the shah himself a year later.

Muhammad Mirza (1808–1848; r. 1834–1848), the eldest son of the late crown prince, inherited the throne along with the bitter legacy of capitulations and epidemics from his grandfather. Domestic unrest, empty government coffers, and the new monarch's own inertia in enacting reforms decreased Tehran's appetite for technical and intellectual knowhow from the West. The

fate of progress suffered another setback when Britain, fearing Russia's growing influence on the doorsteps of its possessions in India, recalled its advisers
from Iran in 1838 as part of a broader effort to force the shah to abandon his
yearlong campaign to reclaim the seditious city of Herat. The reintroduction
of cholera into Iran in 1845 rekindled Muhammad Shah's interest in Western
expertise.[18] The hypochondriacally inclined shah made an official request to
Paris for a French physician to lead his coterie of native doctors and sponsored
the first Iranian in decades to study the latest medical advances in England.[19]
He also ordered the translation and dissemination of a French treatise on
cholera.[20] The work highlighted the causal role of cholera-producing miasma: "poisonous vapors coming out of sick intestines, stagnant water, bad
vegetation, or rotting carcasses," the importance of detecting early signs of the
illness, and nonheroic treatments.[21] Despite this newfound knowledge, the
government did not take any concrete measures against cholera, allowing a
new wave of the epidemic to spread unimpeded across the country in the
following year.

Muhammad Shah and his government abandoned the capital "in the utmost confusion" when cholera eventually reached Tehran in the summer of
1846.[22] Regional leaders filled the power vacuum by taking the country's first
tangible sanitary steps against the disease. Bahman Mirza, the shah's brother
and governor of Azerbaijan, forbade the sale of unripe fruit and the slaughter
of cattle within the walls of the provincial capital of Tabriz. He also prohibited
the temporary vaulting of dead bodies and ordered daily street cleanings, as
prescribed in the recently translated French treatise on cholera.[23] While Bahman Mirza's edicts were generally carried out in Tabriz, smaller cities that were
out of the range of the provincial capital proved to be less compliant. When
cholera broke out in the northwestern town of Urmia during the fasting month
of Ramadan in August of that year, the authorities could not stop commoners from breaking their fast with readily available unripe summer fruits.[24]
These conditions allowed the cholera epidemic to spread across the province
and eventually into Tabriz, where it killed more than six thousand, despite
Bahman Mirza's sanitary efforts to protect the city's residents.

Iranian practitioners attempted to treat cholera's victims by administering
sikanjbin, a traditional syrup of vinegar, honey, and water. Western physicians
residing in Iran similarly used drafts of lemon soda jointly with morphine pills
in a desperate attempt to treat the sick. These interventions did little to curb
the mounting casualties. Tehran alone lost anywhere from ten percent to a
quarter of its population by the time the epidemic abated in 1847. Four of the

shah's immediate family members, including his seven-year-old son, also died.[25] The monarch's powerful chief consort and mother of the crown prince narrowly survived the illness, but his foreign minister was not as lucky; he succumbed to cholera shortly after showing the first signs of being unwell.[26] These events moderated Iran's foreign affairs and curbed the shah's rigid stance in negotiations with the Ottoman Empire. The monarch and his premier, "paralyzed by the fear of cholera," agreed to ratify the Treaty of Erzurum in May of that year, settling boundary disputes that for decades had brought Iran and its western neighbor to the brink of war.[27] The epidemic's virulence also affected the course of faith in the country. Spiritual leaders, spanning Iran's religious spectrum, viewed cholera's lethality as a signal of the coming end of times. Those who espoused Shi'ite Islamic eschatology believed the outbreak heralded the imminent return of the Twelfth Imam in occultation who is prophesied to reappear before the Day of Judgment. The heightened messianic expectation increased conversions to the emerging Babi religious movement, which gave rise to the Baha'i Faith in the second half of the nineteenth century.[28]

When Muhammad Shah died of aggravated gout a year after the epidemic, the crown went to his sixteen-year-old son, Nasir al-Din (1831–1896; r. 1848–1896). The young shah's influential premier, Mirza Taqi Khan Farahani, better known as Amir Kabir, immediately began a series of fundamental improvements in governance and education. These changes increased Iran's globalization and laid the foundations for its integration into the emerging transnational sanitary regime in Europe. Amir Kabir had spent four years in the Ottoman Empire, as a member of Iran's boundary commission, during the Porte's military and administrative reorganization known as the Tanzimat. This experience inspired him to enact similar reforms in Iran. He strengthened the central government by reducing its outlays, modernizing its military, and improving its administration and revenue collection. He established Iran's first official weekly, *Ruznama-yi vaqayi'-i ittifaqiya*, which began the process of disseminating newsworthy information to a wider literate audience.[29] He also attempted the first vaccination campaign in more than two generations. The effort failed to stop a smallpox outbreak in 1850 because of poor compliance with the procedure but moved the premier to address pervasive scientific illiteracy and superstition among Iranians by increasing the number of students sent to Europe to study the sciences.[30] He also established the country's inaugural European-styled Polytechnic College (Dar al-funun) to remedy the continuing technical and leadership deficits in the military.[31] Amir Kabir's plans for the new academy, similar to the Imperial Medical School in Istanbul, called for hiring a core faculty from Austria (Habsburg Empire) whose members would develop the school's curriculum in the hard sciences to educate a new cadre of Iranian officers along Western standards.[32]

Amir Kabir, who had been the army's chief administrator in the North during the cholera outbreak in 1846, was particularly sensitive to the devastating impact of diseases on troops and the need for competent medical personnel in the military.[33] He employed Johannes Lodewik Schlimmer, a Dutch private practitioner, as physician and titular colonel in the Iranian army; and Fortunato Casolani, a British subject of Maltese origin, as the army's chief medical inspector (*hakim-bashi-yi nizam*), tasked with vetting the competence of regimental surgeons.[34] He also ordered the Iranian legation in Vienna to engage a physician, among the Polytechnic College's six European faculty hires, to develop and teach a Western medical curriculum at the school. The instructors received a five-year contract and arrived in Iran in November 1851 but would not meet their benefactor. Amir Kabir's policies, not the least of which included the reduction of annuities to courtiers, angered many influential notables, including the shah's powerful mother. The dowager pressured the young monarch to dismiss the premier two days before the faculty's arrival and conspired to have him murdered thirteen days after the ceremonial inauguration of the school on December 28, 1851.

The Polytechnic College opened its doors as the West began making breakthroughs against cholera; at that time, European epidemiologists had begun to correlate outbreaks of the disease with topographic, economic, and social variables.[35] John Snow's investigations, in particular, challenged the dominant miasmatic etiology of cholera. After arguing that its intestinal pathology, in lieu of the lungs, made corrupted air an unlikely culprit behind the disease, he also convincingly connected cholera epidemics in London with the contamination of drinking water by sewage, setting off the industrial era's hygienic movement.[36] Local and national governments in the West began establishing public health institutions to carry out infrastructural reforms to prevent future outbreaks on the basis of this growing body of evidence. In 1848 Britain passed the Public Health Act, which created the General Board of Health as a central sanitary authority for England and Wales. Similar statutes enacted elsewhere in Europe and North America progressively reduced the lethality of cholera by promoting cleanliness, providing potable water, and improving drainage and sewage systems in cities.[37]

Cholera and the Globalization of Iranian Medicine

Cholera, in epidemic form, returned to Iran shortly after the Polytechnic College's commencement in Tehran. The outbreak crossed into Iran's southwestern Kurdish territories from the port city of Basra, in Ottoman Iraq, where the British East India Company's principal factory in the Persian Gulf was located. It spread to Iran's North by the fall of 1852, rapidly killing an estimated twelve thousand in the city of Tabriz.[38] The epidemic reached Tehran in the spring of 1853 as the population readied to fête the investiture of the newborn Muzaffar al-Din Mirza (1853–1907; r. 1896–1907) as heir to the throne. The shah ended the preparations after a member of his court died of cholera and the infant crown prince began showing signs of illness.[39] The mounting fatalities eventually caused more than eighty percent of the capital's population, including the shah, to flee the diseased city.[40] Even so, cholera claimed approximately ten percent of Tehran's 120,000 inhabitants by the end of the summer.[41] The military fared worse. Only one physician, Hajji Mulla 'Ali Tabib, ministered to the capital's six to seven hundred cholera-stricken troops in a temporary field infirmary; most would die without seeing the doctor.[42] Cholera disproportionally affected the Iranian army in other cities as well. The bulk of Tabriz's early victims were soldiers from the shah's elite Nasiriya Regiment.[43] In Qazvin, the epidemic killed approximately thirty troops a day while sparing the civilian population.[44] Soldiers elsewhere deserted their posts or mutinied against the government to escape outbreaks in their barracks.[45] The authorities eventually released Tehran's regiments to avoid a similar revolt in the capital. Jakob Eduard Polak, the newly arrived Austrian medical instructor at the Polytechnic College, witnessed their departure: "All along the road, I saw men who had fallen, abandoned without any assistance, and very rarely did I see a jug of water, left there by a compassionate comrade."[46]

The sanitary and medical deficiencies in the Iranian military, unmasked during the epidemic, informed the work of the Polytechnic College's first medical instructors.[47] Polak, a former surgeon in Vienna's General Hospital, became the Iranian army's chief medical inspector following the death of Fortunato Casolani, soon after assuming his faculty duties (fig. 1.4). From this vantage point, he realized the venality and shortcomings of the few native physicians assigned to regiments across the country.[48] The absence of military field hygiene and sanitation during the epidemic led him to write his first Persian medical treatise on cholera, which included the latest European findings on its etiology, symptoms, pathology, and hygienic interventions to prevent its

Fig. 1.4. Jakob Eduard Polak. Julius Gertinger, *Jakob Eduard Polak,* 1867, Cdv, black and white on supporting cardboard, 10.5 cm × 6.7 cm. Polak autographed this calling card for Eduard Lewy, a prominent Austrian doctor and educator, in 1868. Author's collection.

spread. The manuscript was printed and circulated before the epidemic's cessation.[49] Johannes Lodewik Schlimmer, who was assigned to the Iranian regiment in the city of Kerman, implemented Polak's hygienic approach by advocating that troops maintain cleanliness in the garrison, drink water directly from its source, and cover barracks with twenty centimeters of charcoal powder to reduce the epidemic's severity.[50] Schlimmer joined Polak on the medical faculty of the Polytechnic College several years later, contributing to the Iranian students' growing familiarity with European sanitary ideas. The Polytechnic College's dedicated printing press allowed Polak and Schlimmer to translate and published a number of medical textbooks during their cumulative fifteen-year tenure at the school (Polak: 1851–1860; Schlimmer: 1857–1866).[51] These first books introduced new medical theories from the West, such as the pathological causes of diseases like cholera, and shaped Iran's modern medical vocabulary.[52]

The epidemic's lethality among the troops also paved the way for Iran's first freestanding military hospital, established in 1853 and based on Polak's design. The establishment, located near the parade ground outside of Tehran's Royal Citadel (Arg), would be replaced over a decade later by the larger civilian Imperial Hospital (Marizkhana-yi mubarak). In the interim, it allowed Polak to give practical bedside instruction to the first cohort of medical students at the Polytechnic College while delivering much needed care to ailing soldiers during the outbreak. Students memorized symptoms and treatments by observing cases in its wards, while they learned foundational biological sciences in the classroom.[53] They also became familiar with principles of field hygiene, the emerging science of sanitation, and empirical approaches to predicting the onset of diseases like cholera before the appearance of prominent symptoms.[54] Polak overcame cultural barriers to applied medical education, including religious prohibitions against dissecting dead bodies in anatomical instruction, by using animal models, detailed drawings, and a human skeleton that he brought from Vienna (fig. 1.5). The latter was a source of morbid curiosity for the uninitiated, prompting numerous requests for explanatory showings from the shah and other notables during inspections of the Polytechnic College.[55] The medical students were quick learners, even impressing the shah with their ability to diagnose every clinical case presented to them during the first royal inspection of the military hospital.[56] However, Polak never expected his first crop of students to become "high caliber" physicians by European standards; rather, he hoped to impart enough biomedical knowledge to allow them to continue their studies on their own or pursue further education in Europe.[57] Because of the Polytechnic College's inclination to admit the children of notables, its early graduates played leading roles in the evolving practice of medicine, administration, and public health in Iran. They taught at their alma mater, became influential court physicians, and, in one case, secured the shah's appointment to govern a province.[58]

While the Polytechnic College disseminated Western perspectives and intellectual trends on diseases and therapeutics among medical professionals, Iran's literate public was becoming increasingly aware of the impact of cholera abroad. The proliferation of the printing press after the mid-nineteenth century expanded the circulation of international news in the country, allowing Iranians to grasp the transnational nature of cholera and their place within the larger global path of recurrent pandemics.[59] Iran's nascent government

Fig. 1.5. Medical students at the Polytechnic College posing with one of the anatomical models at the school. Courtesy of the Institute for Iranian Contemporary Historical Studies (IICHS), Tehran, archive no. '4-4121.

weekly informed the public of cholera outbreaks in "Copenhagen, the capital of Denmark" and the "island of Jamaica," along Kingston's "crowded and dirty streets."[60] Even reports of yellow fever in the United States reached the readers of the broadsheet.[61] In 1853 they learned that the cholera epidemic in Iran had spread to Russia and, from there, to St. Etienne, Berlin, Hamburg, and finally England, killing many along its trail.[62] Later reports showed how the disease reached California from New York, following the new railway lines in the United States.[63]

Literate Iranians also began learning about the sanitary revolution in Britain and its government's hygienic interventions against cholera following the Public Health Act of 1848. Readers of the Iranian government weekly were told of efforts by the mayor of London to have his city's streets and sewers cleaned "to lessen the disease" during the 1853 outbreak.[64] They also learned of the British government's initiative to "white-wash" impoverished dwellings because of cholera's association with tight and crowded living quarters, poverty, and filth.[65] In addition, the more comprehensive *Report of the General Board of*

Health on the Epidemic Cholera reached Iran soon after its publication in England in 1850.[66] The report, translated by Fortunato Casolani shortly before his death, stressed that variables such as occupation, diet, and exposures to poisonous effluvia in the atmosphere and water contributed to outbreaks of cholera.[67] It incorrectly asserted that cholera was not transmissible but rather produced and disseminated by local miasma, as demonstrated in the 1848 outbreak in the English town of Taunton where children at a local school succumbed to cholera despite the disease's absence among the rest of the townspeople. The report associated this inconsistency with cramped conditions and limited air volume in the classrooms, which amplified the student population's exposure to the corrupted atmosphere that caused cholera. It explained that boys at the school were less likely to get sick due to their "unruly habit of breaking windows," which inadvertently increased air circulation, whereas girls, victims of their calmer natures, were more exposed to disease-producing stagnant air.[68] Cholera's prevalence in tenements and poorhouses in London was also associated with overcrowding and lack of air circulation in the report, illustrated by the Christ Church Workhouse's predisposition "to cholera epidemics due to the evil vapors" that arose from manure and animal matter used to produce fertilizer by its workers. The report concluded that elevated and well-drained locales were less vulnerable to epidemics.[69] Influenced by these views, urban Iranians for years to come would seek the protection of mountains and highlands to escape cholera, not the more accessible deserts and flat wastelands.

By the mid-nineteenth century, many of Iran's Galenic practitioners similarly attributed cholera to miasmatic origins. They argued that the putrefaction of air caused outbreaks of the disease and "the arrival of wind or good odor" could stop it from spreading.[70] However, other traditional Iranian physicians maintained that astral misalignments, divine intercession, and person-to-person contagion were principally responsible for epidemics.[71] Despite their ideological differences on the origin of cholera, these physicians were consistent in their Galenic approach to treating the disease, using agents and interventions that could fix the disequilibria in a patient's four humors, as witnessed by a contemporary European observer: "The dreadful practice of the Persian doctors is quite enough to drive the fair dames of Tehran to an English physician. I am told that they give the most nauseating draughts, in immense quantities, to their patients two or three quarts at a time. Thus a hot disease is to be combated by a cold remedy. The classification of these last are somewhat fanciful. Pepper, I know is 'cold' and ice, I think, is 'hot.' It can hardly be otherwise than hot, for it is applied to the stomach in large pieces during cholera."[72] They also advised both the sick and the healthy to exercise, avoid rich foods, and abstain from wines at all costs during outbreaks.

Physicians trained in Western medicine had an edge over their Iranian Galenic counterparts in precisely diagnosing and treating contagions even when they shared similar underlying theoretical perspectives. This was evident both in practice and in the European-authored Persian medical tomes on cholera, which focused on the symptomology, pathology, and treatment of the disease rather than the medico-philosophical etiology of epidemics as articulated in traditional Iranian Galenic treatises.[73] The latter often conveyed information in an ornate, circuitous, and sometimes self-contradicting language. European writers, on the other hand, presented their material in charts organized according to anatomical systems or by giving an alphabetical list of symptoms. This rendered the information more accessible for the lay reader and easier to grasp for the medical student. As a result, literate Iranians increasingly turned to the West for medical expertise, and the prestige of European physicians increased in the second half of the nineteenth century. Even Nasir al-Din Shah could not do without an official attending physician from Europe and appointed Jakob Eduard Polak as his personal doctor after his French physician-in-chief (*hakim-bashi*) died of an accidental poisoning in 1855.

Iran's awareness of the emerging hygienic movement in Europe and its growing professionalized corps of physicians did not result in any significant sanitary changes because of shortfalls in the government following the dismissal of Amir Kabir in 1851. Nasir al-Din Shah's new premier, Aqa Khan Nuri, abandoned much of his predecessor's administrative, military, and educational reforms. Rampant corruption, including nepotism, sale of government positions, and misappropriation of state funds, characterized his tenure. The Polytechnic College and the government weekly newspaper were among the few surviving innovations from the previous administration. Five years into his premiership, the shah's army marched on Herat, triggering a war with Britain. Iran's humiliating capitulation after less than four months of fighting contributed to Nuri's dismissal two years later. An older and more assertive Nasir al-Din Shah did not name a replacement and assumed executive control over the country's administration; as a result, he made an already inefficient central government even worse, which hindered any meaningful sanitary reforms.[74] Unhampered, cholera broke out yearly in the decade that followed the 1853 epidemic in Iran.[75] While none of these flares assumed national proportions, they helped erode the country's economy and stability.[76] Food shortages, double-digit inflation, urban riots, and military rebellions were common features of years after the Anglo-Iranian War.[77] Despite these circumstances, globalization continued in Iran. Broadsheets reporting regional and international news multiplied. Tehran alone had four state-sanctioned weeklies by the 1860s, including an official scientific gazette (*Ruznama-yi 'ilmi-yi dawlat-i 'alliya-yi Iran*) that featured articles on medical and sanitary advances in the West.[78] The Polytechnic College also continued to increase the number of professionally trained physicians, six of whom traveled to France for graduate medical education in 1856 and 1859, eventually obtaining their doctorates from the University of Paris.[79]

In 1860 Polak left Iran "without hate or love" and returned to Austria.[80] Joseph Désiré Tholozan, the shah's new French chief physician, took over his teaching responsibilities at the Polytechnic College, enhancing the scientific foundation of the medical curriculum during his tenure (fig. 1.6).[81] Tholozan had been a senior professor (*professeur agrégé*) at the Military Teaching Hospital of Val-de-Grâce in Paris and a distinguished scholar of epidemic diseases who considered major improvements in Iran's medical education to be crucial to its sanitary progress. He advocated replacing widely referenced tomes, such as Avicenna's *Canon of Medicine* and Imam Riza's *Golden Dissertation of Medicine* (*Al-risala al-dahabiyya fi al-tibb*), with translations of the latest European publications on medicine and public health. The foundation of "popular hygiene" in Iran, as he saw it, depended on substituting traditional "Galenic ideas" perpetuated in older Iranian works with newer theories from the West, such as John Snow's water-borne model of cholera transmission, which were informed by "more precise observations."[82]

Tholozan's empiricism extended to his clinical instruction at the Polytechnic, where he introduced the use of the stethoscope to enhance diagnostic precision. Not satisfied with visual inspection alone, he conveyed the importance of auscultating and palpating the abdomen to detect bowl sounds and other signs of cholera. He also wrote updated textbooks that became mainstays of the medical curriculum, including publications on epidemics, modern diagnostic techniques, and therapeutics.[83] The French National Academy of Medicine (Académie Nationale de Médecine) recognized his work on Asiatic cholera during this period of his residence in Iran with a prize of twenty-five hundred francs in 1872.[84] In contrast to his predecessor, Tholozan sought to train high-caliber physicians, attuned to critical inquiry and research, who

جناب کرمو وزان یحکمانی مخصوص کطهرترش یاشائی

Fig. 1.6. Portrait de Dr. Tholozan, vers 1880–1890, Style Qajar. © Musée d'art et d'histoire, Ville de Genève, Cabinet d'arts graphiques, Legs Jean Pozzi, no inv. 1971-0107-0565.

lived up to European professional standards. He also steered the Polytechnic College to meet the health-care needs of the civilian population in addition to producing physicians and surgeons for Iran's modernizing military, educating a cadre of seventy physicians during his tenure.[85] These graduates became central drivers in the development of a modern sanitary culture in Iran.

The completion of the Indo-European telegraph network by the Siemens Corporation in 1865 further increased Iran's global connectivity and opened the floodgates of information from abroad. The privately owned grid, which passed through the country, connected Tehran with capitals around the world.[86] Iranians received immediate news of the broader international impact of cholera, ranging from the cholera-caused evacuation of the British army in Lahore to cholera's role in Algeria's food shortages.[87] The telegraph also transformed Iran's transnational engagements, allowing its diplomats to coordinate their efforts and participate in international conferences, thus ratifying binding conventions in real-time consultation with Tehran.

The International Sanitary Conference of 1866

The technological developments of the Industrial Revolution that facilitated East-West exchanges in the mid-nineteenth century not only transformed Iran's intellectual horizons but also increased the frequency, range, and speed of cholera pandemics. In 1865 Asiatic cholera rapidly spread from Mecca to Egypt, and from there to Europe, owing to the ubiquity of steam-powered ships and rail transport. The following year, the French government convened a multinational sanitary conference in Istanbul to formulate a common policy to prevent future outbreaks from reaching the West. Delegates from Iran, Austria, Belgium, Denmark, Spain, the Papal States, France, Great Britain, Greece, Italy, the Netherlands, Portugal, Prussia, Russia, Sweden/Norway (then politically united), Egypt, and the Ottoman Empire were invited to the meeting. The French government, which did not call upon the Iranians at similar gatherings in 1851 and 1859, had come to recognize that Iran's position between cholera's endemic home in India and Europe made it crucial to the success of any international strategy against the disease.

The conveners first approached Tholozan to represent Iran at the conference. However, he turned down the offer, citing the importance of having native delegates in a position of making decisions with potentially significant implications for Iran's sovereignty.[88] Mirza Malkam Khan, one of the most influential liberal thinkers of Nasir al-Din Shah's reign, and Sawas Effendi, an Ottoman government physician and sanitary inspector, represented Iran in his stead.[89] Jakob Eduard Polak, who had returned to Vienna, also attended as Austria's delegate. The representatives largely agreed that cholera, in pandemic form, always began in the Ganges Delta and could be stopped at one of the "intermediary" countries in "the Orient" before it reached the West. They remained divided, however, on cholera's mode of transmission and the ideal interventions to halt its global spread. The anticontagionist block of countries, led by England, believed miasmatic theories best explained cholera's propagation and deemed quarantines ineffective against future outbreaks, not to mention harmful to international trade.[90] On the other hand, the Mediterranean nations, spearheaded by Italy, accepted the evidence supporting the contagious nature of cholera and the efficacy of quarantines.[91] The participants eventually set aside their ideological differences and agreed that restrictive and prophylactic interventions on maritime and overland routes of the pandemic could protect Europe, provided the measures were rapidly and comprehensively instituted.[92]

Most of the delegates at the conference called for particularly draconian restrictions on the Hajj to prevent Muslim pilgrims from transmitting Asiatic cholera to Europe, including the use of gunships to enforce sanitary regulations in the Red Sea and weeks-long mass quarantines on routes to Mecca.[93] They also advocated equally strict measures along Shi'ite Muslim pilgrimage routes from India and Afghanistan to the venerated shrines in Mashhad, Karbala, and Najaf. The Iranian government was advised to ban pilgrims from crossing its borders during outbreaks in neighboring countries and to adopt a broad range of interventions within Iran itself to prevent the westward transmission of cholera.[94] These included requiring cadavers be embalmed, hermetically sealed, and transported to shrine cities solely in winter months, because outbreaks were often associated with the Shi'ite practice of burying the dead in hallowed ground during warmer seasons. Finally, the conference recommended establishing a national sanitary council in Tehran, composed of equal numbers of Iranians and supervising European members, to implement the country's sanitary obligations.[95]

The proposals did not sit well with Iran's representatives. Malkam Khan felt that restrictions on pilgrimage and exhumations during outbreaks of cholera, which occurred almost yearly in the decade preceding the conference, would bring Iran's religious life to a virtual standstill, causing widespread discontent.[96] He also opposed a proposed maritime ban between Arabian ports and Egypt in the event of an outbreak in Mecca for the same reason: "The ideas, customs, doctrines and logic of Asia were so different from those of Europe that the mere idea that Moslem Sovereigns had come to an understanding with European powers to regulate the progression of the pilgrims would be sufficient to change completely the relations of those Sovereigns with their subjects and would expose them to the attacks of fanaticism all the more violent because in recent times everything had been done to restrain it."[97]

Iran's other representative, an Ottoman subject, also opposed the measure, likely because of the potential loss of internment and pilgrimage tax revenue for the sultan's government. On the scientific front, Malkam Khan argued against the opinion that clothes could transmit cholera, recognizing that restrictions on shipments of linens during epidemics could damage Iran's cotton exports. He also opposed the suggested quota of Europeans on the proposed Iranian sanitary council, stressing that such a measure would erode his country's sovereignty.[98] Despite these objections, Iran's delegates eventually joined their counterparts in ratifying the conference's core recommendations in September 1866, marking their country's formal entry in the modern era's emerging global sanitary order.

Cholera, Sanitary Planning, and Fleeting Administrative Stability

As the conference ended in Istanbul, pilgrims returning from Mecca gradually carried a new cholera epidemic toward Iran's borders. The Iranian government, paralyzed by unrest and gridlock, did not take any steps to prevent the disease's imminent ingress into the country despite knowing of its presence in neighboring Ottoman territories.[99] Nasir al-Din Shah retained his executive monopoly on governance but lacked the administrative capability to undertake the measures recommended at the sanitary conference to mitigate the risks of another outbreak. In the fall of 1866, cholera crossed Iran's western frontier into the province of Azerbaijan, where it killed more than twenty percent of Tabriz's citizenry in a forty-day span; other towns and villages in the region suffered similar losses.[100] A year later, cholera broke out in the city of Gorgan (Astarabad), on the eastern shore of the Caspian Sea, during an Iranian military offensive to quell a major rebellion by the region's Yomut Turkmen tribes. Many of the ten thousand troops garrisoned in the city perished, setting back the government's pacification efforts in the province.[101] When an attenuated wave of the disease eventually reached Tehran in 1867, the shah changed his course on the country's neglected sanitary readiness and ordered Joseph Désiré Tholozan, his trusted physician-in-chief of eleven years, to organize a sanitary council (*majlis-i hifz-i sihhat*) and a medical service to intervene against the outbreak and prevent future flare-ups.[102]

Tholozan convened the Iranian Sanitary Council in the winter of 1867.[103] The gathering quickly sent physicians to twelve major cities as sanitary "observers," but the council's action could not prevent cholera from flaring up again several months later without the resources to implement precautionary medical and hygienic measures.[104] The new wave of the disease began in Mashhad, where it killed the province's prince governor (a son of Nasir al-Din Shah) and thousands of others in the shrine city.[105] Infected pilgrims rapidly spread the epidemic to every corner of the country, including Gorgan, where its impact on the city's reconstituted garrison forced the army to once again disband and emboldened rebellious Turkmen tribesmen to "plunder openly."[106]

The epidemic also reached Tehran after the cash-strapped government refused to establish a sanitary cordon to inspect and quarantine travelers into the capital.[107] Making matters worse, Iranian officials did not even acknowledge the presence of cholera in Tehran and took no steps "to arrest the progress, or mitigate the evils of the malady" for almost a week after the outbreak in the city.[108]

Tholozan stopped convening the Sanitary Council in 1868 after realizing that the assembly could not carry on without resources to support its mandate. In an ensuing report, he advised the shah to establish an annual national health budget, funded by a special tax, to underwrite regular meetings of a sanitary council and its activities.[109] He also recommended forming municipal councils of health, composed of local notables and physicians, to report outbreaks rapidly and enforce sanitary regulations in the provinces. Tholozan felt that almost two decades' worth of medical graduates from the Polytechnic College equipped the government with the needed medical workforce to run the country's sanitary services and proposed relegating foreign physicians to an advisory role in any future Iranian sanitary council.[110]

Tholozan's experiences as a medical officer at the Siege of Sevastopol during the Crimean War convinced him that improvements in hygiene were the foremost way to stop the recurrence of cholera.[111] He believed that land-based quarantines were ineffective in Iran, being both physically and fiscally impractical during an outbreak. Instead, he urged the shah to improve the quality of urban drinking water by safeguarding its purity at the source and using underground channels for its conveyance. He also recommended building public laundries, latrines, and sanitizing existing urban bathhouses to prevent future epidemics. In his view "the Constantinople [Istanbul] Conference, without any precise information on the sanitary condition of Persia, without any knowledge of the habit of cholera in this country [Iran], formulated a set of measures destined to guarantee [the protection] of Turkey and Russia, by calling for restrictive [quarantine] measures in Persia."[112]

Nasir al-Din Shah was obliged to table most of Tholozan's suggestions as he confronted one of the worst famines in Iran's history between 1869 and 1872. The famine was brought on by years of draught, insecurity, epidemics, and poor governance. The resulting shortage was then worsened by senior bureaucrats, large landowners, grain merchants, and high-ranking religious officials, who hoarded foodstuffs and manipulated the agricultural market. Drastic social consequences included massive migration from rural areas into cities in search of food, bread riots in urban centers, and looting of cattle and crops from villages by starving nomads across Iran. People who could not afford staples turned to eating dogs and cats. When the supply of domestic animals ran out, they began eating grass, roots and even dung. Iranian cities were rife with rumors of cannibals unearthing cadavers and kidnapping children for consumption.[113] The hunger in Mashhad illustrated the country's misery in this period: "All the dogs and cats that could be found have been eaten. People have even been driven to eat their own children. Any eatables seen in the shops are forcibly carried away. The Lord have mercy upon us! 500 Toorkomans besieged Sheref-Abad, 5 fursukhs [50 kilometers] from this town [Mashhad], and carried off 24 captives and 50 cattle."[114] Iran's fortunes remained unchanged until the closing months of 1871, when the food crisis began to diminish and the military slowly gained the upper hand over rebellious Turkmen tribes.[115] But by then, the famine had already killed more than ten percent of the country's population.[116]

The calamity convinced Nasir al-Din Shah that he could not lead alone and wholly lacked the expertise to run departments like the Sanitary Council in his expanding governing administration.[117] In November of that year, he appointed Husayn Khan Mushir al-Dawla, his progressive minister of justice, as the country's first prime minister in more than a decade, bringing Iran out of bureaucratic limbo and strengthening the arm of the central government.[118] Like Amir Kabir a generation earlier, Husayn Khan had observed the Ottoman Empire's modernization efforts during his twelve years as the shah's envoy in Istanbul, inspiring him to initiate similar improvements in Tehran's finance, administration, and communication systems.[119]

Husayn Khan expanded the government's telegraph network, linking Tehran with Iran's most important provincial cities, as the earlier Indo-European line had connected the Iranian capital with the rest of the world. The telegraph allowed the shah's bureaucracy to extend its reach to previously difficult-to-access peripheries, rapidly acquiring information and intelligence on the health of the country's provinces. It also enabled the central government to better enforce its hygienic mandate as demonstrated by the Foreign Ministry's orders to cleanse Bushehr in 1871. The government's local representatives initially ignored the command, as they had in prior years, by claiming that the sanitary instructions sent via courier had been lost en route from the capital.[120] However, the newly completed telegraph lines between Tehran and Bushehr allowed the foreign minister to immediately resend the order and to ensure compliance through the threat of punishment and the reliability of rapid communication as the following passage from the telegram indicates:

Should the above orders and the cleanliness of the town of Bushehr not be fully carried out, you will see that, God-forbid, all at once you have laid yourself open to the evils of a most severe reprimand. . . . What must Foreigners think when they see human beings dying at every step and in every street? Has Islam and a sense of honor totally ceased to exist? Why do not you, who are an employee of the government and a Resident at Bushehr, send me any reports? It is evident that you avoid and shirk from business, and that you are lazy and indolent.[121]

The foreign minister's embarrassment with the filthy and insalubrious conditions in Bushehr echoed a growing recognition by Iran's ruling class, including its new premier, of shortfalls in the country's urban hygiene, especially relative to Western cities.

In 1871 Husayn Khan Mushir al-Dawla established Iran's first Ministry of Public Works (*vizarat-i fava'id-i 'amma*) under the direction of the Hasan 'Ali Khan Garusi, who had served as Iran's envoy to France at the height of Georges-Eugène Haussmann's urban reform program in Paris. Following Haussmann's lead, Garusi improved Tehran's cramped and unhealthy environment by initiating a building program that included wider roads, fountains, and green spaces. He also expanded the Office for Public Order (*idara-yi ihtisabiya*), whose municipal responsibilities involved the capital's upkeep.[122] Johannes Schlimmer, who by this time had spent more than twenty-five years in Iran, saw these developments in the context of larger shifting Iranian attitudes to sanitation and hygiene:

I live and I write in a country which we can compare, until 1871, to a Babel of the nineteenth century, which we call the Persian Empire, precluded from all scientific resources, from all encouragement, from all that can help a physician during the difficult time of an epidemic; where more than anywhere else in the world, the proverb: "Everyone for themselves and God for all!" was in full vigor; it is only since 1871 . . . that the Government began being effectively interested in the hygiene of the country and to taking sanitary measures that proved that Persia, under a well-managed administration, can just as well as other countries, elevate itself one day to the rank of normal powers.[123]

Despite this progress, Husayn Khan's "well-managed administration" lacked the capital to upgrade Iran's urban sanitary infrastructure on a national level, as Tholozan had advocated. The years of cholera, starvation, and insecurity had "devastated" Iran's economy and pushed down its exports to a historic nadir, leaving the premier no choice but to seek foreign investments to modernize the country's utilities.[124] In 1872 he convinced the shah to sign a sweeping agreement with Baron Julius de Reuter, a naturalized British subject, to develop Iran's lagging infrastructure in exchange for dividends and exclusive rights over segments of the country's natural resources. The Reuter's concession, as it would come to be known, included plans to modernize Iran's waterworks, including its canals, wells, and other "artificial water courses," satisfying Tholozan's recommended upgrades.[125] However, the improvements would never see the light of day because of opposition from clerics and other conservatives who decried the purported un-Islamic aspects of the agreement, leading to protests that obliged the shah to cancel the concession in 1873.

The Reuter's debacle led to Husayn Khan's demotion, ending a short period of administrative stability and cabinet rule in Iran. Yet the impetus for municipal development in Tehran endured beyond his premiership. Nasir al-Din Shah's appreciation of the urban layout of Western capitals, which he observed during his 1873 European tour, led him to retain Hasan 'Ali Khan Garusi as the minister of public works, supporting his construction projects in Tehran for another nine years.[126] The shah also officially instructed the Office for Public Order to maintain the hygiene and salubrity of the Iranian capital through regular street cleanings, trash disposal, and the protection of the city's vital waterways from garbage, stench, and other refuse that could be the source of disease-causing miasma (see appendix A).[127] However, the office lacked the funding and workforce to carry out the shah's decree outside of the limited confines of the Royal Citadel, where palaces, government buildings, and ceremonial grounds were located. Even after fifteen years of growth in 1888, Tehran had only a paltry 278 municipal workers, ranging from city managers and sanitary workers to police, serving the city of approximately 200,000 at the time.[128] While administrative and economic difficulties contributed to this shortfall, Nasir al-Din Shah's own urban priorities also played a role in limiting the implementation of sanitary services in Tehran. His municipal reforms emphasized the aesthetic manifestation of modernity as it related to the legitimization of his reign rather than the less visible hygienic and sanitary infrastructure of a city in the last decades of the nineteenth century.[129] The shah's vision of transforming Tehran into a modern capital befitting his rule did not include ensuring a clean and healthy environment for the city's masses as Haussmann had done during his urban renewal of Paris's slums and gutters. As a result, Tehran's only secure and relatively untainted source of water came from the shah's own Nasiri qanat (subterranean water channel), which directly supplied the inhabitants of the Royal Citadel.[130] This meant that the most developed city in Iran and the seat of its government remained intensely susceptible to water-borne diseases like cholera at a time when such illnesses had started to retreat from Western capitals.

Europe's Porous Frontier and the International Sanitary Conference of 1874

The return of cholera to Europe between 1869 and 1874 revealed the flaws of the international consensus on the nature of the disease and the approach to preventing its global dissemination. Unlike the 1865 pandemic, which preceded the International Sanitary Conference in Istanbul, the most recent waves of cholera reached the West via Iran and across the Caucasian provinces of the Russian Empire, not by way of Muslim pilgrims traveling through Egypt to Mecca. This caused European nations to reevaluate their sanitary policies, beginning with England due to its dominant position in the global maritime trade network and its rule over India, the endemic home of Asiatic cholera. In 1872 the British Local Government Board, which succeeded the General Board of Health in overseeing public health in England and Wales, completed its investigation on the latest spread of cholera into Western Europe.[131] John Simon, the charismatic leader of the Victorian sanitary movement, forwarded the published results of the study to the Iranian government and to "Dr. Tholozan for the use by the [Iranian] Board of Health."[132] The report, which was translated and brought to Nasir al-Din Shah's personal attention, held the increased trade between Iran and Russia as the main cause of cholera's westward diffusion.[133]

The Russo-Iranian commercial relationship owed its growth to the Ottoman Empire's heavy-handed sanitary policies a decade earlier. In 1863 the Ottomans imposed strict quarantines on people and merchandise coming from Iran based solely on unfounded rumors of a bubonic plague outbreak. The Iranian government, angered by what it perceived as an Ottoman scheme to extort and harass its merchants, began trade talks with Russia and eventually agreed to redirect most of its westward commercial traffic from the Ottoman Black Sea port of Trabzon to the Russian port of Poti farther north.[134] The agreement not only changed the route of Iran's exports but unwittingly altered the course of future pandemics along major trade arteries into the Russian heartland.[135] This development formed the basis for the report's warning that the imminent rail link between Russian territories in the Caucasus and the West would all but ensure that any "contagious current in

Persia will become current in Europe."[136] Henceforth, the health of Europe was inexorably linked with Iran's sanitary condition, prompting the report to conclude: "From this point of view the internal state of Persia, and the recurring famines which afflict its population, will become a subject of nearer interest and greater moment to European nations than was apprehended even by the international sanitary conference of 1866."[137] The British, not surprisingly, underemphasized the role of their own commercial activities in diffusing cholera out of India and instead unscrupulously faulted the sanitary failings and menace of mercantile rivals like Russia.

The findings described in the Local Government Board's report prompted the Austro-Hungarians to convene an international sanitary conference in 1874 to reexamine prior agreements on best practices against cholera.[138] The meeting, held in Vienna, mostly reaffirmed the conclusions of the conference in Istanbul: pandemic cholera originated in India, and outbreaks in other countries were always imported; in particular, cholera in Europe never arose spontaneously or from latent local repositories. One difference, however, was the unanimous agreement among delegates that individuals and contaminated liquids, especially water, could transmit the disease, a decisive shift from the anticontagionist position held by some at the previous conference. The role of food in epidemics remained contentious, however, and some countries, including Iran, voted against the presumption that animals could transmit cholera. National interest continued to have an overwhelming influence on the scientific position of the delegates at the conference. Those that had the most to lose from potential restrictions on trade, a group that included Iran, Great Britain, Russia, and Serbia, withheld their votes on the role of merchandise and other commodities in transmitting the pandemic. Despite these abstentions, most participants agreed that cholera could be conveyed by infected apparel.[139]

The Vienna conference deviated from the meeting in Istanbul on the utility of overland quarantines. The committee charged with assessing their efficacy, which included Jakob Eduard Polak, who had agreed to lead the Iranian delegation at this conference, highlighted their overall impracticality and the negative impact of land-based quarantines on commerce. But the delegates mostly agreed on the benefits of maritime quarantines outside of Europe and recommended a system of on-board medical inspections for ships docking in their ports from non-European territories. Once again, they also called for the establishment of an international health council(*conseil de santé international*) in Iran to stop cholera before it reached Europe. This time, however, Great

Britain and other countries with growing trade interests in the region squashed the proposal, despite the existence of similar councils in Istanbul and Alexandria.[140] The British worried that an internationally mandated council could legitimize Iranian interference with Great Britain's commercial and colonial concerns in the Persian Gulf under the guise of safeguarding public health.[141] Nevertheless, Iran's strategic position on the path of cholera pandemics was unchallenged, and even the British acknowledged the need for appropriate sanitary interventions to stop future outbreaks from crossing through Iranian territory into the West.

Plague and the Return of the Iranian Sanitary Council

The defunct Iranian Sanitary Council reconvened two years after the meeting in Vienna, not because of cholera but in response to an advancing bubonic plague epidemic that began in the Arabian Peninsula.[142] The disease, transmitted by infected fleabites, caused fever, headache, vomiting, distinctively swollen lymph nodes (buboes), and a high fatality rate. It reached Mesopotamia in 1876, where it took a particularly heavy toll among Iranian Shi'ite theologians in Karbala, located about sixty-two miles southwest of Baghdad:

> In Karbala, I am writing in the grips of the Bubonic Plague, in 1876, among acquaintances, Aqa Sayyid Baqir Lahijani died, Mulla Qasim Lahijani died, Mulla Ghulam Lahijani—what a youth he was—died, in one week they all died. In one house in the Madrasa Mulla Zayn al-'Abidin died alone, Mulla Muhammad Khurasani died, the wife of Mulla Murad Lahijani died, the wife of Aqa Sayyid Sadr al-Din . . . with her child died in a day, Mulla Ghulam Rashti died, the wife of Mulla Rashid Rudbari Gilani died, my neighbor, an Arab, died, died . . . died, in one inexplicable week a thousand breaths, that I knew about, died. Among those I did not know, only God knows the terrible toll, may God have mercy on us![143]

The Iranian government learned of the growing menace along its southern frontier, despite Ottoman efforts to hide the scourge from the outside world, prompting 'Aliquli Mirza I'tizad al-Saltana, the minister of public instruction, to reassemble the Sanitary Council on March 27, 1876.[144] The gathering, which met in the Polytechnic College, included Tholozan, physicians from leading Western embassies, the medical director of the Indo-European Telegraph Department, several prominent Iranian doctors, and the superintendent of the Polytechnic College.[145] They discussed the progression of the epidemic and the needed interventions to prevent its spread into Iran, as indicated in the following record of the meeting: "A warm discussion took place regarding the efficacy of restrictive quarantine measures, one of the principal doctors 'Mulk el-a-Tubbah' (Prince of Physicians) insisting on the inefficacy of quarantine along a great line of frontier, and stating also that quarantine was ineffectual in keeping cholera out of St. Petersburg. The advantage of having Persian sanitary officers appointed at several places on the frontier and the Persian Gulf and of establishing a quarantine of observation was recommended."[146]

The Iranian government did not give the Sanitary Council an independent budget or the authority to implement policy to discourage Western powers from meddling with its proceedings for political gain, as was the case at the international sanitary councils in Egypt and the Ottoman Empire.[147] Despite these restrictions, the Sanitary Council met regularly for nine weeks and effectively guided the government's response to the crisis, which included sending three medical graduates of the Polytechnic College to establish and supervise frontier quarantines in Bushehr and Kermanshah. They also began a disinfection process on border crossings, in which they cleansed people with antiseptic showers and fumigated luggage, parcels, and letters from plague-infected areas—a practice that Iranians resumed during epidemics of plague and cholera in ensuing years.[148] On the Sanitary Council's advice, Tehran ordered the governors of provinces bordering the Ottoman Empire to impose a strict fifteen-day observational quarantine on all arrivals without an official clean bill of health from the Ottomans, allowing neither Iranian notables nor Europeans to forgo the process as had been the custom.[149] The government also printed a French pamphlet on bubonic plague, translated into Persian by Tholozan, describing its history, symptoms, treatment, and prevention through hygiene, which they circulated in provincial cities. The Iranians, however, lacked the budget and the navy to effectively conduct sanitary inspections and enforce quarantines along their sizable southern waterway as the council recommended. They persuaded Britain, which had the region's largest maritime force, to carry out the necessary surveillance, establishing a precedent for British control of Iran's Persian Gulf quarantine arrangements that Tehran would later regret. These measures restricted the inroads of bubonic plague to Iran's southwestern frontier regions, where it lasted for several months before finally receding in the summer of 1876.[150]

A renewed outbreak of bubonic plague, this time in northern Iran, coupled with the threat of cholera from the east, prompted the Sanitary Council to reassemble in January 1877. After lengthy deliberations, its conveners agreed that the country's sanitary deficiencies were largely responsible for the plague's inroads in the North and that quarantines, by themselves, would not suffice to stop a similar ingress by cholera. The recommended sanitary improvements, however, remained out of reach for the cash-strapped government unless it received a rapid and sizable fiscal windfall, such as a reprieve from the reparations imposed on it by the decades-old Russo-Iranian War. Tholozan, who had an amicable relationship with the Russians, failed to convince their government to reduce the staggering financial burden, even after appealing to their humanity. Embittered by their intransigence and callousness, he turned down Russia's conciliatory offer to support his candidacy to lead the Iranian Sanitary Council.[151]

When the threat of plague and cholera receded in late 1877, the Sanitary Council once again faded into obscurity, meeting only irregularly, during times of national public health crisis. It "ceased to function, as much as a result of the lack of indispensable material means, as from the pretension of not wanting to give a consultative voice to the foreign participants on the council."[152] Edward G. Browne, famed British orientalist and physician, reported that the Sanitary Council reconvened a decade later, gathering on a weekly basis in the Polytechnic College (fig. 1.7). 'Ali Quli Khan Mukhbir al-Dawla, who succeeded I'tizad al-Saltana as minister of public instruction in 1880, presided over the group of mainly Iranian physicians and conducted the meetings in Persian.[153] The proceedings, standardized by I'tizad al-Saltana in 1877, began with a detailed report on mortality and morbidity in the capital, which was followed by a listing of death rates and prevailing diseases in the provinces based on information supplied by corpse washers and physicians.[154] The epidemiological data informed the council's sanitary recommendations to the government. The physicians in attendance also compared the efficacy of traditional Galenic therapeutics with newer Western approaches to treating infectious diseases. This broadened the impact of the Sanitary Council as a medium for changing conceptions of medical knowledge in Iran, as explained by Browne after attending one its meetings: "I was very favorably impressed with the proceedings, which were, from first to last, characterized by order, courtesy, and scientific method; and from the enlightened efforts of this center of medical knowledge I confidently anticipate considerable sanitary and hygienic reforms in Persia. Already in the capital these efforts have produced a marked effect, and there, as well as to a lesser extent in the provinces, the old Galenic system has begun to give place to the modern theory and practice of medicine."[155]

Fig. 1.7. Iranian Sanitary Council's members in the 1880s. (*Left to right*) Mirza 'Ali Ra'is al-Atibba; Joseph Dickson, British legation physician; Mirza 'Abdullah Tabib; Joseph Désiré Tholozan; 'Ali Asghar Mu'adab al-Dawla Nafisi; Dr. Cherebinin, Russian legation physician; and Mirza Kazim Shimi. Middle East Centre Archive, St Antony's College, Oxford. GB165-0514 Tholozan, Album 3 no. 1.

Conclusion

The Sanitary Council's establishment as a formal advisory body to the government rested on decades of incremental changes in medical and public health paradigms in Iran, instigated by its military setbacks and the first outbreaks of Asiatic cholera. These new sanitary theories and practices, propagated by medical advisers and the proliferation of printed scientific literature from the West, transformed Iranian perspectives on disease prevention from one of passive fatalism to a proactive enterprise. The inauguration of the Polytechnic College played a crucial role in this process. Its faculty conveyed Europe's latest medical theories, helping to shape a new corps of physicians who could serve as sanitary experts to the government. The rise of Iranian newspapers in the second half of the century similarly communicated advances against cholera and Europe's sanitary revolution to a wider literate public. Increasingly, Iran's elite saw Western concepts of medicine and public health as being ascendant.[156] The sanitary conferences in Istanbul and Vienna also ensconced Iran in the major sanitary debates of the era, pushing the Iranian government to recognize its role in promoting the salubrity of its citizens and the need for the Sanitary Council.

The central government's growing involvement in public health was also aided by intervals of political and administrative stability in the last quarter of the nineteenth century. Tehran improved executive control over its provinces through the communication and transportation advances of the industrial era. The telegraph, in particular, permitted the government to stay abreast of emerging epidemics and enforce its commands in real-time during emergencies. Moreover, the combination of rapid communication and dissemination of print matter initiated the demise of theoretical heterogeneity in Iranian notions of health and hygiene, allowing a more uniform and centralized approach to stopping cholera. Ironically, the era's technological advances in transportation also increased cholera's frequency across the globe, challenging Iran's nascent public health infrastructure in the coming decades.

CHAPTER TWO

The 1889–1893 Cholera Epidemics

Iran's growing integration into the steam-driven global marketplace in the last quarter of the nineteenth century increased its vulnerability to pandemic waves of cholera. The completion of the Suez Canal in 1869 reduced the cost of shipping from Europe to southern Iranian ports and expanded maritime trade traffic in the Persian Gulf. In 1888 Nasir al-Din Shah opened the Karun River, Iran's largest and only navigable waterway, to foreign commercial fleets in response to the growing international trade interests in the region.[1] The river stretched from Iran's northwestern Zagros Mountains to the southwestern banks of the Persian Gulf, linking the port of Muhammara in the south to the towns of Ahvaz and Shushtar more than 250 miles inland. British steamships almost immediately began operating on the waterway, increasing the speed and volume of trade between Iran and Asiatic cholera's point of origin in India.[2] Similarly, Russia expanded its maritime trade with Iran's Caspian Sea ports and established extensive railway lines in the Caucasus and Central Asia, along Iran's rugged northern frontier.[3]

These new transportation networks created alternative pilgrimage and trade routes from India across northern Iran to the West.[4] Iran's natural barriers in the north could no longer slow or halt the transmission of cholera pandemics from India as the British Local Government Board had feared in its 1872 report. The growing traffic and the increased velocity of steam-powered ships and trains all but ensured the recurrence and dissemination of the dreaded disease. Iran's central government could not prevent outbreaks from reoccurring in its territory despite the Sanitary Council's precautionary recommendations in preceding years. Its fiscal and administrative deficits impeded any meaningful measures to improve the sanitary infrastructure of its cities and the protection of its frontiers. Lacking potable water and sewage disposal systems, its urban centers became breeding grounds for the disease and multipliers for its casualties.

The 1889 Cholera Epidemic

Sometime in June 1889, two steamers docked at Manameh, the principal port of the island of Bahrain in the Persian Gulf. One of the steamers disembarked a passenger sick with vomiting and diarrhea, and the second unloaded the body of a passenger who had died during the crossing. These ships were part of a larger weekly steamer line that began little more than a year earlier between Bombay and the Persian Gulf port of Basra near the Iranian border.[5] On June 30 a regimental water carrier for the Ottoman garrison in Basra died after a day of violent watery diarrhea and vomiting. Five days later, the garrison's military physician fell ill with abdominal pain, profuse vomiting, diarrhea, cramping in the limbs, thirst, and an undetectable pulse. He died at Basra's Naval Hospital that night after his extremities became ice-cold and his fingernails turned violet, characteristic signs of circulatory shock. Though the symptoms of both victims carried the hallmarks of cholera, the commanding Ottoman naval officer concealed the disease from his superiors to avoid reprimand for his failure to prevent the outbreak. The city's governor officially acknowledged cholera's presence a month later when a young woman from a seminomadic Arab tribe rapidly succumbed to a succession of diarrhea and vomiting. Cholera in Basra multiplied following her death, causing thousands of casualties including the city's British consul and his two children in the ensuing weeks.[6]

In response, the Iranian and Ottoman governments immediately instructed their postal authorities to fumigate parcels from diseased districts and placed a moratorium on passage from Basra.[7] However, a British steamer defied the order, reaching the port of Bushehr with more than ten Iranian passengers on board who were anxious to escape the outbreak. The ship received a clean bill of health from Iran's port authorities despite two cholera-related deaths onboard during the voyage. Soon after the passengers disembarked, cases of cholera began appearing in Bushehr.[8] Mirza 'Abd al-Riza, the port's government-appointed health officer, rapidly established a strict quarantine on all Bushehr bound vessels at Kharg Island, thirty miles off the coast.[9] However, another steamer from Basra succeeded in circumventing Iran's limited maritime patrols and landed more cholera-infected passengers, worsening Bushehr's outbreak.[10]

The growing cholera epidemic in the Persian Gulf added a sense of urgency to the ceremonial groundbreaking of the American Hospital in Tehran, built with funds raised by the Women's Presbyterian Board in Chicago.[11] The United

States Envoy to Iran believed that it was only a matter of time before the outbreak would reach the Iranian heartland from its foci in Baghdad and Bushehr: "In my opinion, if the disease is introduced in the interior [of Iran] it will be through the medium of the pilgrims returning from the holy shrines of Mecca, Medina and Karbala whose route lies through the infected districts about Baghdad. It is for that reason that when consulted on the subject by His Excellency Amin al-Mulk, acting prime minister, I strongly recommended the establishing of a sanitary cordon along the whole southwestern frontier as well as quarantining against Bushehr."[12] His recommendation, though premonitory, came too late. By late August, cholera spread north of the Mesopotamian valley with Iranian pilgrims traveling back from the venerated Shi'ite Muslim shrines of Karbala and Najaf. The epidemic broke out in Khanaqin, an Ottoman frontier transit point for travelers to Iran, and from there spread to neighboring Iranian towns of Sar-i Pul and Kirind, reaching Qasr-e Shirin farther east by the last week of September (fig. 2.1).[13]

The cholera epidemic also made its way into Iran through the port town of Muhammara on the opposite side of the Shatt al-Arab, facing Basra. Muhammara's poor sanitary condition, illustrated by the excrement and garbage that covered its shoreline, made it particularly vulnerable, and by September the city was reporting as many as thirty casualties to the disease each day.[14] From there, cholera rapidly spread to the towns of Ahvaz, Dezful, and Khorramabad, following the northern-bound escape route of panic-stricken residents of the infected Persian Gulf port cities.[15]

Tehran's sluggish response to the epidemic prompted the British embassy's physician to assemble an "informal and private" meeting of European doctors to strategize on ways to halt the expanding outbreak without first informing the Iranian government.[16] The gathering revived the Vienna Sanitary Conference's suggestion to establish an international sanitary council, composed of delegates from "foreign powers," to coordinate the response against cholera in Iran. The proposed council would include only three native delegates, despite the expectation that Iranians alone would furnish "sufficient number of doctors and dispensers" for the group to carry out its mandate.[17] Isma'il Amin al-Mulk, the regent while the shah and the premier were out of the country at the time, understandably ignored the European physicians' advice when it was brought to him, calling instead for the existing Iranian Sanitary Council to reassemble.[18]

The Iranian Sanitary Council, composed of its many native members, was presided over by Ja'far Quli Khan Nayyir al-Mulk, who invited the members

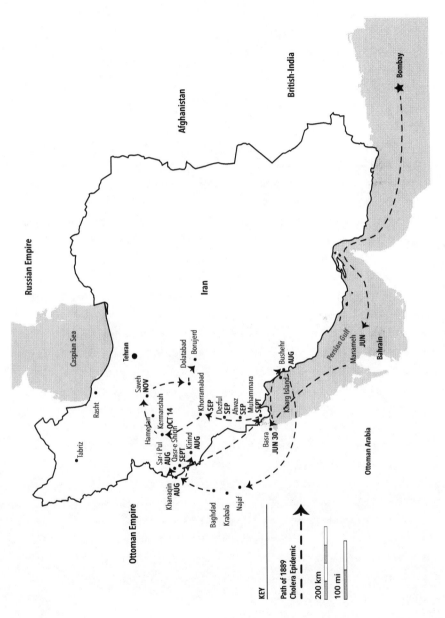

Fig. 2.1. The 1889 cholera epidemic.

of the ad hoc European physicians' committee to attend as observers.[19] The council advised the government to initiate a strict fifteen-day quarantine at Bushehr for those arriving by ships sailing from infected ports and a one-day quarantine for arrivals by land from Muhammara, Shushtar, and the vicinity. It also recommended a ten-day quarantine at Khanaqin and a five-day quarantine at the further inland point of Kermanshah to prevent the extension of the epidemic via pilgrims returning from the shrines at Karbala and Najaf. Mounted patrols would enforce the overland restrictions by turning back anyone without a clean bill of health indicating that they had undergone quarantine.[20]

The Sanitary Council rapidly transmitted its instructions by telegraph to provincial officials. The new quarantine arrangements in Bushehr began on September 13, 1889, owing to an unprecedented level of local collaboration between Sa'ad al-Mulk, the region's governor, and the British-India Company and Marine.[21] Iranian and British authorities also drafted a memorandum of understanding to prevent cholera's introduction by ships arriving from India. Notably, a British gunboat would enforce quarantine measures as outlined by the Iranians, despite a history of British distaste for sanitary restrictions that could affect commerce in the Persian Gulf.[22] Sanitary interventions in other parts of Iran were not as effective. The government lacked the material resources to cope with a disaster of this magnitude, forcing it to either cut back services or address deficits as they emerged. When pilgrims arrived in Tehran in the last week of September, their confinement lasted only two days, rather than the fifteen days advised by the Sanitary Council, due to the lack of tents and sanitary personnel.[23] Similar deficiencies in Kermanshah forced the Iranian government to send Joseph Albu, who had replaced Johannes Schlimmer as the army's chief medical officer, to personally supervise and remedy shortfalls in the city's quarantine measures.[24]

The presence of a countrywide telegraph network allowed the Sanitary Council and the government to receive daily updates on the epidemic's progress. Regular dispatches wired by informants also enabled the regent, Amin al-Mulk, to keep a tight rein on provincial administrators during the shah's absence, holding them responsible for sanitary shortfalls, as his cable to the deputy governor of Kermanshah illustrates: "You say all streets are kept clean, I approve of the quarantine measures you have taken, keep at work in the same way, don't relax your efforts but I must tell you there is a wide difference between your reports and others which have reached me through Europeans. Quarantine is less than you say and you have not provided all things necessary."[25] Despite his best intentions, the lack of resources in Tehran and the reluctance of local leaders in the provinces to establish and maintain quarantine camps and preventive services constrained the regent's ability to effect sanitary improvements in Kermanshah and other districts in the country. The Qajar system of awarding land and political position in the provinces in exchange for cash payments to the crown (*tuyul*) disincentivized regional leaders from spending the needed local funds to implement Tehran's writ.[26]

Without a professional bureaucracy capable of administering the provinces directly, the central government often gave land grants and administrative authority in a particular area to a private individual in return for cash payments. These high officials recuperated their investment through levies on local populations in their jurisdiction and maximized their profits by limiting public expenditures.[27] Likewise, they established and operated quarantine stations on a shoestring budget, making them inferior to their European and British-Indian counterparts at the time.[28] Because regional officials reduced their costs by not purchasing the required amount of rations and shelters, the quarantined in Iran often experienced starvation and exposure. The facility in Kermanshah, for example, had only seventy to eighty standing barracks made of wood or reeds, each able to hold approximately twenty persons. Such accommodations were clearly inadequate for the four thousand pilgrims traveling through the region at the height of the epidemic, even with the addition of fifty or sixty smaller tents to meet the shortfall.[29] Making matters worse, the population at Kermanshah's quarantine facility received a shower of corrosive carbolic acid, ordered as an additional measure to disinfect people, belongings, and animals in the overcrowded camp.[30]

Cholera Spreads and Burns Out

In September, several Jewish families from Baghdad began arriving in Kermanshah to wait out the outbreak in their cholera-stricken city with relatives in Iran.[31] These refugees conceivably carried cholera across sanitary cordons and quarantine stations following a network of Jewish safe houses and reliable trade routes.[32] By September 16, the epidemic had spread to most of the towns between Kermanshah and Qasr-e Shirin.[33] While the disease continued to exact heavy casualties on its northeastern path, the death toll on its flanks began to abate. By September 20, 1889, cholera was no longer present in Muhammara and Bushehr, even though Basra across the border continued to report thirty to forty victims daily.[34]

The epidemic eventually breached Kermanshah's quarantine defenses and overran the city of thirty thousand on October 14.[35] Cholera reached Hamedan the following day, making its first appearance in the house of the city's chief surgeon located in the Jewish quarter.[36] Jewish merchants were the likely vectors, once again, given that the disease bypassed a neighboring "unhealthy small town" that did not have a sizable Jewish population.[37] The outbreak spread to the Presbyterian Mission's school in Hamedan, most probably transmitted by one of its Jewish pupils.[38] Despite precautionary measures to protect the children from cholera, including boiling water before use and isolating the children from the rest of the town, the sick multiplied.[39] Cholera in Hamedan disproportionately killed the very young and the old, many in less than twenty-four hours, whereas the rest of the population had milder symptoms that responded well to treatment.[40] As a result, overall casualties did not exceed twenty-five deaths per day in the city of fifty thousand.[41]

The epidemic spread further north by November, breaking out in the town of Saveh, about seventy-five miles from Tehran.[42] Its proximity to Iran's Caspian seaports prompted the Russians to establish quarantines at the cities of Baku and Julfa, which were the main destination points for migrant Iranian laborers and merchants. However, the contagion outflanked the Russian quarantines and broke out in Tbilisi as it spread north following caravan routes from Saveh through Kurdistan. It also followed the southern caravan routes, spreading to Dolatabad and Borujerd.[43]

Cholera began to wane throughout Iran in the first months of 1890 as Nasir al-Din Shah returned to Tehran from his sojourn in Europe with his new French physician-in-chief, Jean-Baptiste Feuvrier.[44] On February 3, the Iranian government announced the cessation of the epidemic in its territory, but the Russians, ever skeptical of Tehran's optimistic assessment, waited another two months before giving the Iranians a clean bill of health.[45] The epidemic never reached the capital and mainly affected the southern and eastern portions of the country covering six thousand miles of territory inhabited by 400,000 people (table 2.1).[46]

Cholera, Revolt, and Achieving the Virtues of Civilization

Iranian intellectuals viewed the 1889 cholera outbreak as a symptom of broader shortcomings in governance and values, blaming the ruling establishment for allowing a "frightful" disease to overtake the supposed "well-guarded domains" of their country. They also lamented the inherent weakness of Iranian national character for the widespread indifference to hygiene.[47] These

TABLE 2.1.
Mortality by region during the 1889 cholera epidemic

Region	Duration	Deaths recorded
Kermanshah	September–December	1,195
Kurdistan	December	53
Hamedan	November–January	465
Nahavand	November–January	675
Borujerd	November–December	281
Khorramabad	October–December	421
Malayer	November–December	167
Tuyserkan	November–December	295
Qom	December–January	213

views, articulated in the editorials of Iran's first nongovernmental weekly, *Akhtar*, called for sanitation to figure prominently in necessary structural and administrative reforms: "We hope that following the auspicious entrance of His Imperial Majesty's retinue [Nasir al-Din Shah] into the splendor of the royal resting place, that he includes amongst the many reforms, which is of concern to His Imperial Majesty, an ordered sanitary infrastructure, that it be assembled in Tehran and that in view of this [proposal] being accepted it will also be included in His Majesty's exclusive endeavors for the introduction of the good [hygienic] practices of this era's civilization."[48] The newspaper used examples of successful sanitary interventions in neighboring Ottoman provinces as a pointed tool to criticize Iran's leadership.[49] It lionized the "honorable" and "glorious" Ottoman governor Asim Pasha for braving cholera in Baghdad while others ran away and described him as an "assuager of the frightened hearts" for assigning physicians, organizing treatments, cleaning his city, and even nursing those who were ill.[50] Exaggerated or not, this portrayal continued the "mirrors for princes" genre in Iranian literature, which offered guidance on good leadership to monarchs and other high-ranking officials through laudatory examples.

The importance of quarantines and sanitary councils as a civilizing tool was an essential theme within the larger debate over the need to establish an ordered and accountable system of governance in the country. Intellectuals used the term "civilization" (*tamaddun*) to describe the advances in Western sanitary science and, by association, to malign the authorities, implying that Iran's lagging sanitary state was a sign of its uncivilized administration. At the same time, gaining the "fruits of civilization" became a utopian aspiration through which intellectuals in Iran sought to establish a more just society wherein the welfare of its citizens would be paramount.[51] What emerged from this discourse was the definition of an ideal civilization seen through a Western lens. This perspective stood at odds with Iran's religious leadership, which increasingly viewed the West as a threat to the established social order and as the cause of epidemic outbreaks.

In Europe, cholera was associated with political upheavals throughout the nineteenth century.[52] Its class-specific impact on the West's burgeoning industrial population meant that the poor bore the brunt of the disease's casualties. Rising discontent and fear among Europe's underprivileged during outbreaks often resulted in urban riots and insurrections.[53] The Iranian experience with the 1889 cholera epidemic, on the other hand, was class-blind and revealed the singular role of Shi'ite Islam in directing the currents of social unrest that emerged in its wake.[54]

Several features made Iran ready for a clerically led uprising in the context of the 1889 epidemic. First, the latter half of the nineteenth century witnessed the advent of a senior religious jurist (*mujtahid*) with unprecedented prestige and authority in the person of Mirza Muhammad Hasan Shirazi. Acknowledged as a source of emulation (*marja'-i taqlid*) by most Shi'ite religious scholars (*'ulama*), Shirazi surpassed other Iranian clerical leaders of his era in establishing a strong network of disciples and powerful links with the mercantile community.[55] In addition, the growing national telegraph network allowed Shirazi and other high-ranking clergymen to communicate and coordinate their activities rapidly, giving them unparalleled solidarity of action.

European encroachment into the region's commercial sphere was another feature of this period that made Iran susceptible to unrest. The Iranian domestic economy became increasingly vulnerable to global economic currents and fluctuations as its debt load multiplied and its financial institutions fell under growing foreign control. For example, before the inception of the British-owned Imperial Bank of Persia in the late 1880s, native bazaar merchants had a significant role in the lucrative trade of providing loans and circulating currency.[56] With the advent of European-operated banking in Iran, many of these powerful merchants lost their livelihoods. This concession was one of several given to Western powers in the brief span of time between 1888 and 1891.[57] Not surprisingly, clerics were able to direct the wrath of their congregations toward Western economic interests during periods of national crisis. Shi'ite clerics also focused their anger at merchants from minority religious communities, such as Armenian Christian Iranians and Jewish Iranians, whom they regarded as agents of European economic dominion in Iran. These merchants regularly obtained the formal protection of foreign legations, effectively excluding them from the capitulatory tariffs imposed on their Muslim counterparts. This commercial advantage fueled resentments, which culminated in religious minorities suffering the brunt of the popular wrath both during and in the immediate aftermath of cholera outbreaks.

Finally, the clerical leadership itself had a personal financial stake in the beleaguered traditional Iranian mercantile sphere or the bazaar; this interest together with income from religious endowments represented up to forty percent of Iran's budget.[58] For example, Shaykh Muhammad Taqi Aqa Najafi was both a powerful religious jurist (*mujtahid*) and the patriarch of one of Isfahan's most important business families.[59] The Najafis' wealth and fiscal influence had grown to such an extent that they often clashed with the prince-governor, Mas'ud Mirza Zill al-Sultan, over economic and legal control of Isfahan. The primary targets of Aqa Najafi's venomous sermons were the very elements in society that presented the greatest threat to his purse and, by extension, his temporal authority. This group included government representatives, personified by Zill al-Sultan; European mercantile interests; and minorities.[60]

The cholera epidemic worked in favor of Aqa Najafi and other clerics in their campaign against domestic and foreign opponents. It provided them with ammunition for their rhetorical weapon of choice: accusing their enemies of being ritually impure (*najis*) and a source of disease. They exercised strict control over contact and trade with non-Muslims by blaming the outbreak on the desecration of Iran's faithful by "unclean infidels" and restricting or wholly forbidding their congregations from interacting with Westerners and minorities. They also easily mobilized their followers to punish these groups for their alleged transgressions. Iranians, traumatized by cholera, were more likely to comply with these religious edicts, fearing that any deviation could bring divine retribution in the form of a returning epidemic.[61]

The first of these convulsions involved riots against the Jewish communities in the cities of Isfahan and Shiraz in 1889. Aqa Najafi instigated the violence by calling for the application of a series of humiliating restrictions on Iranian Jews, preaching that they caused the epidemic by corrupting Muslims and inviting divine retribution. At their worst, his fiery sermons led to scenes of violent rampage and the murder of Jews and other minorities in Isfahan. Cholera's pattern of transmission, following the path of Jewish refugees from Baghdad into southern Iran, lent more credence to Najafi's sermons.[62] The second upheaval, famously known as the Tobacco Protest, occurred in the aftermath of the cholera epidemic. The first sign of discontent appeared on January 1891,

Fig. 2.2. Shi'ite Muslim clerics leading protestors to the royal palace. *Graphic*, February 20, 1892.

when Nasir al-Din Shah received an anonymous letter lambasting him for granting the monopoly on the manufacture and sale of tobacco in Iran to an English company. Six months later, dissident Iranian newspapers, immune from royal censorship, openly condemned the terms of the concession.[63] The previous year's cholera epidemic played an important role in heightening xenophobia among the Iranian populace. As early as February 1890, rumors of an anti-British riot in Tehran were rampant.[64] By the spring of 1891, the rumors materialized in the form of massive strikes and violent demonstrations in major Iranian cities. The protestors demanded the repeal of the concession and the expulsion of some British tobacco company employees, whom they accused of propagating un-Islamic practices and beliefs.[65]

More than anything else, the general discontent with living conditions in the aftermath of the 1889 cholera epidemic helped clerics mobilize Iranians en masse. For this reason, the initial accusations against Nasir al-Din Shah leading up to the Tobacco Protest focused not so much on the concession itself but on the pitiful state of the country following the outbreak.[66] Secular dissidents used Western norms of public service, including interventions against cholera, as a yardstick to evaluate and criticize the government.[67] They called on the shah to reform Iran's sanitation, hygiene, and disease prevention efforts by implementing urban infrastructure projects and national vaccination campaigns along European standards.[68] While on the surface this clashed with the anti-Western agenda of religious activists, Iran's weakening economic position vis-à-vis the West had forced some clerics to think critically about Iran's future and join hands with secular intellectuals who called for these reforms.[69] The tobacco concession was the spark to the powder keg of discontent with the government and provided a unifying framework that brought these two groups together in protest (fig. 2.2).

Influenza, Cholera, and the Rise of the Germ Theory of Disease

One of history's worst global outbreaks of influenza, which began in Siberia, reached Iran in 1890 on the heels of the receding cholera epidemic and growing social tensions in the country.[70] Iranians perceived pandemic influenza as an unfamiliar illness (*nakhushi-yi gharib*) and sardonically referred to it as "another [dreaded] gift from Russia."[71] The suddenness and severity of the new outbreak even baffled younger physicians trained in Western medicine, who often incorrectly diagnosed and treated the disease as the common cold.[72] In Tehran alone, fatalities attributed to complications of influenza numbered between fifty and seventy daily.[73] The outbreak was especially deadly among the young, killing more than six thousand children across the country.[74] In an attempt to make sense of the rapid diffusion and lethality of the new illness, Iranian publications began to print and translate articles from the West linking epidemic diseases to microbes in terms that lay people could understand. These disease producing "microscopic creatures" (*mujudat-i zara bini*) were described as "a multiplicity of white and transparent worms" that infected people through ingestion or inhalation. Treatment recommendations, which included safari-like fumigations with naphthalene, camphor, and tobacco smoke, reinforced the microbes' zoomorphic depictions.[75]

Although the Sanitary Council had deliberated on Robert Koch's microbial theory of cholera as early as 1884, few outside of Tehran's progressive medical circles framed the etiology of epidemic diseases in the context of microorganisms until the latest outbreaks of cholera and influenza.[76] This was a testament to the longevity of traditional views of illness in Iranian society.[77] As late as December 1889, the weekly *Akhtar* wrote on the work of a prominent German physician who asserted that unhealthy air from confined and poorly ventilated quarters played an important role in causing cholera, as evidenced by the preponderance of cholera-like symptoms in burrowing animals.[78] Another article several years later in the widely circulated Iranian government newspaper *Ittila'* reported the role of demons (*jinn*) in cholera outbreaks.[79] This persistence of religious and humoral views of diseases and epidemics in Iran added to the challenge of developing effective measures to prevent a recurrence of cholera, despite the growing outcry for sanitary improvements from the country's reformists.

The 1892–1893 Cholera Outbreak

Although Iran's cholera epidemic ceased in 1890, the contagion continued in the Arabian Peninsula and the Red Sea region. Mecca saw a dramatic upsurge in cholera related deaths by mid-July 1891.[80] Alarmed, European and American newspapers reported that cholera would not remain contained in the Middle East and that the Italian military, garrisoned in several Red Sea port towns, would transmit the disease to the West.[81] They similarly predicted the inevitability of cholera's diffusion to Europe and North America via the growing traffic of Muslim pilgrims through the Suez Canal, which also served as the main avenue of Western trade with the Indian Ocean. This apprehension exposed a mounting prejudice against Muslims, increasingly portrayed as the main culprits in the global dissemination of Asiatic cholera. The front page of the *New York Times* in 1891 described Mecca as a place of "fervid fanaticism" and a "vast breeding tank of disease, from which samples are sent to every country in the world."[82]

The expected cholera pandemic did not start in Mecca, as predicted in the Western press, but broke out instead among Hindu pilgrims in the town of Haridwar located in the Ganges Delta region of India in 1891. The disease then followed the caravan routes from Kashmir into eastern Afghanistan, reaching Jalalabad and Kabul by 1892. The western Afghan city of Herat, with its strong commercial and religious ties to Iran, reported its first victim on February 22 of that year. Its casualties reached three thousand, almost six percent of the city's population, two months later.[83] Panic-stricken refugees from Herat disseminated the disease in two directions: westward to the towns situated on the main avenue to the Iranian frontier and directly northward along the Imperial Russian border districts of Turkistan. Cholera also broke out among the nomadic Afshar tribe whose migrations between infected regions in Afghanistan and Iran's eastern province of Khorasan introduced the pandemic into Iranian territory. The first Iranian fatality occurred in Torbat-i Jam, near Iran's

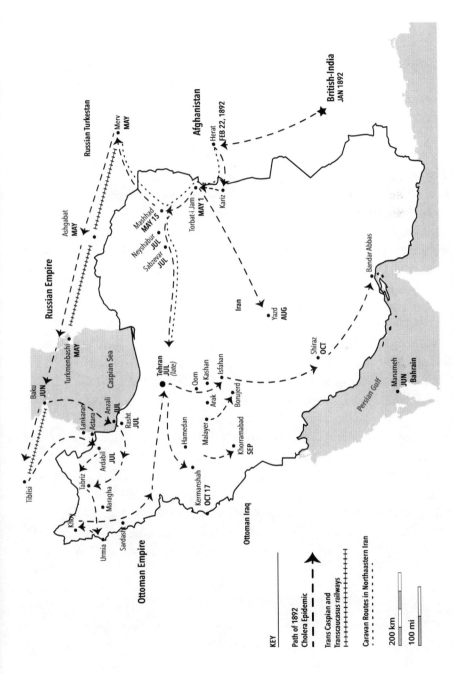

Fig. 2.3. The 1892 cholera epidemic.

border with Afghanistan, where the outbreak lasted from May 1 to the beginning of July. From this epicenter, successive waves of cholera spread north and south following Iran's main caravan routes (fig. 2.3).[84]

The Iranian government reassembled the Sanitary Council once news of the outbreak in Afghanistan reached Tehran, but its officials, fearing the shah's reprimand, kept the ruler in the dark on the westward progress of the pandemic and the fact that the contagion had crossed Iran's frontiers. Meanwhile, Nasir al-Din Shah left the capital on a long tour of the countryside in May 1892. His twenty-thousand-man caravan took a southeasterly path, bearing toward the incoming epidemic from the west, unaware of the danger.[85] The shah's absence from the capital deprived the Iranian government of the executive royal sanction, which it crucially needed, to implement the Sanitary Council's recommendations. Making matters worse, a nominal regent could not assume his duties, as Amin al-Mulk had done during the 1889 epidemic when the shah was in Europe, because his extended sabbatical was within Iran itself. And if the shah's absence was not enough, the ongoing Tobacco Protest had placed Iran's government in a state of administrative paralysis that even an approaching cholera epidemic could not mitigate.

Not until April 1892 did the government's quandaries with the tobacco concession to Great Britain ostensibly reach an end. Nasir al-Din Shah yielded to clerical demands, canceled the concession, and paid a large indemnity to the British for breeching contract. The fiasco weakened his premier, 'Ali Asghar Khan Amin al-Sultan, whose main preoccupation henceforth was to regain his favored position with the shah rather than the effective administration of the country.[86] The shah, hoping to reconcile a defiant clergy with his government, appointed Amin al-Sultan as chief administrator of the Fatima Masuma Shrine in Qom, the second holiest Shi'ite site in Iran. This made Amin al-Sultan vulnerable to even more clerical criticism if he imposed a ban on Shi'ite pilgrims coming from infected regions in Afghanistan and India, and so he did little to stop cholera's ingress. Unhindered, the epidemic followed Shi'ite pilgrims bound for the Imam Riza Shrine in Mashhad across Iran's frontier.

Cholera in Mashhad

The residents of Mashhad braced for disaster as the epidemic crept closer, and alarming news of mounting fatalities in the villages and towns lying on the caravan highway from Torbat-i Jam circulated.[87] By the time cholera reached Mashhad on May 15, it was a deserted city. Most of its terrified inhabitants had left in search of healthier refuge in neighboring mountains with the epidemic in close pursuit.[88] Those on the staff of the British consulate in Mashhad experienced this firsthand when they lost several members to cholera, despite taking precautions against the disease and encamping ten miles from the city. They later discovered that their campsite, located next to a bucolic stream, acquired the infection from an upstream village that used the water to wash its dead before burial.[89] Meanwhile, Mashhad's unsanitary conditions provided a nurturing environment for the cholera bacillus to proliferate. An open water canal (*jub*) that bisected its main boulevard simultaneously served as a public drinking fountain, as well as a place for pre-prayer ablution, washing clothes, and the disposal of animal matter and sewage.[90] Additionally, as a holy city, its numerous and massive cemeteries made it the country's graveyard capital; a place where the dead and the living crowded together and where the pungent aroma from shallow graves and rotting carcasses permeated the atmosphere.[91]

Mashhad's rapid depopulation initially held the number of cholera victims in check, with only fifty cases reported in the first twelve days of the outbreak; but the epidemic resumed its full fury, killing thousands on a weekly basis in the ensuing months, as the city's complacent inhabitants began returning from the mountains and the countryside.[92] Mashhad's hallowed status, which made it a sought-after final resting place, made matters worse. As soon as cholera showed signs of receding, cadavers brought into Mashhad for burial near the Imam Riza Shrine would reintroduce the disease from the surrounding areas (fig. 2.4).[93] These factors allowed the epidemic to grow and eventually invade the whole province of Khorasan by mid-July, including the neighboring cities of Neyshabur and Sabzevar on the Mashhad-Tehran caravan route.[94]

Cholera in Gilan and Azerbaijan

While cholera raced west toward Tehran, another offshoot of the epidemic from Mashhad made its way north across Iran's border with Russia's Central Asian territories. In May 1892 infected Iranian passengers on the recently inaugurated Trans-Caspian Railway nearly simultaneously introduced the disease into Merv, Ashgabat, and several other cities dotting the vast expanse of Russian Turkistan. Instead of stopping internal rail travel, the Russian government in St. Petersburg erroneously ordered all dealings with Iran's northern seaports to cease, allowing the epidemic to reach the railway terminus of Turkmenbashi (Uzun-Ada) on the eastern shore of the Caspian Sea in record time.[95] In June 1892 steamship lines from Turkmenbashi carried the epidemic

Fig. 2.4. Mashhad's sprawling burial ground in the Qajar era. National Geographic, *A view of the Tombs of Mashhad a holy site in the country,* circa 1900, picture id: 600376.

across the body of water into Baku, prompting the Russians to expel the port city's large population of migrant Iranian workers on the misplaced notion that they were responsible for introducing and disseminating cholera.[96] This action unwittingly helped spread the epidemic westward following the recently finished Transcaucasus Railway to Tbilisi and southward following reopened steamer routes to the Russian Caspian seaports of Astara and Lankaran, bordering Iran.[97]

On July 2, 1892, a steamer left Baku for the Iranian port of Anzali carrying infected Iranian workers, nine of whom died and were thrown overboard during the daylong crossing.[98] The ship introduced cholera into Anzali and the neighboring city of Rasht, the capital of Iran's Caspian province of Gilan. Women in both cities had significantly higher casualty rates than men due to their larger presence in the province's rice farming workforce, which obliged them to work long hours in rice paddies, knee-deep in potentially infected water with their children in tow.[99]

From Gilan, the epidemic radiated to the neighboring Azerbaijan Province in the country's Northwest, reaching Tabriz, the provincial capital, and the cities of Maragha, Urmia, and Khoy. It also spread south, following caravan routes, to the town of Sardasht and onward to the Iranian province of Kurdistan.[100] Tabriz's governor, Crown Prince Muzaffar al-Din Mirza, and his courtiers hastily left the city for the perceived safety of a sparsely inhabited outlying village upon hearing that cholera had broken out in nearby Ardabil.[101] However, the prince denied the existence of an epidemic in the region, even after cholera broke out in Tabriz, to avert the deleterious economic impact of a panic-driven exodus from the city. Similarly, the Qa'im Maqam, who was the nominal ruler of the city in the absence of the crown prince, obstinately refused to acknowledge the epidemic until his granddaughter dramatically died of cholera within four hours of showing signs of the disease. Several weeks later, Muzaffar al-Din Mirza also gave in, after mounting casualties in the city made concealment impossible.[102]

The epidemic in Tabriz caused more than 5,572 cases of cholera with a fifty percent mortality rate in the first ten days of the outbreak, which began in late July.[103] By the time the epidemic abated several months later, about 26,000 of the city's 165,000 inhabitants had succumbed to cholera.[104] The severity of the outbreak led to a complete breakdown of public order that began with the flight of the city's leaders and worsened with the self-imposed sequestration of the remaining population as described in the following British diplomatic dispatch:

> I am informed by two reliable persons who remained in Tabriz during the epidemic that for many days the streets were entirely deserted except by those who were compelled by necessity to be out of doors and those engaged in burying the dead. The greatest difficulty possible was experienced in finding people to bury two of the Europeans and the Persian dead were frequently carried to the grave practically as they died and cast into a hole, being covered with a thin layer of earth. Some of the graveyards are in a very bad condition and the smell arising from them is frequently very offensive.[105]

Only the American Presbyterian missionaries seemed to hold the line against cholera, caring for the sick "under the awful circumstances of the battle with the epidemic" and saving hundreds in the process.[106] Basic sanitary services, law, and order completely collapsed with the attrition of the city's municipal workforce and leadership. Theft and looting were common, and clan rivalries, previously held in check, boiled over, resulting in gun battles across Tabriz. Muzaffar al-Din refused to return and resume his duties until he was convinced that the epidemic and lawlessness had completely subsided several months later. The proximity of the outbreak was not enough to motivate the lethargic crown prince, once he had returned, to cover Tabriz's shallow graveyards with the recommended layer of lime or earth to avert another epidemic.[107]

Cholera in Tehran

As cholera ravaged Tabriz, the second wave of the epidemic from Mashhad reached villages near Tehran, obliging the embattled Amin al-Sultan to overcome his reservation about quarantines and other preventive measures recommended by the Sanitary Council. The premier rapidly established inspection posts on the main roads leading into the capital to isolate the sick before they could enter the city.[108] However, plagued by inefficiency and corruption, the posts became places where "orders were issued by everybody, but obeyed by no one."[109] Travelers easily avoided scrutiny by bribing sanitary officials, who even in the best of circumstances quarantined only individuals with prominent physical signs of cholera, allowing those in the early stages of the disease to go on.[110] Making matters worse, Tehran's sanitary inadequacies had remained mostly unchanged since Nasir al-Din Shah's urban renewal efforts almost two decades earlier, prompting George Curzon, Britain's future foreign secretary, to note that the city "has yet much to do before it realizes the full aspiration of its royal Hausmann."[111] Despite the growth of Tehran's Office for Public Order in the 1880s, the municipality lacked enough sweepers and garbage collectors to maintain the Royal Citadel, much less the rest of the city.

The absence of a municipal abattoir in Tehran worsened the sanitary situation by obliging its households to slaughter their daily meal in the open, flooding the capital's streets and alleyways with the drained blood of ritually sacrificed animals. People discarded carcasses, inedible entrails, and other forms of garbage and sewage onto streets and unoccupied lots throughout the city. Since most of Tehran's roads were unpaved, they often carried "evidence of yearly accumulation of animal and vegetable matter in various stages of decomposition."[112] The population also abandoned the remains of larger beasts of burden such as camels, horses, and donkeys along roads and open spaces both within capital and outside its walls where "skeletons whitened the plain in every direction beyond the city limits."[113] The exposed carcasses and trash were the principal food supply of Tehran's scavenging wild dogs, whose howling and barking became such a nuisance that an English telegraph clerk attempted to poison one of the packs that frequented his neighborhood at night (fig. 2.5).[114]

Fig. 2.5. Scavenging wild dogs eating the remains of horses on Tehran's parade grounds (detail). Myron Bement Smith Collection: Antoin Sevruguin Photographs. Freer Gallery of Art and Arthur M. Sackler Gallery Archives. Smithsonian Institution, Washington, DC. Gift of Katherine Dennis Smith, 1973–1985, FSA A.42.12. GN.44.01.

Tehran's leadership did not take any steps to improve the capital's hygiene as a bulwark against cholera on the notion that the approaching epidemic would stop short of the city walls, as it had during the 1889 outbreak. Both officials in the government and the general population looked to God for protection. They mourned and self-flagellated with unmatched fervor during the Muharram observances, which in 1892 began in the fourth week of July. Processions carrying the traditional banner of Husayn ibn 'Ali, the venerated third Shi'ite Imam whose death at the Battle of Karbala in AD 680 represented the climax of the ceremonies, lined the streets of the capital. The deafening public displays of lamentation and religious exhortation gave the city an unusually somber atmosphere, even by the standards of Muharram. But the funerary mood did not dampen the population's appetite for overripe "rotten" fruits, washed and fertilized with potentially infected sewage water, which the government had refused to ban. This seeming insouciance continued until the last week of July when suddenly, in the words of a British observer, "the blow fell—the cholera was in Tehran."[115]

Fig. 2.6. Caravan of pilgrims to Mashhad. *Illustrated London News,* March 28, 1885.

The timetable of the epidemic's progress makes it very likely that Abul Qasim Khan Nasir al-Mulk, an Oxford-educated aristocrat, unintentionally carried the contagion into Tehran when his caravan made its way through the sanitary cordons around the capital without submitting to the required quarantines. On July 25, 1892, cholera broke out about seven miles from Tehran's walls at the golden-domed Shah Abdol Azim shrine.[116] Pilgrims often stopped at the shrine's mosque, situated on the Tehran-Mashhad thoroughfare, to give thanks for their safe homecoming before entering the capital. The outbreak at Shah Abdol Azim coincided with the arrival of a large caravan from Mashhad, probably that of Nasir al-Mulk, which had lost several members to cholera (fig. 2.6). Once the caravan reached its destination in Tehran's Sarcheshmeh area, relatives of the victims washed soiled clothing from the dead along the neighborhood's qanat, whose downstream tributaries supplied the Udlajan and Bazaar quarters (*mahala*) of the capital with drinking water. The first cases of the disease occurred in Sarcheshmeh, less than a day after the outbreak in Shah Abdol Azim, then rapidly disseminated downstream to the adjoining settlements of Sartakht, Pamenar, and Nasiriya (fig. 2.7).[117]

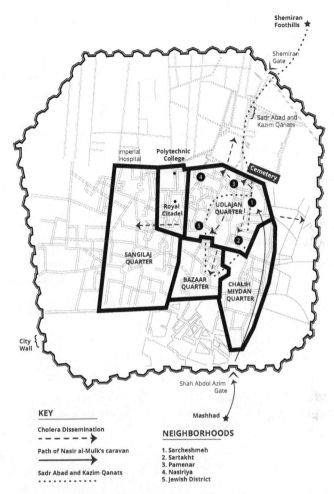

Fig. 2.7. The cholera epidemic in Tehran based on 'Abd al-Ghaffar Najm al- Mulk's 1892 survey of the city. Royal Geographical Society, London: Map Room Collection, mr Iran S.28, Najm al- Mulk, 'Abd al-Ghaffar. *Map of Tehran.* Scale ca. 1:4000. Tehran: Polytechnic College (Dar al-funun), April 1892.

The epidemic in Tehran owed its virulence and longevity to the densely populated makeup of its epicenter in Udlajan quarter. Situated in the heart of the capital's old citadel, Udlajan had 2,125 residential houses, 1,314 shops, 27 caravanserais, 35 public baths, 16 mosques, and 52 barns. It was also religiously diverse, being the only ward with a sizable Jewish population, which meant that the outbreak did not spare segregated minorities in the city. In contrast to the European urban makeup, Tehran's districts were heterogeneous and socioeconomically intertwined, facilitating the indiscriminate spread of cholera across class lines. By way of illustration, the house of Nasir al-Mulk, an aristocrat and high government official, was located next to the caravanserai and tavern of a merchant named Haji 'Ali Shal Furush. The initial outbreak in Nasir al-Mulk's household could have easily infected the large number of itinerants in the adjoining caravanserai building through their shared water source. The residence of the chief servant in charge of the royal palaces and grounds, Mashhadi Hasan Beg Farrash Bashi, was also located in Udlajan.[118] Once infected, this trusted servant may have been responsible for introducing the disease into the royal household where it killed one of Nasir al-Din Shah's wives during the epidemic. Udlajan, which meant a "place of water distribution" in the local vernacular, had been a major watershed in Tehran since the founding of the capital in the early Qajar era. This characteristic also lent itself to the rapid, citywide dissemination of a waterborne disease like cholera once a sufficient load of the microbe had contaminated the quarter's water mains.[119]

Two principal qanats carried drinking water into Udlajan from the Shemiran foothills of the Alborz mountain range, north of Tehran's boundary.[120] To minimize evaporation, they conveyed water through underground horizontal channels connected every twenty or thirty yards to vertical shafts to the surface (fig. 2.8).[121] Once the qanats entered the city limits, people often obtained water by moving stone slabs that covered the conduits that ran under streets and alleyways, making the water vulnerable to pollution. Sinkholes and seismic activity also made the qanats susceptible to microbial contamination by exposing segments of the underground channels to the surface.[122]

After reaching the Udlajan quarter, the qanats branched off among important streets, with tributaries either coming to the surface or flowing directly into one of thirty-three residential cisterns (*ab anbar*) located below the ground level of more affluent households.[123] These cisterns were susceptible to pollution by mud, garbage, and the occasional person who fell into the storage tank in the process of collecting water.[124] In poorer neighborhoods, the water traveled in surface channels (*jub*) that served the dual purpose of conveying drinking water and draining runoff.[125] This scheme of water distribution magnified cholera's impact in Tehran, as described in the following passage by an American resident of the city at the time: "In Tehran there is not a meter of sewer pipe and the open ditch often supplies the lower class

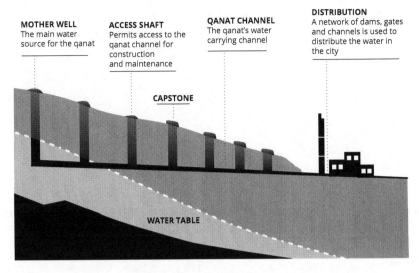

Fig. 2.8. Cross section of a qanat.

with drinking water. Three years ago, when cholera was carrying off from three to four thousand people a day in Tehran, we went out and removed a dozen dead and dying from these streams. The thirst of a cholera patient is terrible and these cool open streams acted as a magnet for the suffering."[126] People often washed laundry in exposed areas in the network of qanats and surface channels, contaminating drinking water with rice-water diarrhea from the sullied clothes and sheets of cholera patients.[127] City water was also susceptible to indirect pollution from adjacent cemeteries and other types of contaminants seeping through the porous soil.[128] Making matters worse, the Islamic ritual of cleansing the deceased before burial led many to wash cholera victims in streams that supplied the city's qanats, thereby infecting the municipal drinking water at its source.[129]

Tehran's system of water husbandry and use created the right conditions for the outbreak to spread from the eastern districts to the west of the city in a short period.[130] Cholera eventually overran the whole capital, causing large segments of its population to panic and seek refuge in the countryside:

One blind impulse seized alike upon rich and poor—flight! flight! All who possessed a field or two in the outlying villages, all who could shelter themselves under a thin canvas roof in the desert, gathered together their scanty possessions, and, with bare necessities of life in their hands, crowded out of northern gateways. The roads leading to the mountains were blocked by streams of fugitives, like an endless procession of Holy Families fleeing before a wrath more terrible than Herod: the women mounted on donkeys and holding their babes in front of them wrapped in the folds of their cloaks, the men hurrying on foot by their side.[131]

More than one-fifth of Tehran's population of approximately 200,000 sought refuge in the surrounding mountain villages.[132] Those who could not afford the journey pitched their tents outside the city perimeter, seeking the sanctuary of even a short distance from the unwholesome urban environment. But escape offered scant protection, as evidenced by the corpses that littered the roadsides leading away from Tehran. Large caravans carrying notables and their retinues were often more susceptible to cholera than their smaller, less privileged counterparts because of the increased opportunities for infection in the crowded convoys of wealthier citizens. Sick travelers received perfunctory medical treatment and were relegated to the back of wagons until they spontaneously remitted or, more commonly, died. Burials, for those who could afford them, had to wait until their transport reached what was perceived as a safe distance from Tehran. The stench of decaying bodies was a hallmark of these escaping convoys owing to the summer heat and the growing number of corpses onboard.[133]

Within the city walls, panic and desperation pervaded. People abandoned sick friends, masters dismissed servants exhibiting symptoms, and caretakers deserted wards on their deathbeds.[134] To many, the outbreak heralded the end times as described in the following passage by an Iranian physician: "The fear of death was such that fathers forgot their sons and mothers no longer worried for their children. Families were torn apart, and the last booming cry of the oppressed signaled Judgment Day."[135] Even the Westerners living in Tehran, who were typically collected and pragmatic on matters of disease, fell prey to the pervasive hysteria. This included the French director of the newly established tramway corporation who ordered the company's physician to escape into the mountains with him, abandoning his employees to their fate during the outbreak. The epidemic even terrorized the chief British physician of the telegraph company, prompting one observer to speculate that "he may persuade himself into cholera unless he can be got to think himself quite well."[136] On the other hand, several enterprising native and Western "rascals" exploited the situation by selling quack remedies for cholera at exorbitant prices, making a significant profit from the collective state of terror in the capital.[137]

The prince regent, Kamran Mirza Na'ib al-Saltana, who administered Tehran while the shah toured the country, abandoned the city like most other notables. In his place, the minister of police and associate governor, Mirza Isa, was left in charge—only to become one of the epidemic's early victims.[138] With the government in disarray and the shah out of the capital, rumors abounded that his eldest son, Mas'ud Mirza Zill al-Sultan, had left his seat in Isfahan with a sizable military force to seize the throne in Tehran. In reality, Zill al-Sultan was also running from the looming cholera epidemic in his city and did not intend to lead a coup.[139]

The breakdown of Tehran's service infrastructure further complicated the plight of its inhabitants. Essential municipal workers either left the capital or sequestered themselves. Even the traditional corpse washers and gravediggers who remained alive refused to work.[140] Fearing contamination, the city's three-hundred-strong police force disobeyed the police chief's order to remove the dead from the city's hospitals and public spaces. This prompted the American vice-consul to drop off bodies from the American Hospital at the doorsteps of the police headquarters.[141] At the height of the epidemic, if someone succumbed to cholera, there was no one to "take him, put him on a stretcher, and bring him someplace to be buried."[142] As a result, corpses littered city's streets where they became food for the scavenging wild dogs along with the rest of the carrion and garbage.[143] Making matters worse, merchants declined to enter Tehran, resulting in severe shortages and increasing food prices. Rumors of impending riots and looting were rampant, and while the anticipated disturbances never materialized, insecurity and criminality were widespread.[144] Highwaymen regularly held up the postal stagecoach in proximity of the capital, and more than thirty-six inmates escaped from Tehran's prison when most of its guards left the city at the peak of the epidemic.[145] Burglars took advantage of the insecurity and the population's flight by robbing the capital's unoccupied properties, including the home of the shah's French physician-in-chief, Jean-Baptiste Feuvrier, while he attended to the monarch in the countryside. After his property's caretaker died of cholera, thieves broke in and stole anything they could remove, even attempting to take a metal truss solidly attached to the ceiling of his living room.[146]

While Tehran's leadership waited out the epidemic in the mountains, cholera continued to take its toll on the capital. At its height, from August 15 to 25, seven hundred people died every day. The dead were haphazardly stacked in wagons and thrown into unmarked pits or the ditch surrounding the city walls.[147] Those who were lucky enough to obtain a proper burial were carted off to cemeteries so overcrowded with shrouded victims waiting to be buried that "it seemed like a large white sheet had been pulled over the burial grounds."[148] Survivors, often with advanced stages of cholera, appeared lifeless; they were dehydrated, were severely hypothermic, and had an undetectable pulse and blood pressure.[149] Tehran's overwhelmed physicians and lay caregivers probably pronounced many dead before their time, and some of these cases would undoubtedly experience a "resurrection" during their ritual washing or shrouding prior to their actual interment—although some unlucky souls could have been buried alive under these conditions.

Tehran did not experience the class-driven mass hysteria, lynching of physicians, and burning of hospitals that characterized reactions to rumors of early demise during cholera epidemics in European cities.[150] Because nobility and commoner died in proportional numbers, Iranians did not harbor class-based resentments during the outbreak. In addition, Iran's deficient health care infrastructure and the dearth of physicians actually protected the profession. Unlike Europe, where countless victims were taken to hospitals never to be seen again, in Tehran the sick died in their homes, some even abandoned by their fleeing families. Therefore, people's fears and anger did not have the same distinct targets.[151]

The Shah on the Run

News of the epidemic finally reached Nasir al-Din Shah in August, causing the panicked monarch to order his retinue to return to the capital after months of absence touring the country.[152] However, cholera had already breached Tehran's walls, and to the great dismay of the shah's physician-in-chief, the twenty-thousand-man royal caravan traveled at breakneck speed toward the very disease the monarch wanted to avoid.[153] The shah's caravan reached the capital's gates on August 15. Inevitably, the large contingent mingled with the populace standing outside the disease-ridden city while simultaneously witnessing cartloads of cadavers driven out for burial. News reached the shah that one of his wives and a governor had died of cholera in Tehran. Fearing a similar fate, he abandoned his large escort and circumvented the capital, with the remaining fifty members of his household, for the safety of the mountains to the north. As the caravan raced around the city walls, village after village succumbed to cholera at a distance of only several hundred meters from the shah's camp. The monarch encountered "convoys of death" and mass burials everywhere, and the typically bustling northern villages of Tajrish and Rostamabad were deserted, their inhabitants either killed or driven away by the scourge.[154] When the epidemic finally caught up with the royal convoy, Feuvrier successfully treated people in the monarch's retinue, including one of the shah's wives, but could not prevent cholera from spreading into the picturesque mountain village of Shahrestanak, where the monarch had decided to camp on September 3. Cholera there lasted only a little more than a week, thanks to the antiseptic measures undertaken by Feuvrier, while the disease continued in the surrounding areas and in the capital.[155] After a short lull in deaths, returning refugees from the Shemiran foothills brought a more violent upsurge of the epidemic into the capital that killed thousands from September 10 to 17.[156] By October, cholera in Tehran began dissipating in earnest, and the city quickly returned to its normal bustle, despite losing more than eight percent of its population to the disease.[157]

Cholera in the South

Cholera took a southerly course from Tehran, breaking out in the cities of Qom, Kashan, Arak, and Isfahan. It also reached the southeastern Iranian city of Yazd, imported there by camel-driven caravans crossing the Salt Desert from Torbat-i Jam.[158] The epidemic's impact on Yazd was mild, killing only an estimated four hundred; but it did not stop the governor and more than half of the city's remaining inhabitants from fleeing. The governor's brother, who took over the city's administration, attempted to reduce the growing hysteria by ordering the immediate burial of the dead and forbidding public mourning. These measures did little to halt the panic-fueled exodus and resulting breakdown of civic authority, which left Yazd in the hands of toughs (*luti*). The city became a shooting gallery where someone was shot and killed on a weekly basis by the brawling *luti*. When a gang of toughs gunned down a man after a perceived affront, the local authorities were unable to arrest the perpetrators, who had "barricaded themselves in a house, shooting at all those who would come near."[159] When another *luti* shot and killed a peasant over a very small loan, the hapless authorities were once again powerless to make an arrest. The ongoing lawlessness brought trade and commerce in Yazd to a screeching halt.[160]

The epidemic's southern wave from Tehran eventually reached Hamedan, spreading from there across the Lorestan Province to the towns of Khoramabad, Malayer, Nahavand, and Borujerd.[161] It also worked its way to the southern city of Shiraz from Isfahan, leaving a trail of insecurity and bandit-infested roads between the two cities.[162] On October 17 cholera broke out in Kermanshah, where it lasted until January 1893.[163] Unlike the 1889 epidemic,

TABLE 2.2.
Mortality in various cities during the 1892 epidemic

City	Approximate population	Estimated total casualties	Casualty ratio
Torbat-i Jam	6,000	725	1:8
Mashhad	100,000	5,206	1:19
Neyshabur	30,000	1,319	1:23
Kashan	30,000	1,319	1:23
Abbasabad	3,000	350	1:9
Mian Dasht	1,000	30	1:33
Damghan	6,000	115	1:52
Shahrud	30,000	400	1:75
Semnan	25,000	360	1:69
Ahmadabad	1,500	30	1:50
Tehran	120,000	12,087	1:10
Anzali	4,000	311	1:12
Rasht	4,000	1,133	1:4
Qazvin	40,000	283	1:141
Rostamabad	2,000	50	1:40
Menjil	2,000	160	1:13
Ardabil	15,000	884	1:17
Tabriz	165,000	26,000	1:6
Khoy	20,000	250	1:80
Qom	30,000	95	1:316
Yazd	10,000	400	1:25
Hamedan	40,000	930	1:43

not a single Jewish resident of Kermanshah got sick, likely because of access to segregated drinking water, obtained from private residential wells sunk deep within each Jewish family's home.[164] Cholera diminished in intensity throughout the country by the time it reached its southernmost point in Iran at the Persian Gulf port of Bandar Abbas, finally ending altogether by May 1893 after it caused tens of thousands of deaths and instability throughout the country (table 2.2).[165]

The Economy of Fear and Newfound Clerical Powers

Westerners and their dependents faced rising hostility from Iran's urban population at the height of the epidemic and in the ensuing months as cholera receded from the country. Locals routinely taunted Europeans strolling through the bazaar in Isfahan, and a mob in Shiraz ransacked the home of a Russian performance group.[166] Even in Tehran, where the central government held the greatest sway, antagonism against foreigners was rampant. Clerics issued anti-Western religious edicts forbidding men to shave their beards and women to wear high-heeled shoes in the capital.[167] A young Gertrude Bell, who would become a renowned orientalist, could feel the growing anti-European sentiment from the confines of the British legation: "Religious fervor grew apace under the influence of fear. . . . The air was full of rumors. It was whispered that the mullahs were working upon native fanaticism, and pointing to the presence of Europeans as a primary cause of evil [i.e., cholera] which must be straightway removed."[168] Growing religious unrest in the northern city of Gorgan around the same time led to looting and destruction of Armenian-owned shops by angry mobs. The town's senior cleric instigated the riots by accusing Armenian wine merchants of breaking Islamic law and causing divine retribution in the form of cholera.[169] Russian troops were obliged to cross the border into Gorgan to restore order and protect their dependents by imposing marital law.[170] In Yazd, harassment and violence prompted more than five hundred of the city's Zoroastrian Parsi community, who were involved in the important gold and spice trade with India, to return to Bombay during the outbreak.[171]

Iran's militant Shi'ite clerics, once again, used the epidemic to enhance their prestige as advocates for the people and to further roll back their economic losses to the West. In Isfahan, Aqa Najafi and his brother Shaykh Muhammad 'Ali, in response to accusations that Europeans were responsible for the scarcity and rising prices of goods during the epidemic, forbade commerce with Westerners, even if it meant breaking earlier contractual agreements.[172] Clerics in Shiraz instigated the population to occupy the government telegraph office in protest against the skyrocketing price of bread. "The mob, on receiving reassuring promises from the governor, demanded that the [former governor]—an enemy of the *mullahs*—should be removed from the mayoralty."[173] Nasir al-Din Shah yielded to the insurgents, which only emboldened them to call for the expulsion of the former governor altogether. The typically compliant shah, unwilling to submit to escalating demands, ordered the telegraph office cleared by force. However, in a last-minute reversal, royal troops sided with the insurgents, which prompted the shah to concede. In Hamedan, a local cleric directed a rampaging mob against the city's Jewish quarter. The shah, having recently promised the British that he would protect his Jewish subjects, immediately summoned the rabble-rousing priest to Tehran. However, in an act of defiance against the shah, a crowd stopped the cleric's escort, conducted him back to Hamedan in triumph, and forced out the city's royal governor. "The shah contemplated sending troops to Hamedan, but the success of such an expedition was doubtful, and the shah accepted a compromise under which the mullah left Hamedan and the governor returned thither."[174]

Clerics also used the epidemic to mobilize resistance against European dominion over local industries. They preached holy war (*jihad*) against the "lowly" and "abject western clothes," urging Iranians to defend life and property (*jan va mal*) from non-Muslims.[175] One of their first edicts during the outbreak in Tehran called for Iranian men to discard their European attire in favor of wearing traditional cloths to atone with God.[176] These pronouncements also called for the boycott of Western products, as described by the French envoy: "They [the *mullahs*] have said that the populace should at no price use European products and should not trade with Christians and that no believing Mohammedan should remain in their service . . . Persian women who wore the *chador* [veil] made out of European cloth have been beaten and their cloths torn apart."[177] While some clerics disallowed only Western goods, others called for ending any association with Europeans, including forbidding Iranian domestics from serving the ritually unclean non-Muslims. As a result, one of the British legation's Iranian domestics promptly resigned to go on "pilgrimage" in the midst of the epidemic.[178] Similarly, the Italian legation lost four of its Iranian servants, and the Germans lost three. The servants who had not left were insulted on a daily basis for waiting on Westerners.[179] These brazen acts of militancy and violence slowly faded with the receding epidemic in the spring of 1893. Yet, the undercurrent of antagonism against European economic and political encroachments in Iran and the status of Shi'ite clerics as leaders of dissent against the government would endure long after the last casualty to cholera.

The International Sanitary Conference of 1894

A year after the epidemic receded from Iran, the French government organized an international sanitary conference in Paris to revisit prior agreements on cholera in light of the most recent global wave of the disease. Sixteen countries participated, including Iran, which sent as delegates two of its native physicians, Zayn al-Abidin Mu'in al-Atibba and Khalil Khan.[180] The meeting focused on the sanitary regulation of the Hajj in view of its role in the pandemic's dissemination from India. It also deliberated on the Persian Gulf and Iran, seen as potential barriers against Asiatic cholera's dissemination to Europe.[181] The gathering criticized the Iranian government for Iran's hygienic shortfalls and for not giving its sanitary council an independent budget to combat the recurrent ingress of the pandemic across its borders:

The absence of all municipal services, charged with cleaning and washing streets, transporting all kinds of garbage, carcasses, etc.; the fact that potable water travels in channels and reservoirs that are not covered, exposed to all kinds of possible defilements, especially when it intermingles with water from public baths and wash houses, with water with which cadavers are washed, the absolute absence of disinfectants and medications for those struck by the illness [cholera], the absence of capable physicians, and finally the fact that, during an epidemic, cadavers are left in homes and even in streets, without coffins—all of these facts and much more . . . explain why cholera [in Iran] propagated so rapidly and took such a terrible toll, as was the case last year.[182]

These shortcomings prompted Austria's delegate to revive a proposal, made at the 1874 conference in Vienna, to place the Iranian Sanitary Council under international jurisdiction. The plan stripped the Iranian government of executive authority over the council but expected it to bankroll the council's estimated 497,000-franc annual budget.[183] Iran's chief delegate, Mu'in al-Atibba, dismissed the proposal as "illusory" due to Tehran's financial constraints. He also highlighted the broader injustice of expecting "Oriental countries" to "instantaneously" make the sanitary improvements that took affluent Western countries decades to accomplish.[184]

Mu'in al-Atibba, who trained under Joseph Désiré Tholozan at the Polytechnic College in Tehran, shared his mentor's pragmatic approach to Iran's sanitary shortfalls. He also had years of experience with Iran's halting public health initiatives dating back to 1867, when the newly established Iranian Sanitary Council sent him to coordinate cholera surveillance efforts in Kermanshah (see chapter 1). In contrast to his predecessors at earlier meetings, Mu'in al-Atibba's experience taught him that fundamental improvements to Iran's public health depended on material assistance from the West. He used this knowledge to sway critics by portraying Iran as a victim of the recurring pandemics from India rather than a propagator of the disease to Europe and characterized Iran's vulnerability to cholera as a global problem that required distributing the costs of its sanitary amelioration proportionally among "interested nations."[185]

While the investments in Iran never materialized, the conference did intervene in the region by standardizing maritime inspection and quarantine regulations in the British-dominated Persian Gulf. It also backed a longstanding Ottoman proposal to establish an international quarantine station on the Fao Peninsula.[186] The location of the proposed establishment, at the mouth of the Shatt al-Arab waterway, prevented passengers and goods from circumventing the station as they had during the previous cholera outbreaks.[187] The British, on the other hand, viewed the proposal as a thinly veiled hostile takeover of the region by the Ottomans, who would gain unrestricted control over navigation into the strategic Shatt al-Arab and all of its upstream tributaries, including Iran's commercially vital Karun River.[188] They also opposed the proposal on financial grounds, claiming that Britain would be obliged to carry an inordinate amount of the station's costs since it controlled more than ninety-eight percent of the taxable shipping in the Persian Gulf. Although a majority of delegates overruled Britain's objections, the plan never materialized because of various regulatory disagreements, which led to the Ottoman refusal to ratify the conference's final convention.[189] This did not dampen the growing international frustration with Britain's seeming lassitude toward the flow of Asiatic cholera out of India and the Iranian government's equally apathetic response to the disease, foreshadowing a new era of assertive European sanitary interventions in the region.

Conclusion

The 1889–1893 cholera epidemics revealed the enduring weakness of executive authority in Iran and the persistent inability of the Qajar polity to contend with national outbreaks in the final decade of the nineteenth century. Economic and administrative obstacles stifled the Iranian Sanitary Council's attempts to improve the country's hygienic infrastructure, allowing Asiatic cholera to reoccur despite Tehran's growing sanitary awareness and medical expertise. On an international level, the government's failure to halt the epidemics heightened European interventionist policies to stop cholera in both Iran and the region as a whole. On a local level, Shi'ite clerics successfully filled the vacuum of authority created by the breakdown of civil administration during the outbreaks. They exploited tensions along Iran's religious and economic divides and mobilized widespread discontent against the government's ineptitude and foreign commercial interests, cementing their status as the people's champion.

The disturbances that followed the epidemics harmed Iran's economy and political stability. The Tobacco Protest led to the cancellation of the British concession, which saddled the country with a £500,000 penalty; and the export of major cash crops such as cotton fell by more than fifty percent after the outbreaks.[190] By 1894, Iran was in the midst of a "serious economic crisis," forcing Nasir al-Din Shah to seek the country's first foreign loans.[191] This opened the floodgates of borrowing and dependence on nonproductive income streams by the Qajar state and its notables.[192] Law and order in the provinces were never fully reestablished after the epidemics. The central government was paralyzed by intrigues, and reformers were largely sidelined by the shah, who was less interested in the affairs of the state. This instability culminated in the assassination of Nasir al-Din Shah on May 1, 1896, by a revolutionary whose stated motive was to bring an end to the growing subservience to the West, oppression, and poverty in the land.[193]

CHAPTER THREE

Epidemics and Sanitary Imperialism, 1896–1904

The 1889–1893 cholera epidemics and ensuing upheavals widened Iran's trade deficit and weakened the Qajar regime. Spiraling inflation, depreciation of silver, and a global recession added to the country's fiscal woes in the years leading up to Nasir al-Din Shah's assassination.[1] With Iran's reserves depleted, his successor, Muzaffar al-Din Shah, sought more foreign credits to keep the government and royal household afloat upon ascending the throne in 1896. This made Iran vulnerable to political pressure from its Western creditors at a time of growing Anglo-Russian military and diplomatic brinkmanship in the region. Britain's efforts to counter Russia's expansion in Asia, known as "The Great Game," intensified with the extension of the Trans-Caspian Railway, which allowed Russia to rapidly deploy its troops on the Indian frontier, threatening the crown jewel of Britain's colonial empire. This escalating rivalry played out in Iran as the two powers vied for political and commercial concessions from the Iranian government in exchange for the loans that Tehran desperately needed to offset its deficit.[2]

The struggle between Russia and Britain occurred against the backdrop of a growing bubonic plague pandemic, which eclipsed cholera as the contagion of concern for the international community and, like cholera, unmasked political and scientific rivalries between the great European powers. The disease made headlines when it spread to the densely populated British colonial port of Hong Kong, killing thousands in 1894. The period's biomedical revolution, which enshrined laboratory-based bacteriology in the fight against epidemic diseases, allowed researchers to isolate the plague's causative bacterium but not before the contagion spread to other parts of the British Empire. When the plague reached Bombay (now Mumbai), two years later, rats were confirmed as key vectors, and shortly afterward fleas were also hypothesized as important in conveying the plague to humans. Despite these discoveries, many physicians and sanitarians continued to hold miasma and direct person-to-person contact responsible for the disease's transmission. European powers prejudicially imposed older plague-control practices, such as the incineration of personal effects and forced quarantines, on natives and colonized subjects.[3] These themes played out in Iran, where Russia and Britain's power struggle shaped the interventions against the approaching epidemic. The Iranian government's inability to pay for the necessary personnel and materials to prevent the ingress of bubonic plague from India extended "The Great Game" to Iran's sanitary realm as the two powers fought to dominate the administration of the country's quarantines.

The Bubonic Plague Panic

Six months into Muzaffar al-Din Shah's reign, in October 1896, the Iranian government learned that a bubonic plague epidemic had broken out in Bombay, killing thousands on a weekly basis.[4] The Iranians recognized that the terrifying outbreak would reach their southern ports, on the heels of India's expanding maritime trade with the Persian Gulf, unless they rapidly implemented preventive measures. Tehran, however, lacked the men, materials, and ships to police its extensive coastline, and so it turned to London, which had the largest navy in the area, for assistance.[5] The British maintained five gunboats on the island of Qeshm, in the Strait of Hormuz, to support their political resident's mission in the Persian Gulf. For more than seventy years, the resident had administered Britain's interests in the entire Gulf region from his headquarters in Bushehr; using the regular patrols of the Royal Navy gunboats to protect commercial shipping lanes against piracy and to enforce London's various policy directives, including the interdiction on slave trading and gunrunning.[6]

The political resident initially refused Iran's appeals to add quarantine enforcement to his navy's list of responsibilities, fearing the detrimental impact of the solicited sanitary measures on India's commerce in the Persian Gulf. After extensive negotiations, however, he agreed to order his ships and men to establish quarantines with "minimal injury to [British] trade" by detaining healthy vessels for no more than five days at Muhammara and Bushehr and seven days at other Iranian ports located closer to India.[7] In return, the Iranian government appointed the British residency surgeon in Bushehr as its principal health officer in the Persian Gulf with jurisdiction over its quarantines.[8] The Iranian governor of Bushehr and other local notables perceived this as yet another concession to foreigners and opposed the arrangement. They tacitly pursued a policy of noncooperation with the residency surgeon and avoided charges of insubordination thanks to the absence of formal orders from Tehran. The British, on the other hand, saw Iranian venality as the principal driver for local opposition because "quarantine arrangements worked by [Iranians] themselves would offer an exceptional opening for extortion."[9]

The Russian government also objected to Tehran's arrangements with London on grounds that Britain's commercial interests in the region would cloud its sanitary responsibilities. In its discussions with the Iranians, St. Petersburg stressed its concerns with preventing an outbreak in the Persian Gulf, especially at the ports of Bandar Lengeh and Bandar Abbas, which adjoined caravan routes that led directly into Russia's Trans-Caspian territories. It proposed to establish its own strategic medical presence at those ports and further inland in Yazd and Kerman—ostensibly, to supervise Britain's sanitary operations while gaining a long-coveted diplomatic foothold in the Persian Gulf. The Iranian government, at London's urging, rejected the offer. The British feared that the Russians could use their oversight to highlight shortfalls in the British-run quarantines and revive the call for an international takeover of quarantine measures in the Persian Gulf made at the International Sanitary Conference in Paris two years earlier.[10]

The spread of plague into neighboring Afghanistan, however, convinced Tehran to accept Russian participation in other aspects of its sanitary affairs not long after it rejected Russian entreaties in the Persian Gulf. The Iranian government invited the European members of its Sanitary Council to the shah's palace to meet with the anxious monarch and his Council of Ministers upon receiving news of the epidemic on the country's eastern flank.[11] In his opening statement at the meeting, the Iranian interior minister advised Muzaffar al-Din Shah to endorse the strict preventative measures suggested by the Europeans. Similarly, the shah's attending Iranian physician, Mirza Mahmud Khan Hakim al-Mulk, called for halting all communication and trade with Afghanistan by closing the roads from Herat to Mashhad and Kandahar to Birjand and Yazd. The foreign minister and the Sanitary Council rapidly formulated a series of regulations to prevent the plague's transmission across Iran's eastern frontier, which they circulated among the medical and administrative personnel in Tehran and key Western legations as an official proclamation from the new shah (see appendix B). The Russians, seeing their opportunity to control trade routes from British India to Iran, agreed with Tehran's request to send two doctors and a contingent of 150 Cossacks to enforce the new measures on the Iran-Afghanistan border.[12]

The Iranian government and its advisers did not achieve the same consensus on quarantines in the Persian Gulf as they had on Iran's frontier with Afghanistan. The majority of the Sanitary Council, including Jean-Etienne

Justin Schneider, Muzaffar al-Din Shah's new French doctor, favored sending additional European physicians to expand the southern quarantines.[13] Joseph Désiré Tholozan, the late shah's influential physician who continued to serve on the council, opposed this suggestion, arguing that the local population would never obey an army of "pretentious" Western medical practitioners.[14] He instead recommended sending a number of Iranian constables (*farrash*) to reinforce the British-run quarantines, casting doubt on the value of modern medical diagnostics in favor of linguistic fluency and cultural competence: "Is it not shameful for our science that all the physicians at Calcutta for a whole month were unable to say whether the illness at Calcutta was the plague or not. First the blood of the convalescent people was placed under the microscope, and they said: 'These people have the plague.' What plague has its victims walking, sleeping, eating, and drinking, in short, able to carry themselves like people who are well! If we send a European physician to the Fars [province] would he be better equipped to diagnose these people than the great English physicians of Calcutta?"[15] Tholozan's diagnostic fatalism was a symptom of the wider malaise with the ascent of the germ theory of disease in the fin de siècle. Older sanitarians like him, who had trained before this era, did not see microbes as the sine qua non of maladies.[16] Schneider, on the other hand, embodied a new generation of medical scientists who were convinced that the bacteriological revolution gave them the capacity and obligation to intervene globally to fight infectious diseases. Veteran sanitarians felt that these emerging disciples of Koch and Pasteur focused on new microbial discoveries for the sake of fame at the expense of time-proven actionable measures to prevent epidemics. This perspective fueled Tholozan's antipathy toward Schneider, whom he portrayed as nothing more than a vainglorious speculator who wanted to make a name for himself in Iran, secure a pension, and return to France to rest on his laurels.[17]

Blinded by his decades-long residence in Iran and disdain for Schneider, Tholozan did not realize that France's growing political and economic interests in the Persian Gulf also influenced his younger colleague. Three years earlier, France established a consulate in Muscat and posted naval cruisers off the coast of Bandar Lengeh, challenging Britain's preeminence in the region.[18] It also ratified the Franco-Russian alliance, which increasingly aligned the two countries against London on a broad range of issues, including sanitary matters in the Persian Gulf. This became apparent at the International Sanitary Conference in Paris when the French delegate, backed by the Russians, attacked Britain's cholera prevention policies and supported the Ottoman proposal to establish an international quarantine sanitation on the Fao Peninsula.[19] France and Russia leveled similar criticisms at Britain's plague prevention efforts in Hong Kong and India in the ensuing years.[20] This new alignment gave Schneider the diplomatic muscle to erode Tholozan's backing in the Iranian Sanitary Council. The gathering eventually turned against its seventy-seven-year-old founder, voicing its outrage at his diminution of the medical profession by suggesting that "simple servants" could replace physicians. Schneider, who appeared to be on a private vendetta against the older doctor, ridiculed Tholozan's proposal to use a limited five buckets of sulfur to disinfect ships as "a mockery of science."[21] Tholozan in actuality called for employing the clearly insufficient resources in the Persian Gulf to respond as soon as possible to the looming epidemic rather than postponing disinfection until the procurement of additional supplies. His attempts to shed light on the controversy failed to win the approval of the shah, the Council of Ministers, or the Sanitary Council for that matter. He died six months later, in July 1897, disillusioned and ignored by the new regime in Tehran.[22]

The British residency surgeon in the Persian Gulf dismissed the Iranian Sanitary Council's call for stricter quarantine and disinfection standards as outcomes of French and Russian machinations to harm his nation's interests in the region.[23] The British knew that Iran could not afford the large sanitary and military staff needed to implement the new directives.[24] Its deteriorating fiscal situation had already impeded the execution of the Sanitary Council's other recommendations, including cleaning provincial villages and towns. The Iranian government was also unable to pay the crew of its cruiser *Persepolis*, which stopped policing the entrance of the Persian Gulf against quarantine-breaking coastal boats.[25] Either way, London was wholly skeptical of the efficacy of longer and more rigorous quarantines in Iran as conveyed in the following dispatch by its envoy in Tehran to the governor general of British India:

> In 1893, during the cholera epidemic the quarantine regulations were an unmixed evil—at certain points, ten days quarantine was ordered—a traveler arrived and freely mixed with the arrivals of the previous nine days and the succeeding nine days so that should he be suffering from cholera, the Quarantine Station actually served as a means of propagating the disease and this I fear may be the result again should Plague in Epidemic form reach Persia. A traveler who is willing to pay can usually escape quarantine regulations but they enable those in authority to levy black mail in a variety of ways.[26]

The British legation's physician serving on the Iranian Sanitary Council failed to "moderate" the assembly's eagerness to strengthen and expand the quarantines in the Persian Gulf.[27] The council dismissed his opinions as thinly veiled attempts to uphold Britain's financial interests. Growing international pressure to halt the epidemic eventually forced London to close Iran's Persian Gulf ports to maritime traffic from India, except at Bandar Abbas, Bandar Lengeh, and Muhammara, where the British residency's assistant surgeons reluctantly enforced the Sanitary Council's mandated fifteen-day quarantine.[28] The Iranian government in turn formally instructed its recalcitrant governor in Bushehr to obey the residency surgeon on all quarantine matters and defer to the British in the event of conflicting orders until the government resolved the problem.[29] Tehran also pledged to inform the British legation of any new sanitary instructions sent to the governor. The Iranians further accepted to transmit their sanitary orders to the residency's surgeon and his subordinates via the British legation in Tehran, forgoing their sovereign right to issue such commands directly. This "indispensable step for the good conduct of the quarantine" assured Britain's supremacy over quarantines and other sanitary interventions in southern Iran, while they continued fighting with the Russians for the rest of the country.[30]

The International Sanitary Conference of 1897

In 1897 the International Sanitary Conference met in Venice to deal with the emerging global threat of "Asiatic plague" and the scientific controversies concerning the disease.[31] The nineteen countries that participated, including Iran, unanimously agreed on the plague's microbial etiology. They also acknowledged the bacterium isolated in Hong Kong in 1894 by Shibasaburo Kitasato and Alexandre Yersin as its causal organism during the month-long proceedings that began on February 16.[32] Differences remained, however, on ways of stopping the pandemic and treating victims of the disease. Most of the Europeans at the conference feared that it was only a matter of time before Muslim pilgrims from India would carry the plague to Mecca; creating a focus of infection that could spread to the West. Austria, for example, worried that its Bosnian-Muslim subjects could assist this westward transmission, and France was concerned with the risks of an outbreak from its North African Muslim colonies. They recommended restricting travel to Mecca and placed similar limitations on Christians traveling to the Holy Lands in Palestine to appear evenhanded to their restive Muslim populations.[33] Britain, on the other hand, had greater political constraints in curbing the Hajj than other European countries due to its history of cultivating Muslim favor in India, as a counter-weight against growing Hindu nationalism, by avoiding prohibitions on pil-grimage to Mecca.[34] Imposing stricter quarantines or banning the Hajj from India ran against London's imperial strategy and risked provoking sectarian disputes and violence. Britain eventually gave in to the growing chorus of crit-ics at the conference, which faulted the Indian government for the smoldering outbreak in Bombay, and banned pilgrimages for the remainder of 1897.

The French had led the attack against Britain's lax sanitary policies, partic-ularly its continued opposition to stricter quarantines in the Persian Gulf. Their delegate at the meeting, Adrien Proust, was a veteran of the international sanitary conferences and a longtime intellectual rival of Joseph Désiré Tholo-zan.[35] His expedition to Iran in 1869 shaped his long-standing opinion that the absence of adequate quarantines along Iran's overland and maritime routes to India threatened Europe's ability to keep cholera and plague pandemics at bay, particularly in view of Russia's growing rail network in Central Asia. This put him at odds with Tholozan, who doubted the efficacy of quarantines with-out overall hygienic improvements in the region.

As in prior meetings, Proust and other European delegates criticized Iran for its inadequate sanitary provisions in view of its position on the "frontline of the [bubonic plague] invasion."[36] Proust believed that Tehran's permissive-ness on sanitary matters reflected a general oriental disregard for human life and the obscurantism of its Shi'ite clerics, which prevented the shah from im-plementing necessary hygienic reforms.[37] The Iranian delegate to the confer-ence, once again, had to deflect a suggested international takeover of the Iranian Sanitary Council by highlighting the number of competent European doctors and Western-educated native physicians serving on the existing council who were capable of gradually improving the country's "deprived" sanitary infrastructure. Moreover, he pointed out that Iranians, unlike other Muslims, were more tolerant of isolations and disinfections when a "Persian" sanitary council mandated the measures.[38] The Russians, contrary to the past, shared this assessment. They expressed their satisfaction with the Iranian gov-ernment's willingness "to provide all the explanations they were asked for" and for accepting "all the interventions of foreign physicians that were proposed to them."[39] St. Petersburg's desire to maintain the existing sanitary status quo, which recently handed quarantine arrangements on Iran's northern frontiers to its military and medical staff, explained its favorable statements. The Brit-ish, on the other hand, refused to acknowledge the conference's consensus declaration that recognized the right of every sovereign country to close its frontier in the event of an outbreak in a neighboring state.[40] They characterized land-based quarantines as futile and "a barbarous attack on personal freedoms, unfit for civilized countries," and supported their position by pointing out that cholera, though prevalent in the rest of Europe throughout the 1880s and 1890s, had been absent from Britain without the use of quarantines.[41] The British were particularly concerned that the Iranian government, encouraged by Russia, would initiate large-scale restrictions on its overland border with India, which they did not control. The Iranian representative, however, endorsed only limited overland interventions against the plague due to shortages in workforce and materials.[42] The conference therefore mandated quarantine and disinfection for ships and goods docking in Iran's Persian Gulf ports and called for only close supervision of Iran's overland frontiers with Russia, India, and Turkey.[43]

Plague, Quarantines, and Anglo-Russian Antagonism

With the absence of plague in the Persian Gulf by the summer of 1897, the Iranian government reduced the duration of quarantines at its open ports but continued to turn back British-Indian caravans at the Sistan frontier for sanitary reasons, causing widespread discontent and "killing trade."[44] Russian patrols also preemptively closed the Herat-Mashhad road, making matters worse for British commerce. Although the Sistan closure agreed with the shah's earlier edict on border control to stop the ingress of plague from India, shutting the Mashhad-Herat road countermanded Tehran's orders. The Iranian government had insisted on leaving open this important pilgrimage route, the only requirement being that travelers and goods face disinfection and a fifteen-day observational quarantine (see appendix B). However, Tehran was unable to provide the required tents, equipment, and medical personnel needed to establish the quarantine stations. The deployed Russian troops used this to justify their blockade, allegedly to prevent the transmission of plague into Iran and Russia.[45]

The Iranian government later discovered that the roads from Afghanistan to Russia remained open with only a cursory three-day quarantine sporadically applied, despite comparable risks of transmission. Russia, it appeared, was using the cover of prevention to divert lucrative trade from Iran to its own territory.[46] The Iranian foreign minister, after informing the shah, ordered the governor of Khorasan Province to open the road, but Russia's Cossack troops refused to lift the blockade until the Iranian government built appropriate quarantine stations.[47] The British urged the Iranians to send at least some tents to the border area to appease Russia's consul general in Mashhad, who controlled the Cossacks and appeared to be "the master of the situation."[48] London's envoy in Tehran was concerned that St. Petersburg was seeking to permanently divert India's trade with Iran to Russia's territory:

> In spite of the Agreement arrived at between the Persian Government and the Russian Legation the Herat-Meshed road still remained closed to traffic of any kind owing to the objections raised by the Russian Consul General at Meshed, which were evidently inspired by the desire of the Russian Authorities to divert the trade with Afghanistan from Persia into Russia. . . . it appears that with this object in view of the Russian Consul General at Meshhed tried to persuade Afghan traders to export their merchandise towards Panjdeh and Merv assuring them that if they did so no passport fees to which they are liable in Persia would be levied from them, that remissions would be granted to them in customs dues, and that articles of manufacture for which there is a demand in Afghanistan would be supplied to them at Merv at the same prices as they are sold in Askabad.[49]

The Russians failed to grasp the importance of the Mashhad-Herat road for pilgrimage to the shrine of Imam Riza and were blindsided when Shi'ite clerics telegraphed the shah directly to complain about the blockade. They reluctantly reopened the highway after several weeks, yielding to the growing anger and the erosion of their political standing in the region.[50]

Notwithstanding this setback, the Russians opened another quarantine station at Torbat-i Heydarieh, seventy-seven miles south of Mashhad, against arrivals from Sistan and the Persian Gulf. The British legation believed that the Russians intended to use the station "to harass and stop legitimate trade between Khorasan and India and the Persian Gulf Ports," especially the export of wool from Afghanistan, which went through Torbat-i Heydarieh.[51] The Russians could also use travel restrictions to divert this important commodity away from Iran into their own markets and work against British commercial interests from the south without incurring the wrath of Shi'ite clerics. The British envoy in Tehran informed his superiors in London that confining caravans that had already undergone the process and been declared disease free in the Persian Gulf was pointless. Moreover, they would have traveled in Iranian territory for at least twenty-five days by the time they reached the Russian camp in Torbat-i Heydarieh, giving the caravans ample time to spread the plague in the event of an undetected infection.[52] Yet the British, facing international denunciation for their negligent sanitary policies in India and persistent opposition to quarantines, could do little to curb Russia's behavior.[53] London realized that any official condemnation of Russia's intervention in eastern Iran would be seen as another example of its self-serving policy of placing trade interests ahead of pandemic prevention. Moreover, since Tehran depended on material and personnel from Russia for the sanitary defense of its eastern borders, the British could not compel the Iranians to order the Russians to lift their quarantines without definite proof of their ulterior motives. This meant that by the closing months of 1897, the Russians were masters of the sanitary regime on Iran's overland frontiers with India and Afghanistan.

In addition to material needs, shortfalls in Iran's medical workforce also increased Tehran's dependence on Russia and Britain to establish quarantines and carry out the rest of its sanitary obligations to stop the plague. This deficiency resulted from the late shah's misgivings with Western education during his last decades in power, fearing its potential radical and subversive influence. Khalil Khan Saqafi, one of Iran's representatives to the 1894 International Sanitary Conference, recalled that Nasir al-Din Shah initially refused his request to study medicine in Europe on grounds that it would corrupt him.[54] The late shah also reduced his patronage of the Polytechnic College, which together with the overall economic downturn, eroded domestic medical education in Iran. Despite continuing financial difficulties, the new shah's government in 1896 began sending large cohorts of graduates to study at medical faculties across Europe to reverse the country's dependence on the Great Powers.[55] The bubonic plague pandemic itself also indirectly helped enhance the prestige of medicine in Iran, thanks to the emerging field of bacteriology, popularized by the plague's microbe hunters in Hong Kong. This precipitated an unprecedented surge of scholarship in the new biomedical sciences in the country. Even several younger Iranian diplomats in foreign postings began pursuing diplomas in medicine during their free time.[56]

The surge of interest in medical scholarship did not bear fruit for several years, forcing Iran to rely on the Great Powers when the plague resurfaced in Calcutta in 1898. During the yearlong outbreak, the British-administered quarantine of ships from India hardly met the internationally mandated ten-day period of observation at Iran's Persian Gulf ports.[57] Even the rumors of a plague outbreak in Bushehr in May 1899 elicited only a muted response from London and Tehran, which assumed that a genuine epidemic would remain limited to coastal areas and be unable to cross into Iran's high-lying central plateau.[58] The Russians, on the other hand, sent a medical mission to investigate the situation in a thinly veiled effort to collect intelligence on ways to establish a strategic foothold in the Persian Gulf.[59] They also proposed to send a regiment of Cossacks to help form a sanitary blockade around Bushehr.[60] However, they needed the formal consent of the Iranian prime minister, Mirza ʿAli Khan Amin al-Dawla, before engaging in the extraterritorial intervention.[61] The prime minister, under British pressure and with Iranian commercial concerns in mind, refused to allow the blockade. He asked the British government in India instead to send additional physicians to the Persian Gulf to investigate and assist in quarantine matters.[62]

The requested contingent from the British Indian Medical Service landed in an unexpectedly tense Bushehr.[63] Its population, distrustful of foreign sanitary measures, opposed their attempts to examine potential cases of bubonic plague and treat the sick:

> The natives have been so much frightened by the sanitary measures that they only show their sick to the doctors when they are sure that the malady is an ordinary case; they never apply to English or Russians [doctors]. They think that the Europeans poison the sick. If the malady breaks out, no sanitary measures can be taken. On the one hand, the people will resist, and, on the other hand, as Bushire receives its supplies from other ports . . . it will be impossible to establish a Sanitary cordon round Bushire without bringing on a famine.[64]

The Iranian prime minister dismissed the British request for more authority to investigate and implement preventive measures as excessive in the absence of verifiable cases of bubonic plague. In truth, the premier was unlikely to empower the British given that a confirmed outbreak in a major commercial port such as Bushehr could set back Iran's already battered economy and undo his premiership. Despite these concerns, the weak position of the prime minister and his government could not withstand the demands of European powers for action to shed light on the rumored cases. On June 22, he mustered the Sanitary Council to launch an inquiry, which determined that the disease was indeed present in Bushehr and the absence of new cases was a temporary hibernation. The council, fearing a flare up of the plague during the cold winter season, instructed the Iranian government to improve the city's hygiene and reinstate its effective former governor.[65]

The British rapidly instituted the precautionary interventions recommended by the Sanitary Council, stirring the restive populace into open revolt.[66] Throughout July, the people of Bushehr, Shiraz, and even Isfahan, which had close commercial ties to the southern cities, engaged in a coordinated protest movement against what they perceived as harsh and unnecessary sanitary measures.[67] A Bushehr-based merchant, whose firm was in a long-running dispute with the British, reportedly instigated the demonstrations and supported the protests by feeding rioters and striking members of the bazaar. The Russians also fueled the anti-British movement to bolster their influence and curry favor with Iranian notables in the region. Immediately prior to the riots, the Russian consul in Isfahan, who was in Bushehr as a guest of the rabble-rousing merchant, declared that "there was no need for the restrictive plague measures" in the absence of widespread illness.[68]

On July 24, stone-throwing demonstrators shattered the windows of the Indo-European Telegraph Department office, the British consulate, and other symbols of British presence in Bushehr. They eventually occupied the government telegraph office and asked Tehran for the suspension of the antiplague measures.[69] The following week, Bushehr's bazaar merchants went on strike and engaged in an extensive sit-in (*bast*) in the city's mosques. By August 3, the situation in Bushehr deteriorated further when rioting townspeople urged tribes from the countryside to join them. The British consul in Bushehr requested troops from nearby gunboats to safeguard British lives and property against the growing violence, and the Iranian government, realizing the gravity of the situation, decided to intervene militarily. The new governor, at the head of one hundred soldiers, restored calm by ordering the protestors home and threatening to shoot whoever broke curfew. The bazaar reopened the next day, the ringleaders of the protests were bastinadoed, and other dissenters were exiled from Bushehr at the request of the British consul.[70]

The residency surgeon declared Bushehr free of bubonic plague by summer's end. However, the Iranian Sanitary Council, fearing a renewed outbreak, refused to reduce the quarantine measures because of the ongoing plague epidemic in India and the approaching winter cold. It advised the Iranian government to not "sink into a false slumber" and prepare for a resurgence of the disease in the near future by stationing soldiers and cavalry in the environs of Bushehr to rapidly establish a sanitary cordon in the event of a recurring outbreak.[71] The Sanitary Council also recommended stocking disinfecting chemicals and ovens, as well as organizing a caravanserai and a lazaretto for the isolation of suspected cases of plague.[72] Meanwhile, the Iranian government sent an agent to confirm the plague's disappearance; what he found instead was an uneven application of the quarantine in Bushehr by the British: "Before my arrival, ships coming from India were not subjected to quarantine, but since I have come to Bushire, quarantine has been reestablished. European passengers have not to undergo quarantine, nor the crews of ships, but only Persian passengers and pilgrims. The Mollahs and the people complain against this partial treatment. The English doctors say that it is not necessary to put persons of consideration into quarantine."[73] Muzaffar al-Din Shah personally demanded an explanation on the discrepant quarantining from the British ambassador, who blamed the "defective" Iranian-built lazarettos as unsuitable for the predominantly European first- and second-class passengers. The residency surgeon had no choice but to remand these passengers to their homes under periodic medical observation rather than confining them in substandard quarters with commoners.[74] Ethnic discrimination in the application of quarantines mirrored prevalent Victorian biases against non-European populations, seen as constitutionally and culturally predisposed to epidemic diseases. This view persisted among British physicians and sanitarians even after the bacterial etiology of the plague was put forth; and Iranians bore the brunt of draconian sanitary practices.[75]

The British continued to apply quarantines unevenly, despite the shah's protests and the rising chorus of criticism from the Iranian Sanitary Council, insisting that changes to their detention policy required Tehran to improve its lazarettos.[76] This gave rise to anti-British conspiracy theories, particularly around disinfections and other prejudicial sanitary interventions:

> In a former memorandum an account was given of the ordering of a disinfector by the Bushire authorities, on the recommendation of the sanitary board. This disinfector has since given rise to grave apprehensions which were communicated by the Sadrazam to Sir H. Durant. The people of Bushire appeared to believe it was an engine designed for the purpose of boiling children who were required in that condition for certain occult purposes by the European doctors, and after earnest warnings from the governor and other authorities, it was finally decided to place it on the Persian guard ship for safe keeping.[77]

Local distrust of quarantines, the persistent intrigues of the Great Powers, and financial difficulties significantly eroded the Iranian government's sanitary authority. Tehran increasingly relied on loans from Britain and Russia to keep afloat even as they strengthened their commercial stranglehold on their respective spheres of influence by the closing months of 1899.[78] Meanwhile, the Sanitary Council, led by its French and Russian delegates, pushed for a greater voice in the administration of quarantines in the Persian Gulf.[79] London, which correctly viewed the council as having "anti-British" undercurrents, responded by compelling the Iranian government to suspend the group altogether.[80] Iran's acquiescence, on the other hand, was emblematic of its government's weakness at this time. Struggling to affect sanitary provisions on its borders and pursuing a policy of balancing concessions between the Great Powers, it essentially yielded to Russia and Britain in their respective spheres of influence. The Russians had a contingent of interpreters, officers, and soldiers at Torbat-i Heydarieh that gave them sway over the quarantines and flow of traffic in that region, and therefore they did not oppose the Sanitary Council's demise since they gained unchecked control over sanitary policy there.[81]

Deprived of the Sanitary Council, the Iranian government exercised its authority in the Persian Gulf through a representative who supervised British-led sanitary arrangements and reported directly to the Iranian Foreign Ministry until May 1901.[82] The British agreed to conduct sanitary inspections, and impose periods of isolation when necessary, on ships docking in the main ports of call in southern Iran.[83]

The Russians Have Come to Stay

The British continued to view Russia's sanitary activities on Iran's northeastern border as an extension of St. Petersburg's expanding military and economic foothold in the region. Russian-administered quarantine stations allowed Iranian caravans to pass freely, while merchants of British-Indian origin were subjected to fumigations lasting several hours, as well as other forms of "harassment" meant to disrupt Indo-Iranian trade.[84] The sanitary cordon, which extended from the Iranian Kavir Desert on the west to the Hari River on the east, placed the Russian military at India's doorstep.[85] From this vantage point, they could observe the main eastern and southern roads from India into Iran and monitor British military and commercial activity in this strategic eastern quadrant. British fears grew with reports that Russian medical officers at Torbat-i Heydarieh lacked supplies to carry out any "serious" disinfections other than cursory fumigations, making their function more military than sanitary. These officers cross-examined caravans and travelers approaching Torbat, asking them for their numbers, origins, route, merchandise, and other details as part of their sanitary investigation. The British suspected that these encounters provided the Russians with important intelligence on their trade with Iran and valuable information on India's northwestern frontier:

> The supervision of the sanitary cordon forms only the minor part of the duties of the Russian officers at Turbat-i Haidari, which is regarded as a post of military observation where information of military and commercial interest relating to

Afghanistan, Balouchistan and Southern Persia can be obtained by the cross-examination of travelers. Their enquiries are largely about roads, water, and food supply, and they are assisted by a large collection of maps including English official maps in one which the Quetta-Nushki route is given in great minuteness of detail. The officers employed are generally members of the staff or special service officers from Transcaspia.[86]

Britain's own decades-long record of using military officers in the region to gather intelligence under the cover of benign diplomatic missions probably fueled its fears.[87] British physicians and health officers also moonlighted as spies throughout the Persian Gulf, particularly along pilgrimage routes to Mecca in the latter half of the nineteenth century.[88]

Adding to British concerns, the commanding officer in charge of Russia's quarantines, Captain Alexander Iyas, had all the qualities of a Victorian spymaster (figs 3.1, 3.2).[89] He was a member of the elite Russian Imperial Guard and a polyglot who spoke French, English, Persian, Urdu, and some Pashtu. He personally cross-examined many of the travelers and merchants brought in for questioning at the quarantine stations. His command of languages, his knowledge of local customs, and his high rank lent him authority with Iranian notables and government administrators. Iyas, supplied with highly detailed maps of the area and reinforced with several detachments of Cossacks, had an upper hand in the event of hostilities with the British. Not only did this immediately threaten India's northwestern frontier; the solid foundations of Russia's quarantine stations meant that they had "come to stay."[90] British authorities began to recognize the need to counter this long-term strategic threat.[91]

The appointment of George Curzon to the Viceroyalty of India in 1899 influenced Britain's increasingly confrontational policy in Iran. Curzon called for an aggressive posture against Russia's expanding presence on India's northwestern frontier. He was also in favor of maintaining Britain's supremacy in the Persian Gulf by blocking Russia's sanitary inroads and countering Russian influence in Tehran.[92] As early as September 1901, the British protested the mistreatment of their subjects by Russian quarantine workers to Iran's foreign minister, Nasrallah Khan Mushir al-Dawla, and unsuccessfully asked for their removal.[93] The British foreign secretary continued to advocate for the elimination of the Russian quarantines with the Iranian premier, 'Ali Asghar Khan Amin al-Sultan, during the shah's visit to England the following year. He dismissed St. Petersburg's contention that its quarantines in Iran protected Russia against plague from India by showing how potentially infected

Fig. 3.1. Alexander Iyas between two Iranian dignitaries at Torbat-i Heydarieh in 1902. Photo courtesy of John Tchalenko.

Fig. 3.2. Alexander Iyas. *Iyas and his Cossack escort, Turbat-i Haydari, 1901–1911,* digitalized original negative, negative size 3.5 × 12 inch (89 × 305 mm). Iyas is dressed in his ceremonial uniform of the Hussars of the Russian Imperial Guard. Photo courtesy of the Finnish Museum of Photography.

individuals from Bombay or Karachi could reach Moscow in a shorter span of time by rail through Central Asia, which continued to lack any sanitary precautions, than it would take them to reach the cordon in eastern Iran by land. The foreign secretary also reminded the Iranian premier that he could compel the Russians to lift their quarantines, as stipulated by the Venice Convention, since India's frontiers with Iran had a clean bill of health and even offered to take over their plague prevention efforts on Iran's eastern frontier.[94]

Amin al-Sultan initially agreed with the British "and intended to take steps for bringing about the suppression of the Russian quarantine."[95] But, fearing reprisals from his Russian benefactors, the premier quickly changed his position, going so far as to deny the existence of a Russian cordon in eastern Iran. On the other hand, the Iranian foreign minister acknowledged their presence and blamed his predecessor for allowing the Russians to get a sanitary foothold there in the first place without the shah's permission.[96] He also complained to the British that the Russian government consistently refused Iran's pleas to disband the quarantines and agreed that their object was most likely to obstruct trade between India and Iran.[97]

Both the Iranians and the British knew that it would be difficult to get the Russians out without causing a serious diplomatic incident.[98] The Iranians eventually lodged muted complaints with St. Petersburg, which led to the replacement of some troops at the quarantine stations with native Iranian Cossacks. These soldiers were under the command of Russian military instructors on loan to the Iranian government. This made the British realize that they could not diplomatically challenge the Russians by invoking the Venice Convention's statute against unsanctioned extraterritorial sanitary interventions since Russia could simply replace its soldiers with obedient native Iranian Cossacks and claim that the cordon was an Iranian enterprise.[99] The Russians could also retaliate by invoking the Venice Convention against British quarantines in southern Iran.[100] As a result, London opted for a more muscular approach by basing a contingent of Sepoy Indian guards, under British command, and consular representatives at Torbat-i Heydarieh and at Birjand.[101] By doing this, the British not only counteracted Russia's military presence in the region but also emulated their way of "gaining intelligence by means of quarantine outposts," which they found to be "a clever method of studying a country for military ends."[102]

Despite the enhanced British military presence, violence against Afghan and Indian merchants in eastern Iran continued unabated into 1903. The Russians used harassment tactics, such as violent interrogations and detentions, to erode British prestige and divert Indo-Afghan trade away from Iran into Russia.[103] They clearly wanted "to make their presence felt to the Afghans [merchants]" and to punish them for being the protégés of the British.[104] The Russo-British rivalry in eastern Iran reached a boiling point in April 1903 when

Mashhad's citizens rioted against the region's British-supported governor, Mirza Husayn Nayyir al-Dawla.[105] The Russians provoked the unrest, hoping to cause the downfall of the governor, who opposed their quarantines in the region.[106] They also planned to use the riots as a justification to occupy Mashhad to protect the lives of Europeans and their protégés. The British, aware of this plan, warned the Iranian government that if calm was not restored and the Russians did invade Mashhad, they would occupy the Persian Gulf ports of Bandar Abbas and Chabahar in turn. The Iranian government quickly dispatched troops to reestablish order in Mashhad and, to appease the Russians, forced the governor to resign.[107]

The events in Mashhad convinced the British that their position in Iran was in danger unless they directly and forcefully intervened to counter the Russians. As a first step, the British foreign secretary, explained his country's grievances with quarantines in eastern Iran directly to the Russian government in a July 1903 memorandum: "His Majesty's Government would certainly be pressed, unless the Russian cordon were withdrawn, to station a British post in its immediate neighborhood for the purpose of protecting British interests. Lord Lansdowne added that he greatly deprecated such exhibitions of rivalry in Persian territory, and that he trusted that the Russian Government would relieve us of the necessity of resorting to the measure which he described."[108] Besides threatening retaliatory measures, the memorandum pointed out that the Venice Convention did not require the enforcement of quarantines on merchants from India since the Indian provinces bordering Iran were not infected.[109] In response, the Russians unapologetically maintained that the articles of the Venice Convention applied solely to Europe and not to Asia.[110] The tsar's government also asserted its dissatisfaction with the measures taken against the ongoing plague epidemic in India, which made it "impossible for them to withdraw a sanitary cordon which is so weak that, if the danger increased, it would probably have to be strengthened, and which is in no way intended to hamper British trade."[111] The British countered by sending a military officer as the new diplomatic consul to Torbat-i Heydarieh with an escort of twenty-five soldiers to serve as a "deterrent against flagrant irregularity," to enhance British prestige among the native merchants, and to monitor Russian military presence on the Afghan frontier.[112]

Asserting Iran's Sovereignty

Britain firmly controlled the sanitary and commercial arrangements in the Persian Gulf while it struggled with the Russians over the quarantines in eastern Iran. The British foreign secretary, in a speech to the House of Lords in May 1903, underscored the importance of protecting and promoting "British trade" and resisting "at any cost" the presence of another power in the Persian Gulf.[113] His doctrine, however, was tested not by another European power but by Iran itself three months later. In August, the British learned that the Iranian government assigned the quarantine in Bushehr to its Customs Department.[114] During their five-year tenure, the Belgian administrators of Iran's customs had increased government revenues, giving Tehran the fiscal capacity to manage its own quarantines (fig 3.3).[115]

Joseph Naus, the successful overseer of the Customs Department, emerged as an influential actor in Iran's government and, as a member of the Supreme Council of the State, personally advised the shah.[116] His attempt to manage the quarantines in the Persian Gulf was a natural extension of the Customs Department's mandate, which included the inspection of ships for merchandise and contraband. A takeover would end Iran's payments to British quarantine administrators and foreseeably increase proceeds through the collection of sanitary tolls and disinfection fees, consistent with the government's effort to expand its revenues from the Persian Gulf.[117] The British, however, doubted the Iranians were acting alone and saw Russia's hand behind Naus's moves. Yet, when the British envoy in Tehran confronted the Iranian prime minister, he was told that the shah himself had given the order for the takeover: "The shah

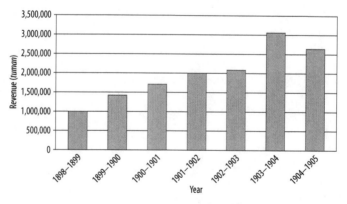

Fig. 3.3. Customs revenues in Iran from 1898 to 1905. The tuman was a unit of Iranian gold currency equivalent to ten silver qiran, the standard denomination used in Qajar and early Pahlavi periods. In 1892 1 tuman was equivalent to 29 pence (5 shillings 9 pence) or approximately 1.4 dollars at the time. Data used in graph taken from Joseph Naus's last report in Destrée, *Les fonctionnaires beiges,* 145.

had been alarmed at the appearance of plague at Bahrain, and having received information that the Indian doctors at the Gulf ports were not taking efficient precautions to prevent its spreading to Persia, had ordered the European officials of the Customs, as more versed in such matters than Persian local authorities, to ensure that the necessary measures should be adopted. His Majesty had also wished to bring from Europe a doctor who should act as Sanitary Inspector General."[118]

Joseph Naus knew that seizing the quarantine in Bushehr needed to be subtle to avoid British opposition until it was a fait accompli. Hence, the government made no official declarations on the impending takeover. Instead, Naus quietly reached out to the French embassy in Tehran to have Jean Bussière, one of their colonial medical officers, assigned to the French consulate in Bushehr to scrutinize the British quarantine arrangements and gradually assume the role of sanitary inspector general in the Persian Gulf.[119]

Around the same time, the director of the Customs Department in Bushehr began to issue orders on behalf of the Iranian government to the local British medical officers. He also officially asked them for information on the working of the quarantine station, prompting the British to confront Naus about the Customs Department's meddling. Ironically, while the British resisted ceding control of the Persian Gulf quarantines, they had no reservations when allowing the Iranians to shoulder their expense, as reflected in the following telegram by the British representative in Bushehr: "It is in my opinion very undesirable that the Customs Department should be allowed to interfere in the executive control of the sanitary arrangements in the Gulf. The financial control may be conveniently left in their hands but they should not be allowed to issue instructions to the residency surgeon whose work will be hampered if he is to receive orders from officials who have neither the requisite knowledge nor experience in such matters."[120] Naus underlined his privilege to issue orders using the channels he deemed most convenient since the Iranian Government paid the salary of the British medical officers who ran the quarantines.[121] The British were taken aback by Naus's bluntness and for not being consulted on an issue that could directly impact their "special commercial position" in the Persian Gulf. Faced with Russia's unwavering hold on quarantines in eastern Iran, they regarded the possibility of losing their sanitary authority in the South as a setback that had to be resisted at all costs.[122]

The British view of the Customs Department's staff as antagonists and proxies of Russia predated their differences over the Persian Gulf quarantines. They previously complained of a Belgian customs agent at Torbat-i Heydarieh who fined and flogged Afghans who evaded border tolls, allegedly at the behest of Russian officers. Though punishing recalcitrant merchants was legal, the British viewed enforcement on their Afghan protégés as an affront to their prestige.[123] Britain's envoy in Tehran similarly protested the behavior of a customs employee from Corsica, whom he disparagingly described as being "made of the same stuff as the bandits of his native land." Among other things, he claimed that the agent mistreated him and harassed other British subjects who crossed the Iranian frontier on pilgrimage.[124] Some Iranian politicians, including the deposed premier 'Ali Khan Amin al-Dawla, also felt that the "the Russian Embassy had made the Belgian Monsieur Naus even more compliant than Amin al-Soltan," who was generally seen as being pro-Russian.[125]

The British soon realized that Naus's plans had more ominous strategic implications. In June 1903, their consulate in Mashhad learned that the Customs Department was sending Russian-supplied tents and other military equipment to the Sistan province, bordering India, to build new quarantine stations.[126] The British saw this as Russia's response by proxy to their recent military buildup and threats of establishing their own quarantine stations in Sistan.[127] Through back channels, they threatened to punish Tehran if it did not stop Russia's "militarization" of the area. The British opted to use subtle diplomacy and veiled warnings against the Iranian government to avoid internationalizing their opposition to an internal Iranian sanitary issue, which could harm their interests. They also recognized that reinforcing their existing quarantine authority in the Persian Gulf was the best approach to thwart St. Petersburg's long-term strategic goal of establishing a foothold in Iran's southern warm-water ports.[128] The Iranian Customs Department mistakenly perceived the muted British response as an opportunity for another rigorous attempt to seize control of the Persian Gulf quarantines. On July 1, 1903, Joseph Naus sent a telegram ordering the British residency surgeon to enforce a ten-day detention on all vessels docking in Iranian ports from Bahrain where cases of bubonic plague were reported.[129] The surgeon, doubting the veracity of the reports, refused to "inconvenience" trade vessels.[130] The Customs Department's obstinacy, however, prompted a reversal in British diplomacy in favor of an openly aggressive response to the quarantine tug of war.

International Sanitary Conference of 1903

The international community revisited its guidelines on quarantines as part of a larger project to bring order to the hodgepodge of previous sanitary agreements during the last three months of 1903 at the International Sanitary

Conference in Paris.[131] The meeting consolidated the four sanitary conventions ratified in the prior decade into a 184-article-long resolution that established a unified set of measures to contain both cholera and plague. The updated codes, informed by new discoveries such as the bubonic plague's rat vector, were more lenient on human isolation and quarantines during outbreaks in Asia. Healthy passengers from plague-infected ships in the Persian Gulf were required to have a five-day quarantine or observation instead of the previously recommended ten days; and, as a protective measure, the bylaws proposed eliminating rats on board ships for the first time in international sanitary history.[132] The conference also recognized that merchandise could not transmit plague or cholera unless verifiably contaminated by infectious agents, thereby reducing the number of articles liable to exclusion or disinfection during national outbreaks. This was a victory for British-Indian shipping and exports, which dominated traffic in the Persian Gulf and had suffered since the plague outbreak in Bombay.[133] On the issue of land-based intervention, which was a source of disagreement between Russia and Britain in eastern Iran, the bylaws acknowledged the right of sovereign countries to close their terrestrial or maritime borders against epidemics and to detain and investigate individuals suspected of having a transmissible illness. However, in an apparent nod to the British, the conference participants recognized the superiority of surveillance to detention or isolation by quarantine as the most beneficial way to protect against the overland spread of contagious diseases.[134]

Adrien Proust, who once again represented France, called for establishing a standing international arbitrating body in Paris to mediate future disputes over sanitary matters, including quarantines. He also revived the proposal to establish an international sanitary station in the Persian Gulf, this time on the island of Hormuz, funded and managed by the International Board of Health in Constantinople (Istanbul).[135] Iran, which unconditionally ratified the convention, likely hoped that an Ottoman sanitary presence on its sovereign territory of Hormuz could serve as a counterweight to British hegemony in the region. The same desire probably motivated Russia to support the measure, despite its delegate's doubts on the "international" nature and competence of the Board of Health in Istanbul.[136] The British, on the other hand, filed their objections to the proposed Hormuz scheme and made it understood that the convention did not apply to their colonies, possessions, or protectorates.[137] Unbowed, the French continued to lobby other dissenting European powers to support the plan. Their government even sent an unsuccessful sanitary mission to the Persian Gulf after the conference to study the efficacy of the proposed station, hoping to use the on-site evidence to convince London to withdraw its objections.[138]

The Iranians used the Hormuz deliberations to highlight their sanitary achievements since the 1894 gathering in an attempt to persuade the international community that they no longer needed Britain's assistance in the Persian Gulf.[139] One of their delegates, Amir Khan Amir A'lam, newly graduated from the French Military Teaching Hospital of Val-de-Grâce, argued that Iran had the expertise and material resources to administer its own quarantine establishments.[140] In addition to experienced foreign physicians in Tehran's employ, he pointed out that the Iranian government had developed a cohort of capable native medical graduates, educated in the West, who could replace the British. He also indicated that Tehran was in the process of purchasing equipment from France to establish a bacteriological laboratory in Bushehr and had already supplied the port with the necessary disinfection appliances, such as a sterilizing steam oven. Even more worrisome for London, Amir A'lam announced that his country's Customs Department could immediately undertake the responsibility of administering the quarantines since it already had an inspection system in place for shipboard contraband and the necessary organizational capabilities and staff, including physicians, to enforce the government's sanitary writ: "As I had the honor of telling you, we possess at this time, all that we need to safeguard the sanitary defense of our country in general and our Persian Gulf ports in particular. The government of H.I.M the Shah, to even further safeguard the sanitary defense of its ports, has recently attached its management to his ministry of posts and customs, which is headed by an eminent European, M. Naus, whose name is well known in Europe."[141]

Losing the Persian Gulf Quarantines

The appointment of 'Abd al-Majid Mirza 'Ayn al-Dawla, an alleged Anglophobe, as prime minister in January 1904, tilted the Iranian government against British interests.[142] The residency surgeon's biased practices, including giving clean bills of health to British ships while impounding Russian vessels from the same point of origin, added to Tehran's antagonism. When questioned, the residency surgeon declined to provide an explanation on the issue to Iranian officials on grounds of being answerable to the British government alone.[143] These jurisdictional disagreements not only threatened Iran's sovereignty but risked drawing Tehran further into the Russo-British fray; incentivizing the Iranians to reassert their control over quarantines in the Persian Gulf.

In 1904 Jean Bussière assumed his position as physician to the French consulate in Bushehr and began establishing himself as sanitary inspector general in the Persian Gulf on the Customs Department's payroll. With Naus's support, one of his first activities was to evaluate the state of quarantines in Iran's southern ports.[144] His subsequent report to Tehran likely influenced the Iranian government to assign one of its own physicians to take over the sanitary measures and the quarantine establishment in the Persian Gulf port of Muhammara. The Iranian physician was instructed to bypass the British residency surgeon and report directly to the prime minister and the foreign minister in Tehran, contrary to previous agreements with London.[145] Around the same time, the governor of Bushehr, at the instigation of the Customs Department, called for a sanitary meeting to protect the town against plague and cholera. The gathering challenged British authority by recognizing the governor's "supreme charge" over matters of public health, including quarantines, in Bushehr.[146]

The British viewed these developments as an attempt "to dislodge the Residency surgeon."[147] Their suspicions grew when the government's physician in Muhammara and Jean Bussière in Bushehr began raising funds, through local subscription, to pay for their growing sanitary activities and to purchase antiplague serum from Paris.[148] Making matters more worrisome for the British, the Customs Department directly ordered the residency surgeon to quarantine vessels from Bahrain. As before, the residency surgeon "regretted his inability to accept instructions from the Customs Department."[149] Unbowed, Bussière and his allies initiated a second meeting of the local Sanitary Council to erode the residency surgeon's power. The governor, the director of the town's customs, Iranian notables, and the British residency surgeon himself attended the gathering. By the end of the meeting, the residency surgeon agreed to split Bushehr into three sanitary zones under the pretext of facilitating "medical supervision." The British were given authority in two of the three areas; however, the most strategic zone that included the town's port and quarantine station was delegated to Bussière.[150] In addition, Bushehr's Customs Department was empowered to fine and punish sanitary offenders, further enhancing Bussière's authority in matters of enforcement.[151] By the end of July, the residency surgeon's influence appeared to reach its nadir when Bussière and the director of the Customs Department ordered the British-built quarantine sheds on the landward side of Bushehr removed and all vessels from infected ports turned away.[152] This prompted the British ambassador to instruct Joseph Naus to cease his "interference" with quarantines in the

Persian Gulf, warning that the British-Indian government would forcefully confront any attempts by Tehran to diminish its influence: "It was only natural that the Government of India should take up the glove thus thrown down and should resist amongst other things, any attempt to undermine an institution, such as our control of the quarantine arrangements affecting shipping, mainly British, which constituted a legitimate recognition by Persia of our peculiar and predominant interest in those waters."[153] An unperturbed Naus reminded the ambassador that the residency surgeon was obliged to follow Tehran's writ since the British physicians in the Persian Gulf were "on loan" to his government, which also paid their salaries.[154] To bolster his position, Naus pointed out that the Russian quarantine staff never refused orders from the Sanitary Council in Tehran, despite the fact that the Russian government paid for the entirety of their operations in Iran.[155] Having failed with Naus, the British raised their concerns directly with the Iranian foreign minister and prime minister who "failed to see why the efforts of the Persian Government to protect its own territory and subjects against epidemics in the manner it thought best should be deemed unfriendly."[156] They demanded to see written evidence that Tehran had formally accepted the existing British-run quarantine arrangements, as London claimed. Because prior diplomatic understandings on the issue were informal, the British could only reemphasize their commercial interests in the Persian Gulf to justify prolonging their unconstrained control.[157]

Iran's boldness was short-lived, however, in the face of growing British diplomatic protests and military threats that climaxed the following year when it dispatched a battleship to Bushehr.[158] The shah was also less inclined to confront London, whose concessionary loans he needed more than before, following Russia's shift in policy away from seeking wholesale concessions from Iran. The Customs Department's ability to challenge the British also diminished due to an economic downturn and growing popular opposition to Joseph Naus's policies in Iran. Naus's stringent application of commercial tariffs and his efforts to reform the land tax had antagonized the country's merchants and ruling landowners, accustomed to evading government levies.[159] After three years of popular "agitation," Naus was dismissed from his post and forced to leave the country altogether.[160]

Meanwhile, Britain strengthened its position in the Persian Gulf by resolving its major overseas differences with France in the Entente Cordiale, ratified in April 1904. In the ensuing year, Paris relaxed its demand for an international sanitary station on Hormuz in favor of a British proposal to establish a sentinel quarantine station on Hengam Island, a location where London already had a foothold and which it considered less onerous for the diversion of its maritime traffic from India.[161] Britain's relationship with the Iranian Sanitary Council also improved due to the efforts of Jean-Etienne Justin Schneider, who turned from London's antagonist to its collaborator. Encouraged by the French ambassador, Schneider used his position as president of the Iranian Sanitary Council to help resolve Tehran's differences with the British-run quarantines in the Persian Gulf. This earned him the gratitude of the British ambassador in Tehran and the Indian government, which recommended him for a medal several years later.[162]

By September 1904, the Iranian government formally changed its position on quarantines in the Persian Gulf and ordered its Customs Department to cease interfering with the British residency surgeon.[163] In return, the British acknowledged that the shah, "having sovereign authority in sanitary and quarantine matters, could transmit his orders [directly] to them."[164] The residency surgeon also expanded his activities to include providing healthcare to the local population through the construction and funding of a charitable clinic in the port of Bandar Abbas.[165] Even with these concessions, Britain's supremacy over quarantines in the Persian Gulf was by no means secure. The victory of Japan, an Asian country, over Russia, a European power, in the Battle of Port Arthur in 1904 energized reformers in Iran who called for a more assertive government that was not beholden to Russia or Britain.[166] Iran's lack of autonomy in the Persian Gulf, where the local governor was seen as a "servant of the British," was particularly contentious for Iranian reformers and intellectuals.[167] This rising tide of Iranian nationalism motivated the British to seek a permanent understanding with St. Petersburg on quarantines and, in so doing, hoped to reduce the likelihood that Russia would instigate an increasingly unfriendly Iranian population against London's interests in the Persian Gulf.[168]

Conclusion

By 1904 Russia and Britain assumed an oversized role in Iran's public health. British physicians oversaw the quarantines in the Persian Gulf ports of Muhammara, Bushehr, Bandar Abbas, Bandar Lengeh, and Jask. Russian Cossacks and physicians managed the quarantines at Torbat-i Heydarieh, Soltanabad, and Kariz; and, additionally, St. Petersburg funded and controlled eleven other sanitary posts between those stations in eastern reaches of the country (fig 3.4).[169] The growing grip of both powers around the

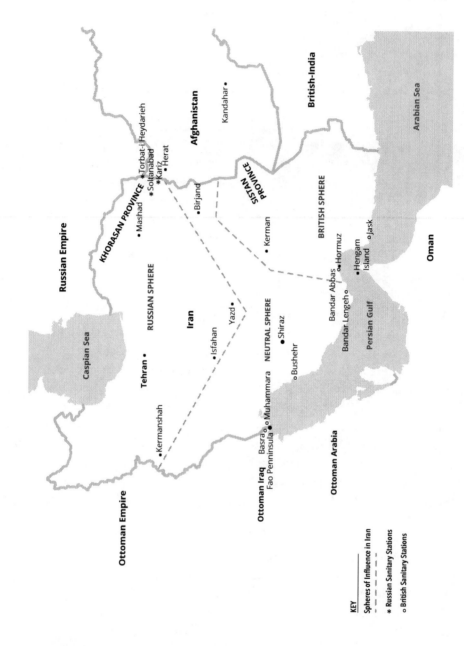

Fig. 3.4. Russian and British quarantines and sanitary interests in Iran.

administration of quarantines in their respective zones of influence forced the Iranian government to increasingly subordinate its sanitary policy to the commercial and political interests of London and St. Petersburg, even after the Anglo-Russian rivalry diminished.

Russia, weakened by defeat and internal unrest, began contemplating a larger negotiated settlement with London over conflicting concerns throughout Asia in the closing month of 1904. The British, no longer perceiving the Russians as their principal strategic rivals and anxious about the rise of German power and Iranian nationalism, also favored a grand bargain with St. Petersburg to resolve outstanding disagreements in the region. The resulting negotiations culminated in the Anglo-Russian Convention of 1907, which divided Iran into three zones: a northern zone under Russian sway, a middle "neutral" zone, and a southern British zone. The two powers pledged not to seek political or commercial concessions in the other's zone of influence, thereby forestalling conflicts in the future.[170] The Russians specifically acknowledged Great Britain's "special interest" in the Persian Gulf and agreed to "consult" London on all sanitary matters, including quarantines, in southern Iran.[171] This settlement narrowed the Iranian government's ability to maneuver around imperial interests by playing off one power against the other to extract concessions on quarantines and other forms of sanitary and medical assistance. Even the Sanitary Council in Tehran assumed a technical role, staying "out of political and financial questions for which the Council declined all competence."[172] Iran's vanishing sanitary autonomy and its growing subordination to this informal imperial control made it increasingly vulnerable to epidemics from abroad.

CHAPTER FOUR

Cholera, Germs, and the 1906 Constitutional Revolution

Biomedical knowledge in Iran continued to advance in the first decade of the twentieth century, despite the country's eroding sanitary autonomy and halting institutional progress in medicine and public health. Iran's literate strata increasingly understood the science of microbes, their pathology, and ways to prevent germ-borne illnesses thanks to a growing number of Western-trained native physicians, as well as the expansion of medical writing in Iranian newspapers and periodicals. The increasingly accepted empirical explanations for the cause of outbreaks progressively dispelled religious beliefs that associated epidemics with divine punishment and pushed back the traditionally fatalistic attitudes that had contributed to the lack of communal disease prevention efforts in Iran.

The emerging field of bacteriology played a key role in broadening the acceptance of public health as a professional discipline guided by scientific principles. Iranians increasingly considered hygiene and municipal sanitation a shared responsibility by every citizen, and reformers increased their demands for the government to invest in the country's sanitary infrastructure. However, political and economic paralysis largely prevented the realization of these new demands. This became apparent when Tehran failed to secure the equipment necessary to establish a bacteriological laboratory at the Polytechnic College in the closing months of 1903.[1] The Qajar administration's persistent inability to translate intellectual progress into lasting institutional and infrastructural reforms for public health perpetuated Iran's vulnerability to pandemics at a time when large-scale outbreaks were increasingly rare in the West. The government's failure to fulfill its sanitary mandate also fueled a growing revolutionary undercurrent that sought to bring social justice to Iran by changing what was perceived as a sclerotic Qajar order bent on self-enrichment at the expense of its people's welfare.

Sanitary and Clinical Conceptions of Cholera in the New Century

The turn of the century was an era when the importance of public health, particularly the science of disease prevention, worked its way to the forefront of urbane conceptions of the "new civilization" (*tamaddun-i jadid*) to which Iranian intellectuals aspired. This was epitomized by a flurry of articles written in the popular weekly newspaper *Tarbiyat* extolling the importance of maintaining a salubrious and hygienic environment for all of the shah's subjects.[2] These pieces framed sanitation primarily in the urban context, as an extension of the broader standards for "municipal maintenance" in the modern era. Tehran was the main site where the government's sanitary performance was evaluated. Its chief administrator was praised for his "tireless toiling to maintain the wellbeing of the capital's inhabitants."[3] Iranian thinkers also increasingly characterized public health maintenance as a civic responsibility.[4] Their emphasis on hygiene as an individual's moral duty grew with the surge in nationalism and patriotism at this time.[5] The new century also witnessed the rise of popular medicine in Iran, embedded in the germ theory of disease, which advocated higher standards for personal hygiene. These views were promoted by a growing number of European-trained native practitioners of medicine and by nationalist reformers. Their statements and editorials in *Tarbiyat* informed readers of the evils of drinking from the communal jugs (*kuzih*), smoking shared water pipes, and washing the dead in municipal waterways. Reselling clothes of the contagious sick was cited as a particularly reprehensible offense, "akin to murder, and such people would have the same fate as murderers on judgment day."[6]

Following their Western counterparts, Iranian newspapers began to rely on the recommendations of experts to shape their readers' notions of disease and prevention.[7] This occurred with the rise to prominence of the first generation of physicians, who were educated at the Polytechnic College and sent to Europe for their advanced medical training. As the numbers of European-trained Iranian physicians grew, so did the interest and confidence in European theories of disease and therapeutics. These native physicians were better able to disseminate and popularize the microbial cause of illnesses.

The biography of Mirza Muhammad Khan Kirmanshahi, an expert contributor to *Tarbiyat* on cholera, exemplifies this cadre. Kirmanshahi began his higher education as a clerical student in the holy city of Najaf. He learned French from two fellow travelers during his passage back to Iran upon completion of his religious studies.[8] He then enrolled in the newly inaugurated Polytechnic College and studied Galenic Medicine under the direction of Hajj Mirza 'Abd-al Baqi Tabib Hakim Bashi I'tizad al-Attiba.[9] He also learned Western medicine (*tibb-i jadid*) from Joseph Désiré Tholozan and purportedly performed the first autopsy by an Iranian.[10] After graduating, Kirmanshahi administered the quarantine near his native city of Kermanshah during the cholera outbreaks of the 1860s. In 1870 he left for Paris, at the urging of Tholozan, to further his medical education. He completed his thesis at the Medical Faculty of Paris under Georges Dieulafoy, a pioneer in gastric surgery, obtaining his doctorate and license to practice medicine in France in 1879.[11] After returning to Iran, Kirmanshahi met and impressed Nasir al-Din Shah, who appointed him to various positions, including attending physician to the shah himself, head of the Imperial Hospital, and physician to the Iranian telegraph department. He also became a professor of medicine at the Polytechnic College, where he taught students to use the microscope and to perform histopathology for the first time.[12] He was the first Iranian physician to demonstrate red blood cells in vivo and integrate "Pasteurian microbiology" into his medical practice.[13] However, Kirmanshahi's radical views on religion, developed during his residence in France, progressively alienated fundamentalist clerics in Tehran, who began calling him infidel (*kufri*).[14] After several years, Nasir al-Din Shah dismissed him from his service, and the government ousted him from the Polytechnic College and the Imperial Hospital due to pressure from conservative notables.[15] This forced him to work from his house in Tehran, which he turned into a private clinic and a proprietary school of medicine.[16]

By the twentieth century, Kirmanshahi's views were more in tune with popular opinion. The growing breed of European-educated Iranian doctors from the provinces, like him, were better able to articulate emerging medical theories from the West and make them accessible—and acceptable—to their lay countrymen. Still, the embattled medical traditionalists continued to oppose him, as recalled by one of his students:

> He [Kimanshahi] taught his students new subjects in the science of medicine and biology, and these subjects did not conform with the views of some, especially the physicians who were in general opposed to new medicine [*tibb-i jadid*]; it did not bode well. As a result, around his [Kirmanshahi's] medical practice in Tehran two groups had aligned themselves against each other: a group that supported Doctor Kirmanshahi, had faith in the new medicine, and had been to the

West, and another group that was as a rule opposed to new medicine. The leader of the latter group was Sultan al-Hukama (Mirza Abul Qasim Na'ini). As this story played out, war between the believers in old medicine and new medicine continued for some time in Tehran (and in provinces as well).[17]

These Galenic practitioners felt besieged by those Iranians, especially among the elite, who questioned their traditional humoral prescriptions and looked to physicians like Kirmanshahi and to the germ theory he promoted for both medical treatment and sanitary strategies to stop the ingress and dissemination of epidemics in Iran.[18]

Kirmanshahi's 1904 article on cholera in *Tarbiyat* described Robert Koch's discovery of the disease's "seeds" in the rice-water stools of advanced cholera patients twenty-two years earlier. He illustrated Koch's bacillus as a one- to two-millimeter organism with a thin body and a thicker head, in the form of a comma, "the same as microbes."[19] Clearly, Iranians had gained a better understanding of microbial entities since the "transparent worms" of the 1890 influenza epidemic. In accordance with Koch's postulates of disease transmission, Kirmanshahi indicated that infective microbes, cultured in vitro, could give rise to illness when introduced into healthy subjects and even described self-experimentation using laboratory-grown microbes.[20]

Public comprehension of the actual pathology by which infectious organisms caused a diseased state also increased with the popularization of the microbial paradigm. More Iranians understood the emergence and course of illness as an organized and rational (albeit less familiar) microbiological process rather than the disorganized actions of the voracious transparent worms of the previous decade. Kirmanshahi's narrative, for example, divided the life cycle of the cholera bacillus into four parts. The first asymptomatic stage lasted from several hours to ten or eleven days.[21] The second "prologue" stage started at the point of cholera's symptomatic appearance, described as particularly virulent due to the large microbial load in patients' diarrhea. The "revelatory" third stage began when the diarrhea acquired white flakes, giving it cholera's hallmark "rice water" appearance that resulted from "the destruction of the intestinal passageways."[22] Worsening abdominal cramps, dry mouth, thirst, "freezing" extremities, slowing pulse, and cessation of urinary output would follow, and the patient became "gaunt" and mostly unresponsive, "like a corpse."[23] In the fourth and final stage of cholera's life cycle, Kirmanshahi explained that the body either cured itself or lost the battle. One of the principal signs of recovery was a rising core temperature. Patients' would then regain much of their color and progressively recuperate within a few days.[24] Kirmanshahi's detailed explanations of cholera's clinical and pathological process reflected the growing etiological sophistication of Iranian physicians. The fact that this undiluted technical information was communicated in a weekly broadsheet to a largely nonmedical audience showed the lay population's rising biomedical literacy as well.

Iranian literature on cholera during the nineteenth century generally described the individual as being either diseased or healthy on the basis of symptoms.[25] The sick would experience a largely linear deterioration, eventually resulting in death or a turnaround toward recovery. The science of bacteriology opened Iranian thinkers and nonprofessionals to the notion that the appearance of wellness and the absence of symptoms did not rule out disease, as conveyed in Kirmanshahi's descriptions of the "latency" phase of cholera. Of even greater significance was an understanding that cholera did not necessarily have an ordered progression; in some cases it could kill a person "within hours," whereas on other occasions it could simply fade into a mild and nondebilitating illness.[26]

The lionization of Pasteur and Koch in Iranian newspapers around the same time reflected the rising status of bacteriology and the scientists who studied it.[27] This went hand in hand with the growing acknowledgment that microbes were the principal agents of disease and, with it, an expanding view that vaccines were their weakness. For millennia, Iranians looked to the great philosophical opuses by the likes of Hippocrates, Galen, and Avicenna for answers on disease prevention and treatment. In the first decade of the twentieth century, they increasingly recognized that the conquest of epidemics did not reside in the writings of worldly philosophers but rather in Western laboratories equipped with instruments of modernity—namely, microscopes, test tubes, and vaccines.

The rising prestige of bacteriology in Iran helped broaden its lay populations' understanding of the pathophysiological process behind the clinical signs of cholera. This included how the cholera bacilli caused rapid fluid loss in patients, an essential feature of the disease's presenting symptoms and the main cause of its lethality.[28] Much of the treatment recommendations for cholera in Iran centered on replenishing body fluids lost to massive diarrhea and emesis, as explained by Sa'id al-Attiba, a prominent physician in Tehran: "Because the patient loses much of his blood-water in a short period owing to this disease, the person becomes so weak that it seems like he has been anemic for years. To remedy this problem, a special synthetic serum, meaning artificial blood-water, is introduced into the body of the patient. This procedure is very significant, and I have treated several patients . . . and obtained astounding results."[29] Iranian households, lacking serum and the technical knowhow for intravenous fluid infusion, resorted instead to administering a cocktail of chloroform, sour-iced lemonade, cinnamon, coffee, and tea "pending the arrival of a physician."[30] The consumption of "lactic lemonade" (*limunad-i laktik*), a mixture of lactic amide, water, and sugar, was considered especially effective in stopping diarrhea in cholera patients. Other treatments included undergoing a hot-water enema, drinking tonic, applying heat wraps to warm the extremities, and ingesting Laudanum to slow gastric emptying.[31]

Nowhere in Iran had the face of medicine so radically changed than in the newly popular philosophy of hygiene. In the 1890s, the use of disinfectants was limited to sanitary authorities and physicians, a practice sometimes ridiculed by the public. A Tehran notable's observation of a doctor's habits during the 1892 cholera epidemic was characteristic of this hesitation: "I realized that the doctor's shirt was almost black with dirt. I inquired, shocked, only to find out that the doctor believed washing clothes would be contamination. He soaked his shirts and pants in Tehran with carbolic acid and refused to change. He began an extensive second lecture on the matter when he realized that I was scrutinizing his clothes. He made me sick of all the modern-minded people. He was the most harmful virus himself."[32] Iranians increasingly accepted the use of diluted corrosives and other disinfectants in tandem with the growing awareness of disease-causing microbes by the turn of the century. Newspapers advised cholera-stricken households to use sulfate or sulfuric acid to decontaminate their yards and other outdoor disposal sites, to sterilize patients' plates and utensils, and even to use the disinfectants in hand washing.[33] Iranian advocates of popular medicine also promoted the use of antiseptics as a preventative before the actual manifestation of cholera. The newly popularized notions of microbialism and antisepsis can thus be credited with establishing stronger individual responsibility to prevent outbreaks as Iranians faced a renewed epidemic of cholera.[34]

The 1904 Cholera Epidemic in Iran

Ottoman Iraq, bordering Iran's Southwest, grappled with a seemingly interminable cholera epidemic in the latter half of 1903.[35] The contagion reached the Euphrates valley from the trading hub of Aleppo in the Levant and eventually spread south to Karbala during the holy month of Ramadan, in November of that year. Shi'ite Iranians made their annual pilgrimage there in this period to pay homage at the revered site of Imam Husayn's martyrdom and to bury their dead in hallowed ground.[36] Their returning convoys, which were purposely overcrowded to dissuade marauding Kurdish tribes, created an ideal setting for cholera to thrive. The cramped and unkempt caravanserais along the way also lent themselves to spreading the disease. In less than a month, the contagion slowly followed the caravans to Iran's frontier, where it wreaked havoc in Ottoman villages on the border.[37]

Unlike Iran's Persian Gulf ports and its eastern reaches, the sanitary defense of its border with the Ottoman Empire remained under Tehran's jurisdiction. In late December, the Iranian foreign minister sent a telegram urging 'Abd al-Husayn Mirza Farman Farma, the region's governor in Kermanshah, to implement a ban on the transport of cadavers across the border for burial in Karbala.[38] The minister also declared a ten-day quarantine on arrivals from Ottoman territory and began preparations to send 150 Iranian Cossack troops and a government physician to enforce his directives along with tents and provisions to establish several quarantine stations at the frontier. The governor was instructed to lay the groundwork for the stations and obtain several large cauldrons, paid for by Tehran, to boil and disinfect contaminated apparel. However, the expedition, which would have taken almost a month to reach the area, did not leave Tehran for three weeks because of the government's inability to gather the necessary funds and supplies.[39] A very limited twenty-four-hour-long quarantine was eventually established by the local agents of the Iranian Customs Department near the border town of Qasr-e Shirin for anyone who had not previously undergone inspection or isolation at Ottoman sanitary stations. But this was inadequate for the considerable pilgrimage and trade traffic that traversed the area, making the epidemic's introduction into the country only a matter of time.[40]

Iran's lethargic quarantine arrangement worried the Russians, whose adjacent Caucasian territories were vulnerable to a potential outbreak in the region. The Russian government rapidly dispatched a medical officer to evaluate the operation at Qasr-e Shirin, threatening to close its entire border with Iran if his suggestions were not adopted. The British viewed Russia's threat as an attempt by St. Petersburg to blackmail the Iranian Customs Department with the financial consequences of a blockade to obtain control over the quarantine.[41] Despite their recent de-escalation, the two powers were still three years away from finalizing the Anglo-Russian Convention, causing the British concern that Russia could use the quarantine to harm their growing commercial interests in Iran's western districts, particularly along the strategic Baghdad-Kermanshah road.[42] This prompted their ambassador in Tehran to send his own medical representative to the area "to balance" Russia's sanitary influence.[43] The Iranian government, fearing further erosions to its sovereignty, followed suit by rapidly hiring and sending the Ottoman embassy's contracted physician, Hamid Riza Vaume Bey, to the border region to neutralize any justification for an international takeover of the quarantine.

On the surface, Vaume was an unlikely choice to represent the shah's interests. He was an elderly Frenchman who openly harbored radical republican views that bordered on communism and had spent his entire professional life in Egyptian and Ottoman employ. He also caused a scandal in Tehran's European colony a year earlier by converting to Islam, which allowed polygamy, to wed a "fat middle-aged" Frenchwoman of ill repute before divorcing his wife in France. Despite his quirks, Vaume's extensive experience with similar epidemics in the region and his Ottoman credentials made him an ideal candidate to coordinate arrangements at the border with the sultan's government.[44] Moreover, the Iranians likely realized that the European doctor would have a better chance of reining in his Russian and British colleagues than would a native.

While the physicians prepared to leave the capital, the feared epidemic slowly crept across Iran's frontier in the direction of Kermanshah with returning pilgrims, who had learned to circumvent quarantine by veering off the main roads as they approached a sanitary post (figs. 4.1, 4.2).[45] They carried the epidemic northward where it killed an Iranian muleteer on his way back from Baghdad and spread to pilgrim encampments around Kermanshah. An official declaration of three casualties in Kermanshah on April 8 marked the start of the outbreak in Iran.[46] The epidemic quickly spread to the city's previously intact Jewish quarter through the sale of tainted clothes recycled from victims of cholera by Jewish rag peddlers.[47]

Kermanshah's urban landscape helped disseminate cholera, much like Tehran's during the 1892 outbreak. Most of its households were overcrowded and unhygienic, with as many as seven or eight people living in a single-roomed dwelling. Its water system, supplied by springs located three miles away, reached the city through open channels that passed through several villages. This made the water vulnerable to pollution even before reaching the city. Just as in Tehran, the water from these channels flowed from house to house to fill the residential storage cisterns used for both drinking and household washing needs. To make matters worse, Kermanshah's interconnected cistern

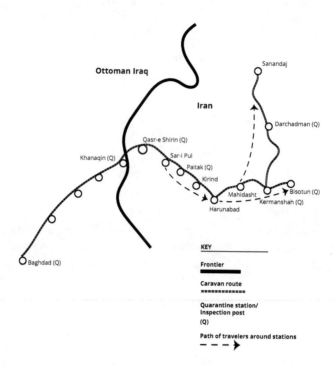

Fig. 4.1. How sanitary stations along the Ottoman frontier were evaded. Data based on Scott, "The Recent Cholera," 620.

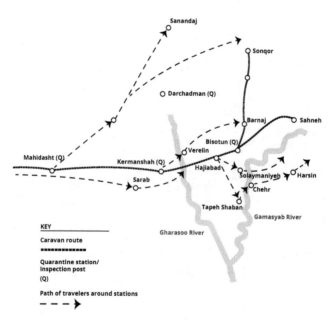

Fig. 4.2. How sanitary stations near Kermanshah were evaded. Data based on Scott, "The Recent Cholera," 621.

system eased the spread of cholera through the whole city's potable water system even if one cistern was contaminated. Its underground sewer system, often left uncovered, also exposed the population to infected fecal matter because of its tendency to overflow, particularly in the poorer parts of town where the gutters surfaced. The surrounding agrarian population used a portion of the runoff for the cultivation of vegetables while the rest drained into the Gharasoo River, which ran adjacent to the city. The germ would then be reimported via unwashed fruits or travel downstream, infecting riverside populations.[48]

With a growing outbreak at hand, Tehran ordered Farman Farma to assist the Customs Department with tents and equipment to establish additional quarantine stations around Kermanshah, near Mahidasht and Bisotun.[49] Upon arrival, Vaume Bey took over the management of the region's existing sanitary arrangements. The British, satisfied that their physician's presence in Kermanshah had halted Russia's intrigues, did not oppose his leadership.[50] In addition to expanding the quarantine, Vaume Bey organized a sanitary department to improve Kermanshah's hygiene with four local Iranian doctors and a French chemist, who had accompanied Vaume Bey from Tehran. The new department rapidly disinfected the homes of the sick and regulated burials under the strictest hygienic guidelines.[51] Controlling funeral rites was particularly difficult, as it challenged established religious customs. The Jews of the city, for example, refused to bury their dead on Sabbath, keeping them exposed for a full twenty-four hours against the department's orders, and Kermanshah's Muslims similarly resisted restrictions on ritual washing and rapid internment of the dead. The sanitary department's use of tact, dialogue, and coercion ensured the short-term success of these efforts, but the tacit opposition foreshadowed a growing religious resistance to sanitary interventions in the region.[52]

The response to cholera in Kermanshah was mostly a local affair owing to Tehran's distance and its financial difficulties. The region's Customs Department agents administered the quarantines and provided the operational support needed to sanitize the city and treat the ill, while Farman Farma, the governor, provided most of the material resources, including tents and foodstuff.[53] Also, immediately before the outbreak, he established a fifteen-bed hospital, which included a well-stocked pharmacy "with European medications," an operating room, and its own artesian spring water source.[54] Kermanshah's municipal authorities did not flee in panic from cholera as they had during previous outbreaks owing to publications like *Tarbiyat*, which made the disease seem less mysterious and terrifying. Its governor, aware of cholera's fecal-oral microbial etiology, prohibited the sale of fruits in the city's markets. He also required everyone to wash their "hands and face with phenicated water" before being admitted to see him.[55] The city authorities forbade customary visits to the homes of the sick and ordered municipal functionaries to report cholera cases to Kermanshah's newly constituted sanitary department. Other measures taken by the city included burning certain belongings of the sick, such as beddings, cloths, and rugs, and confining itinerant rag sellers to a caravanserai outside the city walls to ensure that the clothing of the deceased would not be "recycled."[56] The local government converted another caravanserai to a cholera hospital for the poor and destitute, where medicines and disinfectants were supplied free of charge.[57] This response contained the contagion and reduced casualties by the second week of April, until Shaykh Muhammad Hasan Mamaqani, a senior Shi'ite cleric from Najaf, and his entourage broke the lull.[58]

Cholera, Religion, and Rebellion

Grand Ayatollah Shaykh Muhammad Hasan Mamaqani was the most senior religious jurist (*mujtahid*) in the seminary city of Najaf in Ottoman Iraq. He was considered a source of emulation (*marja'-i taqlid*) by the Shi'ite faithful, and "his book of prayers was accepted by half the population in Iran."[59] The emotions that the octogenarian elicited far outweighed his diminutive and listless appearance. He had a pale complexion "like that of a corpse" and a prominent nose that protruded from under his "gigantic" turban.[60] His public appearances were usually in mosques, seated aloft on a pulpit (*minbar*) that rose above a sea of adoring devotees who struggled to reach him so they could kiss his hands and feet and hear his pronouncements.[61] Like Mirza Muhammad Hasan Shirazi, his nineteenth-century predecessor who led the Tobacco Protest, Mamaqani was also active in Iranian politics—albeit to a more limited degree. In 1903 he helped bring about the dismissal of the shah's former premier, Mirza 'Ali Asghar Khan Amin al-Sultan, by signing a decree accusing him of being an unbeliever.[62] The shah's new prime minister, 'Abd al-Majid Mirza 'Ayn al-Dawla, therefore, owed his position to Mamaqani and other Shi'ite clerics who had campaigned against the "apostate" Amin al-Sultan.[63] Mamaqani did not miss this fact when he communicated his displeasure at being quarantined at Qasr-e Shirin to Farman Farma and 'Ayn al-Dawla in the following telegram:

He [Hasan Mamaqani] is detained here to-day, and troubles your Highness by asking you to do your best to relieve him of suspense and ease the minds of Moslems. Almighty God is witness that through fatigue, old age, weakness of health, and extremity of uneasiness, he had no sleep all night. Are not the Turkish Quarantine and passports sufficient for these poor helpless Shiahs, that this trouble should be added? This is what I have heard about the ascendancy of unbelief [*tasallut-i kufr*] which has become apparent at the first station. Is God willing that while you are in possession of all the bounties of heaven, his slaves, and the guests of the martyred Seyed, Imam Husain (i.e., pilgrims) should remain in the extremity of misfortune and distress and perhaps even hunger?[64]

The 1903 decree that charged Amin al-Sultan with apostasy also denounced his policies for contributing to the "ascendancy of unbelief" (*tasallut-i kufr*). By applying the same phrase to his confinement, Mamaqani placed 'Ayn al-Dawla on notice that he could suffer the same fate as his predecessor unless he remedied the situation. The prime minister responded that the quarantine precautions "had been taken in accordance with the indications of Islam" and to protect Muslims in Iran from cholera.[65] By appealing to Mamaqani's religious sensitivities, 'Ayn al-Dawla hoped to deflect accusations of being an unbeliever. However, his explanation that the quarantines helped pilgrims and kept the Shi'ite shrines free of disease did not appease the cleric. Mamaqani saw the restrictions as "child's play" and the "innovation of infidels," which only hampered pilgrims from performing their religious duties.[66] He viewed the Customs Department's European functionaries in charge of the quarantines as "impious infidels" attempting to "defeat divine will."[67] The prime minister had no choice but to instruct the governor in Kermanshah to show Mamaqani "the greatest reverence" and allow his caravan to pass unhindered through the quarantines in the area.[68]

On April 11, the cleric and eight hundred of his followers left their confinement in Qasr-e Shirin and continued toward Kermanshah carrying cholera back into the city.[69] Mamaqani's caravan not only infected the city; it also spread the cleric's opposition to quarantines to Kermanshah's inhabitants, who increasingly resented the government's stringent sanitary measures.[70] In this already tense environment, the city's bazaar was abuzz with rumors of European physicians administering lethal pills to patients so they could use their corpses for medical purposes.[71] This rumor apparently began when a victim of cholera rapidly deteriorated and died after taking antidiarrheal medications prescribed by the recently arrived Russian medical officer.[72] The population's anger reached a boiling point by the third week of April and resulted in city-wide demonstrations against European physicians and the closing of the bazaar in protest.[73] The riots eventually converged on the governor's palace, and protest chants exhorting 'Ali (*ya 'Ali*), the revered cousin of Prophet Muhammad, could be heard throughout. Like the riots in the city of Yazd during the 1892 cholera epidemic, the city toughs (*luti*) became a catalyst for this uprising. The danger to Europeans increased further due to religiously motivated xenophobia fanned by Mamaqani and his followers. This prompted the British to remind the city's governor that they held him responsible if anything happened to their subjects.[74] The governor, lacking the strength to quell the unrest, ordered the cessation of restrictive sanitary measures.[75] This reduced tensions immediately and slowly brought an end to the protests and strikes in Kermanshah.

Mamaqani's caravan continued its northeastern course, picking up "hundreds of beggars, dervishes, and dirty and diseased folk" along the way.[76] At every stage, the cleric challenged the government's policies, publicly condemning quarantines and the tolls levied on the road to Qom by the British-owned Imperial Bank of Persia. He encouraged "believers" to refuse the levies, preaching that payment at the tolls was tantamount to impiety.[77] In Hamedan, Mamaqani spoke out against "the faithful being prevented from visiting the sacred shrines, and described the quarantine measures as [sacrilegious] innovations of the Europeans."[78] He practiced what he preached by personally defying the government when the opportunity presented itself. He disobeyed a travel ban and quarantine on Kangavar, "a most unsanitary village" located on the outskirts of Kermanshah, by accepting an invitation to enter the settlement.[79] A mob of villagers and followers of Mamaqani beat Vaume Bey when he attempted to stop the caravan. Fearing for his life, Vaume Bey lifted the quarantine around Kangavar and fled the area with his French deputy.[80] In the town of Sabzevar, Mamaqani intervened in a dispute between a Muslim and a Christian Armenian over a sale of wine, ordered the Armenian flogged, and decreed that all Armenians should be driven out of their shops. The local authorities carried out his command against Tehran's wishes, causing the town's terrified Armenian community to seek refuge in the British telegraph office until the cleric's departure.[81]

Mamaqani eventually reached the vicinity of Tehran and camped on the grounds of the Shah Abdol Azim shrine, with his army of followers, to meet Muzaffar al-Din Shah. He used the encounter with the monarch to endorse the newly minted premier who had freed him from the quarantine in Qasr-e Shirin, while the hypochondriacal shah spent the entire meeting complaining of his failing health. The shah, who wanted Mamaqani's blessing, derived "real or fancied" relief when the cleric touched the sides of his forehead. He also drank Mamaqani's washings, to which he attributed healing powers, as described in this British diplomatic dispatch: "A basin of water was brought, and the shah himself washed Mamaghani's hands in it, after which his Majesty drank part of the washings and ordered the rest to be taken back to the palace to be consumed by the ladies of the harem."[82] Muzaffar al-Din Shah's interaction with Mamaqani was a quintessential example of the superstitious monarch's belief in faith-based cures at a time when most of his literate subjects were drifting away from traditional medical practices.[83]

Shortly after the meeting, Mamaqani's wife, who was traveling with his retinue, fell ill and succumbed to cholera. The prime minister, fearing the epidemic's dissemination from the cleric's camp to the capital, personally urged Mamaqani to "hasten" his departure to Mashhad.[84] The rest of the cleric's journey in Iran was uneventful, probably because he was mourning his wife's death. He even ignored requests by Mashhad's clerics to issue orders forbidding Muslims to trade with Armenian merchants, as he had done weeks before in Sabzevar.[85] After a short sojourn in Mashhad, he returned to Najaf, to the relief of the city's non-Muslim population.[86]

Mamaqani's high status and reverence in Iran made him an effective nexus for countrywide antigovernment and anti-European protests that could have hastened the Constitutional Revolution. However, he died less than a year after returning to Najaf in 1905, and political quietism and scholarship characterized his last days. He preferred to "dwell in a *sardab* [a basement used for study to escape the desert heat]" rather than engaging with "the cares and affairs of this world."[87] This diminished his chances of leading the growing nationalist movement against Muzaffar al-Din Shah's government and the meddlesome Great Powers. This change of heart was demonstrated by his refusal to grant the request of Iranian clerics to issue a decree (*fatwa*) permitting Shi'ite Muslims in the Caucasus to proclaim holy war (*jihad*) against Russia. When Shaykh Muhammad Taqi Aqa Najafi, the quarrelsome cleric from Isfahan, asked for his support against the government, Mamaqani replied, "Let Aqa Najafi sell all his properties and give the proceeds, as I have done, to the poor, and he will no longer be troubled by the government."[88] Out of fourteen letters Mamaqani received from Qom during his travels in Iran, he answered only four of them, which dealt with subjects of religion and ritual.[89] This left the helm of the brewing Constitutional Revolution in the hands of younger and less ascetic clerics who were actively involved in emerging radical secret societies.[90] Mamaqani's role in weakening Iran's sanitary defenses and facilitating the spread of cholera, however, indirectly played an important role in provoking the political conflagration to come.

Cholera's Spread

The second eruption of cholera in Kermanshah, which intensified following Mamaqani's departure, never became as deadly as the first outbreak due to heavy rains and hailstorms, which "flushed out the streets and sewers."[91] However, it trailed the cleric's caravan and caused more deaths in other locations. First, cholera spread to the village of Kangavar, where it became endemic following the forced departure of Vaume Bey and his staff. From there it spread to Hamedan in early May and killed up to a hundred a day by the end of June (fig. 4.3).[92] By then, cholera had disseminated to the towns of Arak, Malayer, Dolatabad, and Borujerd. The disease also reached Qom where corpses washed in a river that flowed adjacent to the Fatima Masuma Shrine, considered the second holiest Shi'ite site in Iran, helped propagate the epidemic. Residents and pilgrims in Qom used the waterway for ablution and domestic consumption. From Qom, the epidemic spread further north to Tehran and south to Kashan and Isfahan.[93] Smaller cities such as Konartakhteh and Kazerun were also affected, though their sparser populations resulted in fewer casualties overall.[94]

It did not take long for cholera to spread to Iran's southern Fars Province. New roads opened since the 1892 epidemic added speed and range to its dissemination. One major thoroughfare, the Lynch-Bakhtiyari road built in 1897, was critical in transmitting the epidemic from Iran's northern cities to the Center and South of the country.[95] Local authorities absconded their posts, commerce stopped, and citizens took flight when cholera reached Shiraz and Isfahan.[96] At its height, the outbreak claimed more than a thousand victims a day in Shiraz. The staggering number of corpses made it difficult to find help for funeral arrangements, leaving distraught relatives to wash and bury the victims.[97] Murders and robberies were committed "up to the walls of towns."[98] In Shiraz, criminals used a feigned case of cholera to frighten away a corn seller and steal his corn.[99] Other cities in the province witnessed similar opportunism. In a small town near Yazd, the residents paraded several coffins and spread rumors that cholera was rampant to scare off tax collectors. In Yazd itself, a fugitive evaded capture by hiding in a coffin, pretending to be a chol-

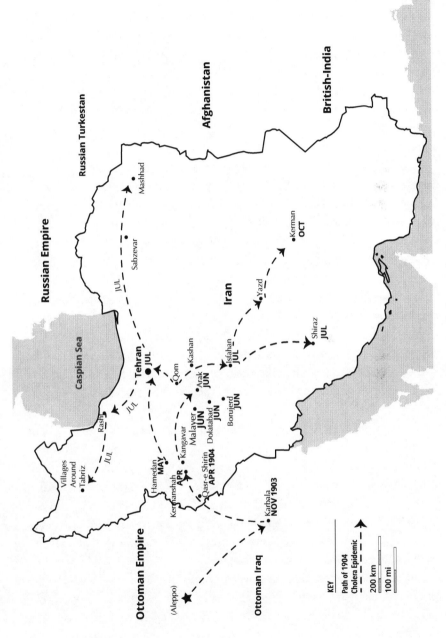

Fig. 4.3. The 1904 cholera epidemic.

era victim and thus deceiving the soldiers sent to arrest him.[100] In another case, nomadic Lur tribesmen, who had come to Yazd to trade, were tricked into believing that cholera was ravaging the city when in reality it had subsided. The duplicitous townsmen were successful in pressuring the tribesmen, anxious to leave as soon as possible, to sell their goods at a vastly discounted rate.[101] Other cities in the South showed familiar patterns of exodus and disorder, even though the outbreak there peaked at no more than thirty cases a day; in Kerman, for example, when the governor and local officials left the city for healthier climes, burglars took advantage of the confusion and made an unsuccessful attempt to rob the city's bank.[102]

With the Anglo-Russian Convention not yet ratified, the Great Powers also used the epidemic to enhance their influence in southern Iran. The British tried to have Shiraz's quarantine placed under the authority of the residency surgeon in the Persian Gulf.[103] The Russians, who had sent a medical commission to Shiraz, wanted to test a new cholera serum there. With prevailing anti-European sentiments in Iran, Shiraz's citizenry was unwilling to accept either intervention, going so far as to threaten violence. This prompted the local authorities to warn the Russian physicians to limit their attention only to individuals soliciting their help.[104]

As with the 1892 cholera epidemic, a large segment of Iranians still saw divine intervention as the only sure defense against the dreaded disease. The din of prayers and loud entreaties seeking the protection of God and saintly figures covered Iran's urban landscape in the South as described in this passage by a British missionary: "Last night a 'rozi khani' [ceremonial lamentation ritual] was held near us—a great noise pleading God & Ali—'oh Ali' about 50 times—'ya Haqq' many times—then 'Ya Haqq Ali'—another 50 times. I should think they say it to keep away cholera—everywhere else it is the same—in town hardly any work is being done—'rowza khanis' are held every day—may our God answer us and not send cholera if it is his will."[105] In the city of Yazd about 250 ceremonial lamentation rituals were held on a daily basis. People also engaged in increased charity work as a way of showing gratitude to God for keeping cholera at bay.[106]

The germ theory continued to make inroads in the provinces despite the tenacity of religious determinants of health and disease among the poorly educated strata of the local population. In Yazd, people erroneously took the science of antisepsis one step further by holding pieces of cotton dipped in carbolic acid near their noses to keep cholera away.[107] This was probably applied within the framework of older notions of miasmatic prevention that emphasized stopping disease causing vapors from being inhaled. Microbes also entered the popular vernacular in the provinces as evidenced by the following fanciful description of cholera by locals in Yazd: "Hajji's people . . . said that the *farangi* [Europeans] had been on the roof looking through telescopes and had seen microbes of cholera in the air. Some said like poppy heads, some like worms."[108]

Cholera in Tehran and the North

As the cholera epidemic took its toll in Iran's southern provinces, it also reached further north, making landfall in Tehran. Its first victim was Vaume Bey, who was infected while attending to a sick soldier in Qom and died shortly after he arrived in the capital for treatment. By the third week of July, cholera was claiming up to two hundred victims daily in the city.[109] The outbreak led to the familiar pattern of general panic and exodus to the surrounding mountains, as described by Muzaffar al-Din Shah's sister, Taj al-Saltana: "We, husband and wife, were so terrified we could not remain at home. Unmindful of the fact that heat and travel would increase chances of contamination we decided to go to Posht-e Kuh. Forming a large caravan with others, we set out for Shemiran. Our traveling companions—men, women, and children—numbered eighty-six. Our first two stops were very pleasant. But for the remainder of the trip we saw sick and dead people everywhere. Most of the contaminated ones had been thrown out of the villages, and these unfortunates were suffering the agonies of death under the hot sun."[110] Tehran's commercial life came to a standstill, and frightened people of all classes fled, leaving a trail of abandoned victims behind. Muzaffar al-Din Shah, kept in the dark on the outbreak in the capital, eventually saw through the unusual precautions and forced his doctors to tell him the truth.[111] Terrified by the revelation that cholera was at his doorsteps, the shah did his best to isolate himself: "The shah lived in the Sahebqaraniyya [royal palace], terribly rattled. All traffic around the royal mansion, even in the village, was forbidden. No one could see the royal person, except a few servants and Sayyed Bahrayni who was constantly reciting from the Holy Book and chanting prayers for protection. The functioning of the state had been suspended totally."[112]

It did not take long for the superstitious monarch to abandon the city, as he sought protection in the Talikhan hills, forty miles away. He based his decision on an augury taken in the Quran, while going against the advice of his physicians, who worried that the altitude could harm his "sick heart."[113]

However, the epidemic caught up with the shah's retinue less than a day after his arrival. Afraid of being blamed for the outbreak, the prime minister, 'Ayn al-Dawala, refused to inform the shah that cholera had infected his camp. Exasperated by the premier's stance that "Moslems believe that they are in God's hand," the shah's physicians took it upon themselves to warn the monarch that an outbreak was at hand and that they would not be responsible for his health if he remained.[114] A terrified Muzaffar al-Din Shah began planning a secret departure to Europe by car, leaving his entourage behind in the diseased camp, but gave up on the idea and reluctantly agreed to return to Tehran after recognizing the disastrous political fallout of a "sudden and undignified flight" during a national emergency. Workers removed the unburied bodies of soldiers who succumbed to cholera from the main road to Niavaran Palace and disinfected the royal compounds prior to the shah's arrival to reduce his chances of panicking and turning back. Once in Tehran, the terrified shah sequestered himself and restricted access, even to ministers, causing the state apparatus to grind to a screeching halt.[115]

Not all Iranian officials opted for the shah's fright-induced seclusion and ignominious flight. In the early days of the epidemic, with "all the native doctors having fled except one," Tehran's municipal chief "labored night and day to organize the sanitary defense of the capital" and performed his duties while others panicked.[116] However, he died soon after the outbreak, one of the earliest casualties to cholera in Tehran.[117] Making matters worse, one of the few Iranian physicians who had not fled also perished after accidentally drinking carbolic acid, having mistaken it for brandy.[118] These conditions paralyzed the government, and Tehran became "an unremitting graveyard and her people the dead."[119] With physicians gone and the city's de facto governor dead, the Customs Department and the American Presbyterian missionaries led the relief work in the capital. The Americans started a voluntary cholera field hospital with a staff of forty people less than two days after the outbreak.[120] They treated the deluge of patients around the clock and educated Iranians on Western sanitary science.[121] The American example of civic service inspired some of Tehran's remaining residents to engage in the care of their fellow citizens through the Cholera Relief Corps. Donations from the Russian and British banks (contributing a thousand tuman each) and the Iranian government (fifteen hundred tuman) funded the group. Iran's foreign minister contributed two hundred tuman from his private estate, and other Iranian notables followed suite with personal donations.[122] The money financed the salary of a nurse, free treatment for the poor, and the operation of the new cholera hospital. The Americans also used the funds to buy a large tract of land outside the city gates to build a cemetery that would allow the burial of cholera victims to meet established sanitary standards.[123] The American missionaries were praised in the native press for "spending their money wisely," and they inspired larger than usual participation by both Iranian and non-Iranian residents of Tehran in their relief efforts.[124]

Samuel Jordan, renowned educator and founder of the American College of Tehran (Alborz High School), was a leading member of the Cholera Relief Corps. He and his wife were completing the first part of their mission in Iran as teachers at the American Boys School of Tehran when the epidemic reached the capital.[125] Many homes in Tehran drove out sick relatives to die on street corners, fearing the risk of contagion from family members with cholera.[126] Jordan, along with some of his students and fellow teachers, patrolled the city in hired carriages to find and transport these neglected victims to the American hospital, which took 215 cases in July alone.[127]

The Americans treated patients in the early stages of infection with mercuric salts and other intestinal "antiseptics." However, most patients in the hospital were at the later stages of the disease when vomiting, cramps, and rice-water stools were already common. These advanced cases received antispasmodics, mustard on the stomach, and vigorous massage of the extremities. The physician-in-chief of the American service also used a remedy popular in his native state of Indiana, called "Hot Drops"—an equal mixture of opium, rhubarb, ginger, capsicum, and wine of ipecac given after each bowel movement.[128] The patients received boluses of saline solution to rehydrate them and prevent circulatory collapse once their acute symptoms subsided.[129]

Despite these efforts, the sick were unlikely to recover, even with treatment. When the disease was caught early, "particularly in the homes of the better class Persians, the death rate did not exceed 35 percent"; otherwise the mortality rate was 60 percent.[130] Relatives often had to care for one another, sometimes with tragic consequences as described by an American missionary physician:

> Near-by the hospital was a young mother who took cholera, and her husband cared for her with great tenderness and thoughtfulness, under the direction of one of the hospital physicians; but, in spite of all that was done for her, she grew rapidly worse. Finally, she declined to take medicine, and her husband, in his extreme anxiety for recovery, remarking that the medicine was pleasant, put the spoon to his own lips which she had been using. In a few hours, he developed

the disease, and in order to hide from his wife his own illness he excused himself by saying he must go to inquire concerning the welfare of his father's household. He died a few hours after his wife's death, without knowing that she had gone before. They were buried, with their newborn babe, a few hours later.[131]

Whole families in Tehran died within a day or two of each other, mostly due to exposure that was avoidable had they understood cholera's fecal-oral transmission. In order to educate the population, the Americans quickly printed a booklet that described simple directions and precautions against the contagion in both English and Persian, which they circulated free of charge in Tehran's bazaar.[132] Even a senior Shi'ite cleric copied the booklet and then sent it out under his name, to the astonishment of the American missionaries.[133] The Iranian weekly *Tarbiyat* also spread "the American Hospital's Recommendations" on prevention.[134] It described cholera as a disease acquired in the digestive system and associated with insalubrious environments and unhealthy lifestyles, views that echoed notions of moral hygiene that predominated among progressives in the United States at this time. They admonished public officials for leaving their posts, discouraged the consumption of fruits and vegetables, and presented temperance, cleanliness, and education as the cornerstones of public health.[135]

Following the American recommendations, the municipality in the northern city of Rasht rapidly cleaned the streets and forbade the sale of fruits and vegetables. By the time the epidemic reached the city, it had lost its potency due to these sanitary measures.[136] Muhammad 'Ali Mirza, Iran's crown prince and governor of the northern province of Azerbaijan, immediately ordered the region's most qualified physicians to protect his territory against cholera from Tehran around the same time. His French-trained physician, Mirza Zayn al-'Abidin Khan Adham Luqman al-Mamalik, established a strict ten-day quarantine on the road from Tehran to the provincial capital of Tabriz.[137] Even royal relatives and notables were obliged to undergo a period of observation before allowing them into the city. These quarantines prevented the transmission of cholera by people from the surrounding infected towns and villages to Tabriz.[138] The willingness of Luqman al-Mamalik and other notables to spend several thousand francs establishing stations and paying the salaries of doctors improved staff moral and performance at the quarantines. It also reduced corruption, paving the way for a successful prevention campaign.[139]

TABLE 4.1.
Mortality in various cities and provinces during the 1904 cholera epidemic

Province	Total deaths	Province	Total deaths
Kermanshah	5,000	Mazandaran	2,000
Hamedan	3,000	Khorasan	5,000
'Iraq-i 'Ajam	1,000	Isfahan	2,000
Tuyserkan	1,500	Shiraz	15,000
Qom	1,000	Arabestan	2,000
Qazvin	200	Azerbaijan	2,000
Gilan	2,000		

Cholera slowly receded from the North and, after extracting a severe toll from Iran's southern provinces, from the rest of the country by the fall of 1904 (table 4.1).[140]

Cholera, Sanitary Reform, and the Constitutional Revolution

The 1904 cholera epidemic's impact on Iran not only altered notions of sanitation and disease but also helped shape the rhetoric of the 1906 Constitutional Revolution. Iran's inability to cope with the large influx of patients during the outbreak provoked calls to correct the deficiencies in its health-care infrastructure. Tehran, which had more hospitals than any other Iranian city, was overwhelmed. Its major inpatient facilities were limited to the decades-old Imperial Hospital, the sixty-bed American Hospital, the Cossack Brigade's Hospital, and a thirty-bed Russian Red Cross Hospital.[141] Radical newspapers, like *Nida-yi vatan,* acknowledged the country's health-care deficiencies, while simultaneously casting a suspect gaze on Western countries that established the much-needed facilities: "Using foreign funds, for the first time European powers have founded hospitals and schools in every city whereby by this method they can spread their language. Otherwise in all frankness the American government would not show us this much kindness and why is it that the Russian government establishes so many hospitals and schools and spends all these funds? Where does this friendship [*dusti*] before being acquainted [*ashna-yi*] come from?"[142] The scarcity of hospitals was not the only handicap in Iran's public health infrastructure. Municipal administration and hygiene had not changed since the 1892 cholera epidemic and did not keep pace with shifting popular attitudes on sanitation and disease. Even all of the religious strata were no longer opposed to Western hygienic practices, though resistance against preventive measures such as quarantines remained.[143] The following 1904 editorial in *Tarbiyat* reflected this departure from earlier clashes between religious notions of disease causation and the Western empirical perspectives on antisepsis and hygiene:

> For the majority of the ignorant population whose sanitary works has no law or logic, most make themselves happy with this saying "God is our keeper [*khuda hafiz ast*]." I also say that God is our keeper, although conditionally, and there is no doubt that the Dear Glorious God has created all creatures and just as he has created the scorpion and snake and ants and locust and fire and plants, He has also created very small creatures that can be seen only with the help of a microscope, and because these creatures enter the bodies of men or animals, they give rise to disease, and the Glorious Creator has given man intelligence, eyes, and hands so that it can keep itself [from disease] as long as it knows that the sting of a scorpion or fire will cause him to have pain and will burn and [therefore] must defend against it instead of sitting idly out of laziness and saying God is my keeper. If you see the scorpion on your hand or if fire is in your surroundings, will you sit idly and say God is my keeper? I don't think so.[144]

The role of infected water had become a sine qua non of cholera among Iran's literate lay population, even though most Muslim clerics still rhetorically maintained that a religiously mandated volume of water (*ab-i kur*) could not be defiled. This secular understanding of infectious diseases had unforeseen consequences. The Iranian population increasingly recognized that outbreaks could be controlled and rendered less deadly by maintaining hygienic standards as mandated by modern science. They slowly marched away from the fatalism and divine determinism that characterized their reactions to cholera in the past. While this empirically based sanitary perspective gained momentum after the 1892 outbreak, the 1903–1904 cholera epidemic brought it to the forefront of contemporary Iranian political rhetoric.

Mounting calls for the government to improve the country's public health emerged from the notion that people could alter their environments to make them more salubrious and resistant to contagions.[145] Newspapers urged Iranians to take responsibility for the hygiene of their communities by keeping their homes, neighborhoods, streets, and waterways clean.[146] Tehran was singled out by the weekly *Tarbiyat* as the "the beating heart of the nation," making its cleanliness and well-being vital to the rest of the country. The newspaper called on the capital's municipality to continue the sanitary undertakings of its recently deceased chief, whom they described as a "martyr" who had given his life in service to his country.[147] Building on this secularizing trend, *Tarbiyat* criticized officials for leaving Iran's sanitary condition in "God's hands." Going even one-step further, it sardonically stated that the most likely reason for the government's prevalent inattention to hygiene was that "God wants us to eat poison and die."[148] The revival of the Iranian Sanitary Council embodied the aspirations to fix the country's hygienic shortfalls and institutionalize public health on a national level to prevent future outbreaks.[149] Joseph Naus helped engineer the council's resumption and maneuvered to have Jean-Etienne Justin Schneider, his personal doctor with whom he was "very intimate," preside over it.[150]

Schneider's selection to lead the council reflected his rise in Muzaffar al-Din Shah's good graces.[151] In 1900 he helped the royal party navigate the social and medical hurdles in his native France, where the shah had come to receive treatment for his chronic gout and renal failure at the Contrexéville Spa (fig. 4.4).[152] During this trip, President Emil Loubet of France personally asked Muzaffar al-Din Shah to issue an edict appointing Schneider as his chief physician in place of his English doctor, Hugh Adcock.[153] Even the *British Medical Journal* recognized the political significance of Schneider's eventual promotion: "Sir Hugh Adcock . . . retains the rank of Honorary Consulting Physician to His Majesty. Dr. Schneider, a Frenchman, has been appointed physician in chief. We understand that the real meaning of these appointments is that Sir Hugh Adcock, who was the Shah's Physician for many years, has practically been supplanted by the French practitioner, a fact which is considered to be of some political importance."[154] This influence allowed Schneider to hold the Sanitary Council's first meeting in the comfort of his own home, on August 6, 1904. The shah's doctors, a Customs Department representative, and physicians from the major Western legations attended at Schneider's request.[155] Discussions focused on public health and "prophylactic" policies to stop cholera from returning to Iran and the necessary municipal measures to safeguard Tehran's population against disease.[156] The British worried that the Iranian Sanitary Council could once again threaten their interests if it gained the same type of international sanction as the Sanitary Council in Constantinople (Istanbul). However, a hypochondriacally inclined Muzaffar al-Din Shah strongly supported reestablishing the organization, against British wishes, to prevent the dreaded cholera epidemic from returning to the capital.[157] Furthermore, as this dispatch from the British envoy in Tehran indicates, London doubted the Sanitary Council's longevity and, by extension, its ability to challenge Britain's control of quarantines in the Persian Gulf over the long run: "To me it appears on the whole more probable that, with the disappearance of the cholera panic which has brought it [the Sanitary Council] into formal

Fig. 4.4. Muzaffar al-Din Shah's consulting physicians at the Contrexéville Spa. (*Left to right*) Eugen Hollander, Khalil Khan A'lam al-Dawla Saqafi, Mahmud Khan Hakim al-Mulk, Ibrahim Khan Hakim al-Mulk, Albert-Emile Debout d'Estrées, Sigismond Jaccoud, Richard Pfeiffer, Georges Dieulafoy, Jean-Etienne Justin Schneider, and Hugh Adcock. Photo courtesy of Bibliothèque interuniversitaire de Santé, Paris.

existence, it will gradually cease to assemble and will eventually be relegated to the limbo in which so many Persian schemes of reform sleep."[158]

The Iranian government reported the Sanitary Council's proceedings to the public for the first time in decades, praising it for ensuring the welfare of the shah's subjects through the prevention of disease. It also distributed the printed transcript of each meeting within the administration and among all major embassies in Tehran. Publications sympathetic to the government highlighted the shah's demand for the council to meet weekly, responding to the growing view that the monarch was accountable for the health of his subjects.[159] The weekly *Tarbiyat* praised the Sanitary Council and officials from the Departments of Public Works (*baladia*) and Public Safety (*nazmia*) for their efforts to address Tehran's hygienic deficits.[160] However, continued obstruction by the Great Powers and the bankruptcy of the Iranian economy hindered the council from instituting its ambitious plans in the near term. This meant that by the dawn of the Constitutional Revolution in 1906, the calls to improve the state of hygiene and public health had gone unanswered, adding to a growing wave of dissatisfaction.

Iranian resentment toward the government had already heightened after the cholera epidemic, as reflected in the following French diplomatic dispatch from Tehran: "The cholera epidemic that ravaged the capital and nearly all the provinces had slowed down, even stopped for a time, all affairs and created a situation of malaise and trepidation that the whole country feels. We can say that the discontent is general. The Sovereign who has never been popular is today less than ever, and the Grand Vazir is even less [popular] than his master."[161] The most recent outbreak in the country not only fueled the population's anger because of the government's apathy but also added to the general dissatisfaction by disrupting Iran's already strained economic fabric.[162] The epidemic "killed the farmer and stopped agriculture," causing widespread shortages in foodstuff.[163] At the height of the epidemic, farmers in the country's northwestern "rice bowl" abandoned their homes for safer highland climes, missing important cultivation cycles.[164] Even the quarantines, meant to halt the contagion, were damaging to agrarian production, particularly during the critical spring and summer months. Travel restrictions on villagers delayed harvests, prevented the cleaning of irrigation channels on farmland, and hindered the essential daily delivery of cattle feed, hay, and logs from the surrounding countryside. One observer remarked that "these interventions [quarantines] are worse than the disease itself. If the disease lasts ten days these interventions will be the cause of a downright famine in the future."[165]

Although the initiatives of the Belgian-led Customs Department had increased government revenues between 1899 and 1904, it proved to be extremely unpopular among merchants who had to bear the burden of new tariffs.[166] They also had to endure the expenses of a poor silk harvest caused by bad weather and drought in 1904.[167] The cholera epidemic made matters worse by bringing Iran's land and maritime trade to a standstill. Porters and muleteers in Baghdad, for example, refused to transport goods into Iran because of the strict quarantine and customs duties on the Iran-Ottoman border. Business in the southern commercial hub of Shiraz came to a standstill due not only to decreased shipping to the port of Bushehr but also to a travel ban between the two cities during the outbreak.[168] Domestic quarantines, such as those around Tabriz, and the prevailing lack of security significantly curtailed commerce along Iran's main arteries in the North as well.[169] Local trade was also at a standstill with the exception of the sale of foodstuffs and wine brandy, a popular prophylactic against cholera.[170] By the end of 1904, Iran's customs revenues had fallen more than thirty percent because of the epidemic.[171]

To meet these losses, the prime minister restricted government spending on pensions and salaries and increased taxes, including a ten percent surcharge on stamps. However, the government refused to use this income to pay back its sizable debt to the merchant population.[172] Tehran's lack of financial accountability magnified the growing social discontent, which began extending to rural areas, as demonstrated by the refusal of several villages to pay their taxes.[173] Even members of the royal family became displeased with the country's plummeting financial situation and draconian tax measures. The shah's brother left Iran to reside in Istanbul, and one of his sisters chose Paris as her tax haven—where she lived "in European style and fashion."[174]

Secret revolutionary societies began to multiply in response to the mounting dissatisfaction with the government. These organizations, whose membership reflected the political, religious, and mercantile elite, sought radical social and political change.[175] The Secret Society (Anjuman-i makhfi), for example, was founded in Tehran in 1905. Its members would become some of the most important leaders and interlocutors of the Constitutional Revolution.[176] After long deliberations about modernization, education, and social justice, the society agreed on a program of constitutionalism through revolution.[177]

The first meeting of the Secret Society drew attention to the poor sanitary state of the country, including its "dirty streets, smelly and tainted baths and waters."[178] The society attacked the government for not setting the foundations for a single hospital or poorhouse in any of Iran's cities and not meeting the country's sanitary needs:

> Is the three-thousand-year-old Government of Iran not strong enough to establish in all its larger cities an office of public health and appoint three or four physicians to those places so that the people of that city are [saved] from the false sense of knowledge that is caused by the ignorance of quacks? Why is it that there are no hospitals to treat a disease like leprosy. . . . in this category, we need more than goodness, we need plans, repairs, and lack of greed under the condition that people see themselves as from the Government and the Government sees itself as from the people.[179]

The Secret Society also reproached the government for high taxes, the exorbitant price of foodstuff, and the lack of investment in teaching facilities.[180] During its third meeting, it also blamed Tehran for the lack of paved roads and factories and especially for the absence of clean drinking water: "Please pay attention that all the civilized and uncivilized countries do not fear for the cleanliness and safety of their water. We Iranians, in abject servility, must drink these dirty waters replete with impurities, [and] even that is limited and not allowed."[181] It even condemned the typically irreproachable shah and the prime minister for charging exorbitant prices for "dirty water that animals would shun," preferring to feed their orchards rather than quenching their people's thirst.[182] This reflected a growing recognition of the government's responsibility to ensure public access to clean water as part of its municipal responsibilities. The role of public health in the meetings of the Secret Society was a clear indication that the concept of reform among revolutionaries included the need to alleviate the poor state of hygiene and sanitation in the country. More importantly, the powerful religious elite, who could make or break the emerging political movement, increasingly accepted the secular notion of sanitation, as articulated by these revolutionaries.[183]

By 1905 Iran's worsening economy resulted in double-digit inflation, a thirty-three percent increase in the price of sugar, and a ninety percent appreciation in the cost of wheat.[184] The country faced near famine conditions, and Tehran could not remedy the shortages by importing wheat from Russia, as it had done in the past, due to the ongoing Russo-Japanese War.[185] The French envoy in Tehran articulated the harmful long-term impact of the epidemic on commerce in Iran as follows:

> The epidemic has now notably decreased: those who had escaped are returning home, but business, completely stopped for six weeks, is not starting back up. The commerce, which had already [experienced] the negative impacts of the Russian conflict [Russo-Japanese War], is nearly annihilated by the deplorable sanitary condition of the country, and the southern customs have seen their receipts decline, mainly as a result of the epidemics of cholera and plague that reign for several months in the region of the Persian Gulf; according to the information that was given to me by Mr. Naus, their receipts have declined in frightful proportions: all the data indicate that this year will be, for Persia, a deplorable year.[186]

This economic freefall and the government's inability to address popular expectations generated growing public protests that would culminate in the Constitutional Revolution of 1906.

The revolution started with disturbances in 1905 when Bushehr's Customs Department began enforcing higher tariffs on imports in response to falling revenues. Merchants refused to clear their goods through customs unless the shah repealed the new levies.[187] Shortly afterward, religious seminary students distributed a picture of Joseph Naus, the director of the Customs Department, masquerading as an Iranian Shi'ite cleric at a costume ball. This offended Iran's religious leadership who called for the dismissal of the Belgian civil servants. A month later, a group of prominent merchants, unhappy with the new tariffs, took sanctuary (*bast*) in protest in Shah Abdol Azim Mosque and once again called for Naus's dismissal. In December 1905, Tehran's bazaar closed in protest, and a gathering of two thousand merchants, clerical leaders, and theology students took sanctuary in Shah Abdol Azim after the government bastinadoed two leading merchants to coerce them to reduce their price for sugar.[188] The shah accepted the protestors' demands to dismiss the governor who had ordered their punishment and to establish a consultative House of Justice (Idalatkhana). Strikes resumed in the summer of 1906 when the shah attempted to arrest some antigovernment preachers and refused to convene the promised House of Justice. The demonstrations turned bloody when on two occasions Iranian Cossack troops and police fired on the crowds. Triggered by the wounding and killing of some of the protesters, Iran's clerics joined the strike, effectively halting the country's judicial system. More than twelve thousand protestors then took sanctuary in the compound of the British legation, demanding that the shah gather a national assembly to draft a written constitution limiting royal power. Facing a sustained general strike in Tehran and waves of opposition in the provinces, Muzaffar al-Din Shah acceded to their demands on August 5, 1906. The country's first National Assembly (Majlis) gathered three months later, bringing a short pause to Iran's long history of absolutism.

Conclusion

Iran's Constitutional Revolution, triggered by the social and economic impact of the 1904 cholera outbreak, could not have occurred without a growing secular view of epidemics. Iranian perspectives on the etiology and prevention of contagious diseases changed in step with the popularization of the germ theory. Mind-sets in the country slowly moved from a fatalistic view of epidemics, as irreversible acts of God, toward an empirical conceptualization of outbreaks as preventable and treatable by worldly authorities. This made the Iranian government accountable for the severity of the 1904 cholera epidemic, magnifying the existing popular resentments that led to the Constitutional Revolution two years later. The secularization of outbreaks also fostered a sense of sanitary responsibility among Iranians, who increasingly viewed municipal hygiene and disease prevention as a social contract. This galvanized regional administrators, professionals, and even local populations to intervene against cholera in 1904, particularly in the face of government shortfalls. The emergence of a civic sanitary spirit would be the lasting legacy of the public health movement in Iran during the period leading to the Constitutional Revolution.

However, the country's evolving culture of public health did not result in any large-scale sanitary improvements under the new constitutional regime. The Great Powers continued to thwart disease prevention initiatives that could interfere with their commercial and political interests, and Tehran's nagging economic and administrative difficulties hampered urgently needed upgrades in Iran's hygienic infrastructure. Despite these limitations, the crystallization of the germ theory of disease and the expectations that emerged from the constitutional movement played a decisive role in institutionalizing public health as a core element of governance in the country's new administration.

CHAPTER FIVE

Wars, Plagues, and Institutional Developments in Health, 1906–1926

The first Iranian National Assembly (Majlis), born out of the 1906 Constitutional Revolution, brought a populist and regional influence into the traditionally Tehran-centered Iranian government. The newly elected representatives from the provinces tried to improve public health throughout Iran, not just the capital. They furnished the Sanitary Council, reconvened during the 1904 cholera epidemic, with a modest budget that allowed it to meet monthly, gather epidemiological data from the country's major districts, and publish its proceedings.[1] However, the council was unable to enact any meaningful national public health reforms due to its limited financial resources and persistent interference from the European imperial powers.

The 1907 Anglo-Russian Convention divided Iran into a Russian zone of influence in the North, a British zone in the South, and a neutral zone in the Center of the country. The agreement ended the political rivalry between the two powers and consolidated their dominance over quarantines in the frontier regions where they held sway. Several months after the convention was signed, Jean Bussière, the French Iranian Customs Department's sanitary inspector general for the Persian Gulf, received strict instructions from the French and Russian embassies in Tehran to cease interfering with the British-run quarantines in the country's southern ports. A year later, the French government withdrew him from Iran altogether after he ignored their order and conveyed his concerns with the sanitary defense of the ports to the Iranian Sanitary Council during a looming plague epidemic. The Russians similarly enforced their sanitary prerogative in the North of the country when they obliged the Iranian government to lift its quarantine against arrivals from their cholera-infested Caucasian territories, allowing the disease to spread to Tabriz in 1908.[2] Unable to establish quarantines or other preventive measures that could threaten Russia and Britain's commercial and political interests, the newly inaugurated National Assembly focused instead on reforming its domestic health-care system through investments in medical education.

Medical Education during the Constitutional Period

Reformers in the nascent National Assembly recognized that continued progress in public health depended, in large part, on increasing the overall quality and capacity of health-care delivery in the country. This prompted the Iranian government to focus its limited resources and political capital on regulating Iran's health professions, beginning with reforms in medical education as recommended by Jean-Etienne Schneider, the shah's French physician-in-chief. Schneider felt that Iranians were remarkable students and practitioners of medicine: "It should be recognized, herein, that Persians have great dispositions for the study and practice of medicine. They are intelligent, precise, delicate, and still have the qualities that we do not develop enough among our European students. They are observing, attentive, and take care of [looking out for] the affect, the decubiti [ulcers], and the [changes] in the pulse of their patient, and in sum they show remarkable clinical qualities. In addition to everything else, the students have zeal for work and frank ardor to educate themselves."[3]

As early as 1894, Schneider obtained scholarships from the French Government and a reticent Nasir al-Din Shah to send Iranian students to the French Military Medical School of Lyon (École du Service de santé militaire de Lyon) and the Military Teaching Hospital of Val-de-Grâce in Paris (Hôpital d'instruction des armées du Val-de-Grâce) for advanced clinical training.[4] The frequency of students sent abroad to study medicine increased with the ascension of Muzaffar al-Din Shah in 1896. During a state visit to France four years after his coronation, Schneider convinced the shah to sponsor twenty students from the Polytechnic College to study medicine throughout Europe, including in Saint Petersburg, Moscow, Berlin, Vienna, Paris, London, and Istanbul. The following year, five additional students were sent to Paris and four to Vienna; followed by another cohort of eight students in 1902.[5] Schneider's leadership in medical education and the Alliance Française, the French cultural center in Tehran, endeared him to the francophone Iranian political elite in the capital. This allowed him to maintain his influence after the Constitutional Revolution, resulting in his appointment to the Supreme Council for Public Instruction (Shura-yi 'aali-yi anjuman-i ma'arif), which directed Iran's national education programs.[6]

Using his position, Schneider convinced the government to increase medical teaching standards at the Polytechnic College by highlighting the absence of scientific foundations in its curriculum. Poorly trained graduates, he pointed out, would routinely prescribe the newest medication on the market "that could be found in newspaper advertisements" without any regard for potentially harmful side effects.[7] Observations such as these led to the elimination of the antiquated post of lecturer in Galenic medicine at the school, relegating humoral medical theories and treatments to the realm of folk practice.[8] At Schneider's urging, the Iranian government recruited new lecturers from France in the basic sciences to enhance the Polytechnic College's foundational medical courses. They included a curator from the Paris Museum and a chemistry professor from the Sorbonne to teach physics, chemistry, zoology, botany, and medical mineralogy. These appointments elevated the first year of the institution's three-year medical program to French educational standards. Schneider also persuaded the government to hire two lecturers, physician majors Georges and Galley, from the Military Medical School of Lyon to improve clinical instruction in medicine and surgery for the more advanced students.[9] In addition to their core teaching responsibilities in clinical medicine, Galley introduced an international public health (*prophylaxie internationale*) lecture series into the curriculum, helping shape a new generation of globally minded Iranian sanitarians.[10]

Six years later, another group of educators from France who taught the natural sciences, physics, chemistry, medicine, and surgery replaced these professors.[11] Iranian physicians would later credit these instructors with giving the Polytechnic College, still the country's only government-supported institution of higher learning, a new lease on life.[12] What's more, these investments were a testament to Schneider's powers of persuasion since the cash-strapped administration had almost disbanded the school in 1904.[13] For the rest of the decade, the government also continued to send students to the Military Medical School of Lyon, where, in addition to developing expertise in the clinical practice of medicine, they gained proficiency in the science of sanitation and hygiene. This stemmed from the French military perspective that their medical corps needed the skill set to implement and administer large-scale programs in disease prevention to avoid the kind of losses to infectious diseases that they experienced decades earlier in the Franco-Prussian War.[14] Moussa Khan articulated the growing commitment to sanitation and hygiene among the Iran's younger medical cadre in his 1906 doctoral thesis at the school: "As Persian students, we came to France seeking, in addition to a solid instruction in medicine, eminently practical notions of hygiene; this science born of the past is already called the 'civilizer of the future.'"[15] Several of these students were unable to complete their studies due to inadequate financial support, and at least one student committed suicide due to depression, apparently triggered by homesickness.[16] However, those that did finish returned to Iran and played important roles in shaping the constitutional government's efforts to improve military medicine and public health in the country.[17]

Amir Khan Amir A'lam epitomized this class of influential medical technocrats (fig 5.1). Shortly after graduating from the Military Medical School of Lyon, he represented Iran at the 1902 International Hygiene Conference in Brussels and the 1903 International Sanitary Conference in Paris. Following advanced clinical training and research at the Military Teaching Hospital of Val-de-Grâce, Amir A'lam returned home in 1906 to teach anatomy and medicine at the Polytechnic College.[18] He was also elected to represent his native Mashhad (November 1909 to December 1911) in the Second National Assembly, joining the powerful social democrat faction (Democrat Party) of the parliament. Using his training and experience, he shaped two landmark public health laws: the Hygiene and Smallpox Vaccination Act (*Qanun-i hifz-i sihhat va abila-kubi*) and the Medical Practice Act (*Qanun-i ṭibbabat*), respectively ratified in 1910 and 1911. The latter established the legal framework

Fig. 5.1. Amir Khan Amir A'lam teaching advanced medical students at the Polytechnic College. Courtesy of the Institute for Iranian Contemporary Historical Studies (IICHS), Tehran, archive no. ʿ1-3935.

for medical education and practice in Iran (see appendix C). A decade earlier, the Belgian consul in Tehran described the practice of medicine as "absolutely free requiring neither exams nor diplomas, from either the European physicians or Persian ones."[19] The Medical Practice Act stopped this unregulated culture of health care by recognizing the government as the ultimate arbiter in sanctioning a physician's authority. Even foreign physicians practicing in Iran were required to obtain a government license, though Westerners largely sidestepped this regulation.[20] The country's Ministry of Education required a diploma either from the Polytechnic College or from a recognized foreign (Western) medical school in granting a license to practice medicine under the law. Its enforcement brought an end to the traditional apprenticeship-based model of medical training in Iran, one in which expertise and professional legitimacy were principally derived from a mentor's reputation and skill.

The Sanitary Council in the Constitutional Period

Jean-Etienne Schneider, in addition to advising Tehran on educational matters, continued to lead the Sanitary Council until 1906 when he resigned to attend to a dying Muzaffar al-Din Shah.[21] Another French military doctor, Charles-René Coppin, who was the newly crowned Muhammad 'Ali Shah's (1872–1925; r. 1907–1909) chief physician, replaced him.[22] The shah, who ascended the throne less than a week after his father ratified the constitution, immediately chafed at the new limits on his rule. Tsarist Russia, which had always been hostile to the democratic aspirations of its southern neighbor, supported his antipathy. Conservative segments of the Shi'ite leadership were also increasingly dissatisfied with the new government's liberal leanings; particularly on jurisprudential matters that typically fell within their purview. Emboldened by these undercurrents, the shah abolished the constitutional government, arrested and executed several leading delegates, and ordered his Cossack troops to bombard the National Assembly in June 1908. Surviving members and sympathizers in the provinces organized a counterattack against the shah and his allies, ushering in a one-year civil war that paused Iran's momentum for social reform until Muhammad 'Ali Shah's defeat. The victors restored the constitutional government and deposed the monarch in favor of his twelve-year-old son, Sultan Ahmad (1898–1930; r. 1909–1925), who assumed the throne under the regency of his great uncle (fig 5.2). The Second National Assembly convened in November 1909, one month after Muhammad 'Ali Shah left the county with his immediate family for a permanent exile in the West.[23]

Fig. 5.2. Amir Khan Amir A'lam preparing to administer an injection to a wounded leader of the victorious constitutionalist forces in Tehran. Courtesy of the Institute for Iranian Contemporary Historical Studies (IICHS), Tehran, archive no. a'-909.

Meanwhile, the increasingly respected European-trained Iranian physicians assumed a more proactive and vocal role on the Sanitary Council. This became apparent when Mirza Zayn al-'Abidin Khan Adham Luqman al-Mamalik, Muhammad 'Ali Shah's French-trained doctor, led the body by consensus for much of 1908 while Coppin waited out the civil war in Europe.[24] After steady improvements in medical education and advanced clinical training in France, native physicians were equipped with the technical knowhow to revitalize the Sanitary Council despite political barriers from the European powers. They were also motivated by the patriotic undercurrents of the Constitutional Revolution as observed in the period's British diplomatic cables: "A marked feature of the proceedings of the council has been the increasing prominence taken by the Persian members in discussions, and their aliveness to Persian interests. This is to be connected with the election to the council of several young natives who have studied medicine in Europe and with the spread of the spirit of nationalism."[25]

Early signs of their boldness emerged in 1907 when an outbreak of bubonic plague in the county's Southeast and Bahrain prompted the Sanitary Council to investigate and identify significant shortfalls in the British quarantine

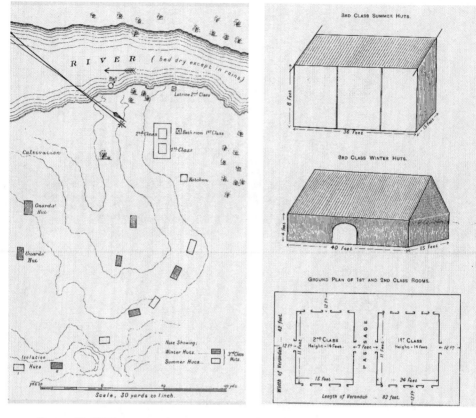

Fig. 5.3. (Left) The quarantine station at Bandar Abbas in 1906. FO 881/8780 Confidential Print (n. 8780) Report by Dr. Theodore Thomson on the Sanitary Requirements of certain places in or near the Persian Gulf, & c. enclosed in Dr. Theodore Thomson to Sir Edward Grey, Local Government Board, Whitehall, July 14, 1906. *Fig. 5.4. (Right)* Layout of the structures in the quarantine station at Bandar Abbas.

arrangements in the Persian Gulf. The Iranians called for upgrades, including additional personnel and equipment at Muhammera, Bandar Lengeh, Bandar Abbas, Jask, and Bushehr (fig. 5.3, fig. 5.4).[26] They also persuaded the British to hoist an Iranian flag over each of the stations instead of the Union Jack and required their quarantine officers to behave as Iranian government officials by employing Persian (or French) for all correspondences that related to their sanitary duties.[27] By 1910, the British completed the called-for renovations to the lazarettos in the five ports under their control and equipped each of them with a disinfection stove and Bushehr with an additional rat-destroying Clayton apparatus.[28] They also posted eight full-time

British sanitary physicians in the ports, with a backup "mobile" doctor and disinfection stove in the Indian port of Karachi that could be deployed to the Persian Gulf in the event of an outbreak.[29]

The Iranians on the Sanitary Council continued to play a key role in growing the organization's mandate and scope of interventions even after Georges, the French medical lecturer at the Polytechnic College, took over its leadership in 1910. They established a commission to study the sanitary arrangements in the North of the country and identified significant shortfalls in Caspian seaports within Russia's zone of influence. As in their negotiations with the British, they suggested splitting the costs of the needed improvements with the Russian government; but buoyed by their recent victory in the civil war and "increasing spirit of nationalism," the Iranians "insisted strongly on the point that the members of the proposed quarantine service should be Persian."[30]

The suggested improvements did not occur in time to protect Iran against the ingress of Asiatic cholera in the summer of that year. The epidemic crossed the country's vulnerable border with Russia's Caucasian territories and, over several months, broke out in its northwestern Azerbaijan province, its Caspian seaport of Anzali, and its northeastern province of Khorasan. From these three points, the disease spread to the rest of the country's major metropolitan areas in a matter of weeks, including Hamedan and Kermanshah, where it caused the greatest number of casualties.

The Iranian government did all that it could to combat the outbreak.[31] The National Assembly agreed to allocate an emergency sum of twenty thousand tuman (about $850,000 in today's currency) from its meager budget to the Sanitary Council to employ doctors, establish lazarettos, and distribute medications to infected areas.[32] The Iranian Customs Department met the remaining costs of the national anti-cholera operation. The emergency funds allowed the council to rapidly organize more than eighteen sanitary stations along Iran's extensive land and sea border with Russia and to send physicians, supplied with medications and disinfectants, to affected districts. The council also convinced the Russians to allow Iranian authorities to quarantine passengers from Russian ships at Iran's Caspian seaports for five days, despite St. Petersburg's protests on the "badness of the [quarantine] accommodations."[33] In the hopes of avoiding a repeat of the Mamaqani affair during the 1904 cholera epidemic, Tehran urged the country's Shi'ite leaders to respect restrictions on travel, particularly the ban on pilgrimage, and quarantine regulations during the epidemic.[34] The Sanitary Council's leadership in the national anti-cholera operation reflected its evolution, in little more than six years, into an effective implementing entity within the government. It improved its executive capabilities by dividing its financial, staffing and sanitary duties among three permanent commissions.[35] This allowed the Sanitary Council to allocate the sums at its disposal with efficiency and speed, limiting the duration of the cholera outbreak to three months, despite the country's infrastructural barriers.[36]

Following the cessation of the epidemic, the National Assembly agreed to expand the scope and power of the Sanitary Council. The council began reporting directly to the Ministry of Interior (instead of the Foreign Ministry) and received an independent budget derived from the tax levied on pilgrims and burial-bound cadavers traveling to Shi'ite shrines. This freed it from the political pressures, applied through discriminatory funding, that hampered the objective application of sanitary interventions in the past. The Sanitary Council used these initial funds to build a standing quarantine station at Qasr-e Shirin, on the Ottoman border, and begin construction on previously proposed quarantine facilities at the Caspian Sea ports of Astara and Anzali.[37] Additional financial backing from the National Assembly, which appropriated fifteen thousand tuman annually from its customs receipts in the North, paid for the completion and eventual operating costs of the Caspian stations.[38]

The Sanitary Council's independent budget did not stop internal rivalries between its European members that mirrored their political antagonisms on the international stage. Anglo-German disputes replaced the traditional Russo-British quarrels that receded with the 1907 Anglo-Russian Convention.[39] Germany, which had emerged as Europe's growing military and industrial powerhouse, continually attacked the British quarantine services for their deficiencies. Like the Russians before them, the Germans based their arguments on investigations and fact-finding missions in the Persian Gulf conducted by their delegate at the Sanitary Council and other governments aligned with them.[40] The Germans even sought voting privileges from consenting European legations not represented on the council to increase their power and eventually obtained the right to vote for Norway and Sweden. The British, not wanting to be outdone, acquired Belgium's vote, and the French secured the Greek vote.[41]

Germany's growing clout in Iran was increasingly worrisome for the other historically influential Great Powers. The newly appointed French professor of medicine at the Polytechnic College, for example, loathed the German physicians running the government hospital, accusing them of using their position and influence to spread pro-German sympathies among Iranians.[42] As a result, the French also sought a large hospital to administer as a "French institution."[43] This was emblematic of efforts by Western powers during this period to establish or control hospitals throughout the country to enhance their influence and prestige. In Tehran alone, the Germans controlled the Imperial Hospital, American Presbyterian missionaries founded and ran the American Hospital, the Russians controlled a hospital dedicated to the Iranian and Russian Cossack Brigades, and the Russian Red Cross managed an ambulatory clinic in the city. In addition, the French and British had active legation dispensaries serving the indigent population of the capital.[44] Despite political rivalries, the Sanitary Council was free of the theoretical bickering that had characterized its past proceedings. Its members received regular updates on the health of the country's provinces, including morbidity and mortality statistics, news of outbreaks, and sanitary developments in local jurisdictions. They unanimously agreed that Iran needed wide-ranging ameliorations in urban hygiene, prophylaxis against plague and cholera, and universal vaccination. They also made plans to expand the country's quarantine stations and sanitary services even under worsening financial and political circumstances in the second decade of the twentieth century.[45]

Pervasive insecurity and political interference by the Great Powers had plagued Iran since the inauguration of the Second National Assembly in 1909. Many of the armed factions that converged on Tehran during the civil war refused to leave the capital after the cessation of hostilities. Prominent figures across the country's political fault lines were either executed, assassinated, or forced into exile. Urban riots, brigandage, and warlordism had become the norm in the provinces, stifling commerce and the ability of the central government to collect taxes. To address these problems and counterbalance the Russian-influenced Iranian Cossack Brigade, the National Assembly formed a Swedish-trained gendarmerie force in 1910 and appointed Amir Khan Amir A'lam to lead its health services. The Iranians also hired a group of American advisers, led by the thirty-four-year-old William Morgan Shuster, to put the country's financial house in order. Schuster's efforts to improve and enforce the Iranian government's revenue collection collided with St. Petersburg's political agenda in Iran. Russia began to attack Shuster directly after the American played an instrumental role in defeating a Russian-supported countercoup by the country's deposed monarch, Muhammad 'Ali Mirza, in July 1911. Matters finally came to a head when Shuster formed a dedicated "treasury gendarmerie," led by a British officer, to enforce security and tax collection in Russia's northern sphere of influence. The Russian government issued an ultimatum to Tehran calling for Shuster's dismissal, the replacement of his gendarmes by Cossacks, and an apology for a trumped-up affront against one of its vice consuls. When the National Assembly refused to comply, Russia landed troops in Anzali and threatened to march on the capital. This precipitated a coup by the regent and the conservative members of the cabinet, bringing an end to the Second National Assembly and Shuster's seven-month tenure in December 1911.[46]

The dissolution of the Second National Assembly did not stop the Sanitary Council's work on the Caspian quarantine stations from moving forward. Jacques Philippe Gachet, a French naval physician, who replaced Georges as president of the council earlier that year, used his maritime sanitary expertise to guide the project through the crisis.[47] The Iranians gained Russia's political and financial support after Gachet promised to place the management of the finished stations in the hands of a Russian "chief sanitary officer for the Caspian Sea." The stations, which began operating in the spring of 1912, possessed a disinfection stove, an infirmary, and residential buildings that accommodated first-, second-, and third-class passengers for extended periods of observation.[48] The Sanitary Council also began to build similar establishments at the ports of Bandar Gaz and Babolsar (Mashhad-i Sar) on the eastern shore of the Caspian.[49]

Improvements to the country's sanitary stations and the Iranian Sanitary Council's expanded ability to respond to epidemics almost immediately diminished the impact of cholera outbreaks in the country. In December 1911 Asiatic cholera reached Kermanshah from the Ottoman Empire, but its spread was prevented after the prompt institution of inspection posts and quarantines around the city.[50] The Sanitary Council's takeover and effective management of Kermanshah's hospital helped lower the number of casualties and limit the duration of the outbreak within the city itself. The new quarantine stations also prevented the reintroduction of cholera from the Ottoman Empire and Russia, where the epidemic lingered.[51] This became apparent several months later, when aggressive medical inspections at the Caspian sanitary stations stopped Asiatic cholera from being transmitted back into Iran during a severe upsurge of the disease in the Russian city of Astrakhan. A similarly rapid response by the Iranian Sanitary Council and British "mobile" sanitary physician limited the duration and scope of another cholera outbreak in the South of the country around the same time.[52]

International Office of Public Health and the International Sanitary Conference of 1911–1912

Since the inauguration of the Second National Assembly, the Iranian government had attempted to raise its profile and legitimacy on the international stage by increasing its multilateral diplomatic engagements around the world.[53] In 1909 it joined twenty-two other countries as a full member of the International Office of Public Health (Office Internationale d'Hygiène Publique).[54] The office, established a year earlier in Paris, was the first permanent global health organization created to oversee international rules on preventing epidemic outbreaks, including maritime quarantine standards and procedures. It also collected and disseminated epidemiological data on infectious diseases, particularly plague, cholera, and yellow fever, and exchanged crucial information on the public health regulations of its member states.[55] The Iranian Sanitary Council regularly sent its minutes along with the country's morbidity and mortality data to the office's headquarters in Paris and used the office's best practices, communicated via monthly bulletins, to shape its policies and interventions. This lent international credibility to the Sanitary Council's proposals, especially when they clashed with the interests of the imperial powers in the country. Iran's membership also allowed its Sanitary Council to participate in cutting-edge international sanitary research projects, such as a 1911 study on the efficacy of the rat-destroying Clayton device in Bushehr on bulk grain containers.[56]

The International Sanitary Conference, held in Paris between November 1911 and January 1912, also relied on data and reports from the International Office of Public Health. The conference focused on recent advances in both the etiology and prevention of cholera and the means of stopping future outbreaks during the Hajj.[57] The Iranian government's financial and political problems prevented Tehran from sending its preferred representative, Amir Khan Amir A'lam, to the conference; however, Physician Major Georges agreed to lead Iran's delegation in his stead, having returned to Paris several months earlier after resigning his commission at the Polytechnic College and the Sanitary Council in Tehran.[58] Georges was accompanied by Iran's ambassador to France and the embassy's first secretary, Ardishir Khan, an enthusiast of bacteriology who had earned a diploma from the Faculty of Medicine in Paris in his spare time.[59] Iran's delegation showcased the country's progress since the last conference in 1903, highlighting the establishment of a regularly convening sanitary council, which effectively intervened against the previous year's

Asiatic cholera outbreak. Georges credited the reestablishment of "popular authority" in Iran after the civil war for the country's recent achievements.[60] He also praised the British for their contributions to improving the quarantines in the five Persian Gulf ports under their control:

> In April 1910, the Persian Sanitary Council, composed—I insist on this detail— of not only indigenous members, but also American and European physicians, delegates from all the powers represented in Tehran, did an inventory of the sanitary situation on the country's frontiers. After expert deliberations, it decided that the question of the [sanitary] defense of the Gulf, which had been the first order of its preoccupations, could and should, because of *implemented improvements*, give way in order of urgency to the question of the [sanitary defense] in the country's north, followed immediately by the sanitary defense of the Kermanshah-Karbala-Mecca pilgrimage highway.[61]

The upgrades in the Persian Gulf enabled sanitary inspectors to rapidly detect and quarantine the Iranian cruiser *Persepolis* off the coast of Muhammara during a potentially disastrous cholera outbreak onboard several months before the conference.[62] Georges's endorsement of the British arrangements also resolved competing demands, made in previous international meetings, for a sentinel quarantine station at the mouth of the Persian Gulf.[63] This allowed the deliberations to focus on more comprehensive cholera control strategies, such as the detection of asymptomatic carriers of the disease, which had baffled international public health authorities. Although routine bacteriological stool examination was suggested as a potential diagnostic solution, Iran's delegation and others at the conference questioned the feasibility and efficacy of large-scale laboratory screenings. They agreed instead that improved surveillance and quarantines were the best approach to detect silent carriers.[64]

The conference's proceedings did not touch on the subject of vaccination as a protective measure against cholera, despite the meeting's endorsement of an immunization program against plague for Muslim pilgrims.[65] Unlike the plague vaccine, the cholera serum, developed by Waldemar Haffkine at the Pasteur Institute in Paris more than a decade earlier, used living attenuated microbes, making it more difficult to prepare and almost impossible to send over long distances. It also had a short shelf life, requiring bacteriologists to frequently manufacture new batches during mass vaccinations. While the vaccine's large-scale use in India had yielded some promising results, the process continued to be controversial and out of reach for Iran and other countries that lacked domestic production capabilities.[66] Moreover, a Russian attempt to administer the Haffkine vaccine to the population of Tabriz during the 1904 cholera epidemic ended in fiasco after terrified locals began assaulting the vaccinators, forcing the Iranian government to halt their activities.[67] Tehran's lack of enthusiasm with the live cholera vaccine did not extend to immunizations against other diseases. Iran's delegation explained that, while cholera killed about sixty thousand every decade, other infectious diseases such as tuberculosis and smallpox killed and maimed even more of their countrymen in any given year. This explained why the establishment of a national vaccination program against infectious diseases had become a top priority for the Iranian Sanitary Council.[68]

The Emergence of National Vaccination

Iran lacked an established immunization program in the first years after the Constitutional Revolution, despite modest attempts at vaccination dating back to the mid-nineteenth century. Jean-Etienne Schneider regularly received tubes of smallpox vaccine from the French Academy of Medicine, which he administered free of charge to his patients during his time in Iran between 1894 and 1907.[69] Most Iranians still resorted to the long-standing tradition of variolation for protection against smallpox. In 1907 the Sanitary Council convinced the Iranian government to fund a "vaccination institute," following a deadly smallpox outbreak in Tehran.[70] The first National Assembly nominally agreed to appropriate yearly funds for its operation; however, the civil war and budget cuts scuttled this initiative.[71] This setback notwithstanding, the Iranian government had grasped the importance of smallpox vaccination on a national level, paving the way for the passage of the Hygiene and Smallpox Vaccination Act by the Second National Assembly in 1910. This legislation helped establish a free national smallpox vaccination program by appropriating one-tenth of the new tax on vehicles and beasts of burden toward the service.[72]

The national vaccination program began in 1911 with the purchase of calf-lymph-derived smallpox vaccine from Paris and the appointment of public vaccinators throughout country.[73] Between February and June of that year, more than 14,400 people were vaccinated in Tehran alone.[74] By 1912, the campaign had succeeded in immunizing a substantial number of schoolchildren, despite lingering resistance by some parents and the occasional lack of access to government vaccinators.[75] In a quest to remove these barriers and increase compliance, the Sanitary Council instituted Iran's first public health advertisement campaign.[76] Posters and brochures written in colloquial language

warned Iranians of the dangers of smallpox and the lifesaving benefits of vaccination. The council also instructed local authorities to "oblige school di-
rectors to vaccinate their students and bar new pupils from enrolling unless
they consented to the operation."[77] In the ensuing two years, the Sanitary
Council added immunizations against typhoid and diphtheria to its vaccination program thanks to sustained government funding and donated vaccines
from the Military Teaching Hospital of Val-de-Grâce and the Pasteur Institute.[78] Ironically, the widespread use of vaccinations and hypodermic needles
also introduced Iranians to intravenous opiate abuse at the threshold of the
First World War. This new scourge would change the country's destiny well
into the twenty-first century. The Sanitary Council's strong response to illegal opiate sales would foreshadow Iran's current laws, which include capital
punishment as a means of deterring illegal opiate sales: "The Commission [on
opiate abuse], after the reading of a letter from the governor of Neishabour
regarding the use of intravenous injection of morphine in cafes and opium
dens, has discussed measures that seem most practical and easiest to apply to
stop this abuse. The [Sanitary Council] proposes to write to the governor of
Neishabour to take the following measures: punish the perpetrators with expensive fines and prison, all individuals who provide morphine without a physician's prescription."[79]

The national vaccination programs could not cure the endemic sources of
infection, particularly in Iran's urban food and water supplies. Tehran's contaminated water, for example, was responsible for deadly typhoid outbreaks,
which endured despite the Sanitary Council's immunization campaigns. Its
cow stables, maintained in crowded and unlit quarters all over the city by the
"milk industry," caused chronic animal-borne diseases. Tehran's garbage, discarded in a "large pit" located near the Arsenal Quarter within the city's
walls, was its main municipal disposal system for the better part of the nineteenth century and a recurrent source of outbreaks.[80] In response, the Sanitary
Council established a hygienic commission, charged with investigating and
fixing deficits in urban sanitation, beginning with the capital. In 1913, it purchased disinfection equipment and produced guidelines for Tehran's municipal employees to improve cleanliness.[81]

In May 1914, the hygienic commission produced a detailed study on the
sanitary and public health needs in the capital.[82] The report highlighted the
lack of hygienic abattoirs and proposed improvements based on the municipal slaughterhouses at La Villette in Paris, including the inspection of animals
for disease and fitness of consumption by on-site veterinarians.[83] The commis-

sion's report prompted the government to decree a set of regulatory measures to improve Tehran's sanitary condition. Owners of ice pits (*yakhchal*) were required to guarantee the purity and hygiene of the water and pits that they used in making ice.[84] The government also enacted new regulations to improve garbage disposal and sanitary standards at bazaars and markets (see appendix D).[85] However, the Sanitary Council's fiscal limitations hampered the enforcement of these edicts. Even the vaccination service, which had an earmarked budget by the National Assembly, periodically lacked the funds required to buy vaccines due to erratic government disbursements.[86] Making matters worse, governors and municipal leaders, unwilling to spend money from their own independent budget for municipal hygiene, often disregarded Tehran's directives.[87] As a result, the head of the Sanitary Council in 1915 complained that he had "been unable to convince the municipality (of Tehran) that it could not ignore its sanitary responsibilities."[88] This inability to address sanitary deficiencies meant that Iran was vulnerable to epidemic diseases on a national scale, despite advances in science, work force, and administration in the postconstitutional era.

Epidemics and the First World War

Iran's nonalignment did not shield its territory from the terrible consequences of the First World War, which broke out in July 1914. Worsening fiscal and administrative difficulties in the wake of the 1911 coup that dissolved the Second National Assembly increased Tehran's inability to police its provinces and borders at the dawn of the global conflagration. This eased the Ottoman Empire's invasion of the country's Northwest in a bid to cut off Russian access to the oil-rich Caspian littoral in December 1914, less than a month after Iran officially declared its neutrality.[89] A counterattack by Russia in 1915 pushed the Ottomans back without achieving a decisive victory, transforming the campaign into a grinding low-intensity conflict that lasted until the end of the war. In Iran's South, the British landed troops to protect oil fields that played a vital role in the Royal Navy's ability to function since its battleships had converted to oil burning engines several years earlier. The Germans, lacking a standing army in the Middle East, responded by inciting Qashqai tribesmen and other sympathetic groups in Iran's southern tribal belt to attack British oil and naval installations. In the summer of 1915, British-Indian and Russian forces led a two-pronged invasion and occupation of the country's eastern provinces to prevent the Central Powers from infiltrating Afghanistan and India through Iran. The following year, the British established a standing military force of more than eight thousand Iranians, known as the South Persia Rifles, commanded by British officers to protect their interests in Iran against further attacks.[90] Around the same time, a revolutionary Iranian paramilitary group known as the Jungle Movement (Jangali) emerged from the confusion of war and took control of the forested northern province of Gilan bordering the Caspian Sea, and nationalist gendarmerie forces began a campaign of resistance against the Allies in the country's western provinces (fig 5.5).[91]

Iran's civilian population bore the brunt of the military intervention and violence during this period, including unprecedented levels of material devastation, economic disruption, and displacement. Forceful requisitioning and looting by occupying armies caused an almost immediate increase in the price of basic food staples.[92] In 1915 the British military purchased the total grain reserves in the southern province of Sistan, and the Russians prohibited crop transfers not destined for their troops in the North. Both militaries requisitioned pack animals, disrupting the cultivation and transportation of foodstuff.[93] Annual anthrax outbreaks decimated Iran's cattle population, causing food insecurity and near famine conditions throughout the country.[94] Urban overcrowding by refugees and peasants escaping violence, brigandage, and scarcity in the countryside increased the incidence of common infectious diseases such as malaria, tuberculosis, gonorrhea, syphilis, typhoid fever, and amoebic dysentery.[95]

The large-scale movement of troops and refugees from neighboring countries and the absence of prewar levels of sanitary surveillance at important junctions leading into Iran also increased the frequency of imported epidemic diseases. British soldiers carried bubonic plague from India into the country's Persian Gulf ports during their buildup at the outset of the conflict (table 5.1).[96] Outbreaks of cholera from Russia, which had become a breeding ground for an array of contagions due to its deteriorating domestic condition, were a far worse calamity.[97] The disease reached epidemic proportions in Iran's Northwest during in the summer of 1915 on the heels of a seventy-thousand-strong Russian offensive in the Caucasus and the influx of Armenian and Assyrian Christian refugees escaping ethnic cleansing operations by the Ottoman military in Anatolia.[98] A separate expeditionary force of twenty thousand Russian Cossacks carried the epidemic to the port of Anzali and the interior of the country as they marched on Tehran to intimidate the pro-German Third National Assembly, which had been reconstituted a year earlier.[99] The Cossacks halted their advance twenty-five miles west of the capital along the Karaj River, giving the Iranian Sanitary Council breathing room to stop the epidemic on

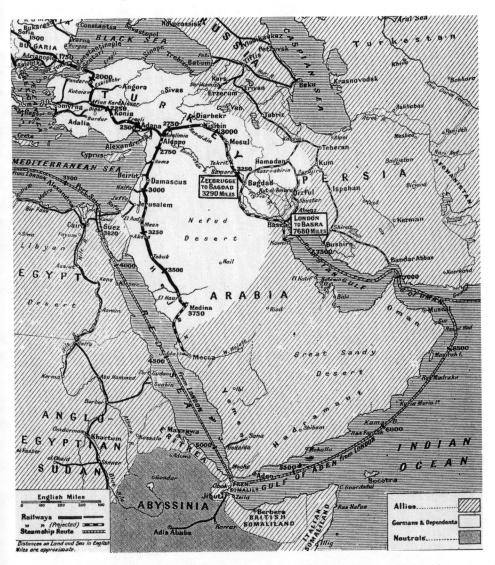

Fig. 5.5. Military occupation of Iran in 1916. Herbert Wrigley Wilson and John Alexander Hammerton, eds., *The Great War: The Standard History of the All-Europe Conflict*, vol. 9 (London: The Amalgamated Press, 1917), 24.

the capital's doorsteps by establishing a six-day quarantine on the highway to Tehran and a provisional sanitary station in one of Fath 'Ali Shah's abandoned palaces near the Russian camp.[100]

The dissolution of the National Assembly under Russian pressure and subsequent government paralysis deprived the Sanitary Council of urgently

TABLE 5.1.

Occurrence of plague, cholera, and typhus in Iran during the Great War (1914–1918)

Date	Plague	Cholera	Typhus
1914			
January			Hamedan (Jewish Quarter)
February	Lingah		Van
March			
April			
May			
June		Anzali	
July			
August			
September			
October			
November			
December			
1915			
January			
February			
March			
April	Bandar Abbas, Bushehr		
May	Muhammara		
June			
July			
August		Azerbaijan Province: Tabriz, Jolfa, Maragha, Khoy, Marand, Salmas	
September		Gilan Province: Astara, Anzali, Khoman, Rasht, Manjil, Talesh	
October		Tabriz, Astara, Anzali, Rasht	
November		Astara, Anzali, Rasht, Manjil, Talesh, Ardabil, Qazvin	
December		Qazvin	
1916			
January		Qazvin	
February			
March			
April		Gilan Province: Anzali, Rasht	
May	Muhammara, Khark	Gilan Province: Anzali, Rasht	
June		Gilan Province: Anzali, Rasht	
July		Rasht, Qazvin	
August		Gilan Province, Hamedan, Qazvin	
September		Gilan Province, Shah Abdol Azim, Tehran	
October		Gilan Province, Jafarabad, Shah Abdol Azim	
November		Gilan Province; Mazandaran Province: Amol, Caspian shore	
December			

Date	Plague	Cholera	Typhus
1917			
January			
February			
March			
April			
May			
June		Mazandaran Province: Amol, Sari, Babol (Barfarush), Bandar Gaz, Astarabad.	
July		Gilan Province: Larijan, Yekshanbeh Bazar; Tehran; Khurasan: Mashhad, Tappeh, Shahrud, Sabzevar; Sistan; Kerman; Fars; Khuzestan: Muhammara	
August			
September			
October			
November			Widespread epidemic
December			Widespread epidemic
1918			
January			Widespread epidemic
February			Widespread epidemic
March		Gilan Province: Astara, Anzali, Rasht	Widespread epidemic
April		Azerbaijan Province: Qazvin, Zanjan, Mianej, Ardabil, Tabriz; Kurdistan Province: Kermanshah; Qom, Sultanabad, Isfahan; Fars Province: Yazd, Kerman.	

needed funds to intervene against another wave of the disease several months later.[101] The new outbreak followed Russian supply lines across the Caspian Sea during Russia's offensive against a second Ottoman invasion of western Iran in the spring and summer of 1916.[102] Cholera entered Anzali, unimpeded by the usual disinfection and isolation procedures, which had been crippled by the Russians earlier in the year when they repurposed the port's sanitary station into a military hospital.[103] The growing insurgency by the Jungle Movement in the forests of Gilan and the Sanitary Council's anemic response allowed the epidemic to spread inland and reach Tehran several months later.[104]

New waves of cholera crossed into Iran from the north during the last two years of the First World War despite Russia's military drawdown after the October 1917 Bolshevik Revolution. The chaos in Russia and worsening instability in Iran created the right conditions for the disease to thrive and spread, as it did in the summer of 1918 when a cholera epidemic reached the embattled northwestern Iranian city of Urmia. Seven months after Russia's withdraw from the region, the city's defenders could no longer hold the breach against advancing Ottoman forces. Fearing a massacre, eighty thousand Assyrians and Armenians from the city and the surroundings began a weeks-long retreat, mostly on foot, to the safety of British-controlled Hamedan, "leaving a trail of death and disease behind them."[105] Approximately half of the marchers succumbed to Ottoman bullets or cholera.[106] Such was the fate of the American consul to Urmia, who died on a "cold desolate mountain road in darkness" with "no medicine, no nourishment, no comfortable place for him lie, and only a limited amount of water," as he directed the columns of men, women, and children through the mountain passes.[107]

A two-year countrywide famine that began in the spring of 1917 magnified the impact of the cholera outbreaks in this period. Rural areas were particularly hard hit by the food crisis, as described by a British military officer traveling through western Iran in 1918:

> The country was in a terrible state and the peasantry was in the last stages of starvation. Every time I was forced to stop my car, I was surrounded by hundreds of near-skeletons who screamed and fought for such scraps as I was able to spare. In a single day's journey of fifty six miles between the towns of Kirind and Kermanshah, I counted twenty-seven corpses by the roadside, most of them those of women and children, and the general condition of life amongst the peasants was so frightful that I was ashamed to eat my simple rations in their presence.[108]

As people migrated to cities in search of food and work, they broadened the dissemination of cholera. Survival strategies such as the consumption of opium, a cheap and readily available crop that served as a "famine food" in Iran, added to their vulnerability. Starvation also increased individual infection rates by lowering salivary and stomach acid levels, which served as primary barriers against the fecal-oral disease. Tehran alone is estimated to have lost thirty thousand to cholera in the summer of 1918, a shocking number when one considers that only two hundred thousand people lived in the city around that time. Observers described scenes of bodies scattered in the streets and mountains of corpses buried in mass graves throughout the capital.[109] Other cities lost thousands as well, although the confluence of various epidemics in 1918, including a deadly typhus outbreak from Russia, made precise mortality data on cholera almost impossible to obtain.[110]

The Russian border was also an entry point for the 1918 influenza pandemic into Iran, transmitted by the anticommunist White Army during its retreat from a Bolshevik counteroffensive in the Caucasus and Central Asia. The speed of modern transportation was a key factor in the rapid propagation of the Spanish flu, as the pandemic came to be known. By September 1918, influenza's first wave overran Iran's eastern and central provinces and, like earlier cholera epidemics, rapidly reached the country's Northwest owing to the Caspian steamship traffic from Baku and the Tbilisi-Julfa railway line. In Tehran, the disease was called the "illness of the wind" (*nakhushi-yi bad*) due to its initial occurrence during a strong westerly wind burst and its rapid spread. As in Europe, Iranian officials and physicians were caught off guard by the deadly outbreak, unable to adequately respond to the growing number of victims. The resulting losses in the capital were staggering, particularly in the slums and poorer districts of the city where the dead were carried by the cartloads to be buried in mass graves.[111]

Several weeks after the outbreak of the Spanish flu in the North, British-Indian expeditionary forces from Bombay introduced a second, more virulent wave of the disease into Iran's southern provinces from their ports of call in the Persian Gulf and military staging grounds in Mesopotamia. This new wave was especially lethal in the cities of Kermanshah and Hamedan because of the large influx of Armenians and Assyrians from Urmia and other areas in the Caucasus. Kermanshah alone had received sixty thousand hungry and diseased refugees, a number equal to the native population of the city.[112] Casualties were higher among these newcomers due to comorbidities like malaria and anemia, which occurred more frequently in this population and made succumbing to the Spanish flu more likely. Similarly susceptibilities in the countryside caused elevated mortality rates throughout rural areas in Iran. Nomadic Qashqai tribesmen, who had been fighting the British in the South, lost more than one-third of their fighting-aged men to influenza. After seeing five armed Qashqai sick by the roadside, a British officer reported finding their dead bodies several days later beside a stream to which they had probably crawled to drink; apparently no one had been left alive to carry them away or bury them.[113]

Influenza's impact on urban areas like Shiraz, which eventually lost a fifth of its population, had unforeseen economic and social consequences. Mounting casualties prompted the city's officials to speculate on the cloth used to make religiously mandated burial shrouds. This caused a shortage of the commodity, leading hundreds of the city's impoverished victims to crawl into mosques in a desperate attempt to die on hallowed ground so as to lessen the ignominy of an inappropriate burial. Municipal services were paralyzed by the large number of medical personnel, transport workers, and telegraph and postal officials who succumbed to influenza.[114] Even the provincial governor, 'Abd al-Husayn Mirza Farman Farma, barely survived his bout with the illness. When the commander of the British expeditionary forces visited the grandee during his convalescence, the governor "explained in his curious French that half of Shiraz was dead"—or, as he put it, "Le demi-monde de Chiraz est Mort."[115]

By the fall of 1919, the Spanish flu ran its course after a third limited wave of the disease struck the province of Baluchistan earlier in the year. The pandemic had an oversized demographic impact on Iran because of the country's wartime conditions. Malnutrition, comorbid illnesses, and rising rates of opium consumption made Iranians more susceptible to dying from complications of the Spanish flu. Exacerbated by such conditions, the pandemic killed from eight to nearly twenty-two percent of the country's total population, a rate higher than any other nation affected by the global scourge.[116]

The Sanitary Council during the First World War

The institutional capabilities of the Iranian Sanitary Council declined significantly with the outbreak of the First World War. Financial shortfalls and political chaos in Tehran limited its capacity to enforce hygienic regulations or carry out the necessary interventions to stop the period's prevailing epidemic and endemic contagions. Many European physicians who held key positions on the council began leaving Iran to join their respective armies in the West. A series of temporary vice presidents, who lacked the consensus power of their predecessors, led the Sanitary Council until Amir Khan Amir A'lam took over as "permanent" president in January 1916.[117] The nomination of an Iranian to head the council on a nontemporary basis was not without controversy. While the French, who viewed the position as theirs by tradition, strongly protested, they could do little to stop Amir A'lam's appointment when their preferred candidate refused to leave Paris. Even the British, who sided with their French ally on most Iranian issues, privately confessed that he was the best choice to lead the council: "The appointment of Dr. Emir Khan is a popular one and is the best that could be made from among the Persian members. He is the most fitted as regards scientific and general knowledge and speaks French well. He holds a French qualification having studied at Lyon and at Paris. He is a protagonist of the '*jeunes Persans*' [young Persians] and was at one time a member of the Democratic party in the Majlis. He has however enough savoir faire to prevent his going to extremes."[118] Under Amir A'lam, the Sanitary Council began reorganizing and professionalizing the vaccination services in an attempt to halt the increasing rates of preventable infectious diseases.[119] Certified physicians were required to staff vaccination posts in Tehran except at the Polytechnic College, where medical students carried out inoculations under the supervision of their instructors at the school's clinic.[120] The Sanitary Council also realigned the vaccination services in the provinces under the direction of a central commission located in each of the provincial capitals. These commissions were composed of a government representative, two Iranian physicians, and a European doctor—when available.[121] They carried out the Sanitary Council's instructions from Tehran and convened training conferences to certify vaccinators along national standards set by the council.[122]

The Sanitary Council's plenary discussions and consensus on matters of prevention during this period highlighted Iran's advances in public health and administration since the 1903–1904 cholera outbreaks. For example, when cholera broke out in the village of Aqa Baba near the northern city of Qazvin in 1916, the Sanitary Council rapidly agreed to take steps to protect the rest of the country without engaging in the type of contentious and slow deliberations that had characterized its proceedings during previous outbreaks.[123] The central government's weakness and ongoing financial difficulties, however, hampered the council's ability to translate recommendation into action. The Iranian government went so far as to wholly deny the existence of the outbreak; fearing that an acknowledgment could trigger a "cholera panic" and bring a torrent of refugees into Tehran, straining the already stretched wartime resources in the capital.[124] Unable to obtain the necessary financial support from the central government, the council took steps to "clear itself of any responsibility should the epidemic reach Tehran."[125] This left local sanitary authorities, where epidemics first broke out, with the burden of protecting the country using limited provincial funds and resources.[126]

Despite the Sanitary Council's setbacks against epidemics, improvements in its vaccination services during the last two years of the war helped double the population of Iranians vaccinated between 1919 and 1924 (fig 5.6). The expansion meant that Iran could no longer exclusively rely on imported antigen and serum to meet its needs. These stocks were not only limited but also often spoiled due to poor refrigeration in route; motivating Iranian sanitarians to seek indigenous means for vaccine production for the broad-based national immunization program needed to address the mounting fatalities.

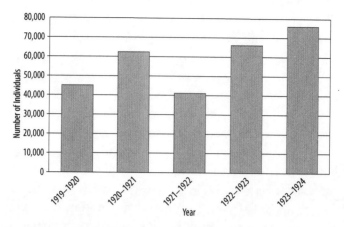

Fig. 5.6. Vaccinations in Iran from 1919 to 1924. Gilmour, *Report on an Investigation,* 29.

The Pasteur Institute of Iran and the Sunset of Cholera

Iran was a failed state by the end of the war, with most of its population diseased or dead, its treasury empty, and its central government unable to project power beyond Tehran's city limits. In the North, the Jungle Movement led a successful secessionist military and political campaign, proclaiming the province of Gilan an independent Soviet Socialist Republic. In the South, the rebellious Shaykh Khaz'al of Muhammara flaunted Tehran's authority and carved out an autonomous sheikhdom. Foreign forces continued to occupy Iranian territory, even after the armistice. Thousands of soldiers from the British-led South Persia Rifles remained in the country and occupied strategic positions stretching from the Mesopotamian border to as far north as Qazvin and Mashhad.

In May 1920 Soviet troops landed at the port of Anzali and marched into Gilan to support the Jungle Movement, vowing to stay until British forces withdrew from Iran. Sultan Ahmad Shah, who had assumed governing duties in 1914 upon reaching majority of age, was a weak monarch with little interest in the kind of governance and diplomacy that could solve these postwar ills.[127] The National Assembly, paralyzed by imperial meddling and an empty purse, could not even cover the cost of operating the government hospital in Tehran, which would have closed down in 1919 had the British government and the Anglo-Persian Oil Company not shouldered the burden of providing the medical manpower and material expenses for the eighty-bed establishment.[128] In the provinces, the dual traumas of war and disease had decimated the rural population and the agrarian economy. This meant that famines, caused by poor harvests, and the collapse of mercantile activity continued even after fighting had stopped. The infrastructure of public health was also severely degraded. The vital steam stoves used on a regular basis for disinfection in Iran's principal southern ports had vanished by the end of the conflict. The machines at Langeh, Jask, and Bandar Abbas had broken down in 1915 and were discarded. The British military in 1916 commandeered the remaining apparatus in Bushehr and rendered it inoperable after two years of wartime use.[129] The quarantine stations in Astara and Anzali in the North had been similarly pillaged and destroyed by the Bolsheviks.[130]

Iran's growing vaccination service, on the other hand, was a bright spot in an otherwise dreadful sanitary picture and inspired the Iranian mission to the Paris Peace Conference in 1919 to seek the assistance of the Pasteur Institute to meet its country's increasing need for vaccines. The institute's leadership had corresponded with Iranian officials in the past and had periodically supplied the Sanitary Council with shipments of vaccines and growth media.[131] The Iranians hoped to persuade the internationally minded Pasteurians to expand on this relationship by establishing Iran's first enterprise in laboratory research and vaccine production.[132] Firuz Farman Farma, Iran's foreign minister and its chief envoy to the Paris Peace Conference, was poignantly aware of the need to improve Iran's capacity to prevent contagions. His experiences during the wartime plague years, including his father's brush with death in the 1918 influenza pandemic, convinced him that Iran needed a strong base in laboratory science and microbiological education to halt future outbreaks.[133] He personally approached the French foreign minister for a list of candidates for the newly created position of professor of bacteriology at the Polytechnic College and led the Iranian delegation to the Pasteur Institute.[134]

On October 23, 1919, Emil Roux, Louis Pasteur's closest collaborator, personally received the Iranian mission at the Institute's headquarters in Paris.[135] After several days of negotiations, the director of the Pasteur Institute agreed to establish a satellite institution in Iran.[136] The Pasteur Institute of Iran, inaugurated in August 1921, was built on a ten-thousand-square-meter private plot outside of Tehran donated by the foreign minister himself, who also covered the initial construction costs.[137] Roux selected Joseph Mesnard, a thirty-four-year-old French microbiologist, to direct the new institute, which included laboratories for histopathology and chemical analysis, prevention and control of animal diseases (particularly rabies and anthrax), and the manufacture of cholera and smallpox vaccines.[138] Mesnard also offered a course in foundational laboratory medicine to senior medical students from the Polytechnic College that included thirty lectures in bacteriology, microbiology, and hygiene at the institute's laboratories.[139] Most of the new institute's microscopic and chemical studies were carried out free of charge at the behest of Iranian government; but private practitioners, who previously sent specimens and patients abroad to obtain professional laboratory analysis, could also use the services for a nominal fee.[140] The institute's vaccine production allowed it to stock 10,000 doses of cholera serum in its first years of operation. It also produced and distributed 186,805 doses of smallpox vaccine during the first nineteen months of its existence and 193,119 doses the following year.[141] However, by 1925 it was still unable to make enough vaccines to meet the needs of the vaccination service, in part because of the government's inability to manufacture or import sufficient amounts of glass ampoules required to contain the vaccines. Its skeletal budget, which limited the purchase of animals needed for the large-scale production of sera, added to the shortfall.[142]

Despite its deficits, the Health Committee of the League of Nations described the Pasteur Institute of Iran as "the most efficient part of the Persian sanitary administration" in 1925.[143] Its manufacture of vaccines and other contributions to public health were instrumental in slowly turning the tide against epidemics in Iran. In 1923, as cholera swept through Mesopotamia towards Iran, the Sanitary Council, with Joseph Mesnard of the Pasteur Institute as its vice president, quickly organized a surveillance station on the Baghdad-Tehran highway, at Qasr-e Shirin, and coordinated the vaccination of the surrounding population with domestically produced cholera sera— forbidding passage without a certificate of vaccination. This intervention prevented cholera from spreading beyond coastal towns on the Persian Gulf and contained an otherwise virulent epidemic that killed 911 out of 961 sick individuals in the city of Abadan alone.[144] The Pasteur Institute was able to prepare six to seven thousand doses of cholera vaccines daily by 1927, when an incursion of cholera from Baghdad once again threatened Iran's frontier.[145]

The Qajar political establishment's successful negotiation with the Pasteur Institute stood in stark contrast to its inability to address Iran's fracturing regionalism and unfettered financial free fall in the years following the war. In a fit of desperation, the foreign minister and his political allies attempted to give Britain control over Iran's financial and military affairs in 1919. By making Iran a British protectorate in the style of Palestine or Iraq, they hoped to halt further economic deterioration and the country's breakup. The agreement ultimately failed, paving the way for Riza Khan (1878–1944; r. 1925–1941), a general in the Iranian Cossack Brigade, and his nationalist political allies to mount a successful coup d'état, six months before the Pasteur Institute of Iran began its operations. Sultan Ahmad Shah quickly approved a new postcoup cabinet under Riza Khan's de facto military leadership in 1921 and left the country two years later, never to return. The new government began stabilizing Iran through a series of shrewd military and political maneuvers that neutralized internal dissent, strengthened the central government in Tehran, and reestablished its control over the provinces. Riza Khan personally led the troops that crushed the Soviet Republic of Gilan, incarcerated the independent-minded Shaykh Khaz'al of Muhammara, and smothered the flames of tribal rebellions and petty brigandage that plagued the country's highways.[146]

The new regime also consolidated the administration of public health. It absorbed the Sanitary Council, which continued to be led by Amir A'lam, into the Ministry of Health, formed by the postcoup cabinet in February 1921. 'Ali Asghar Nafisi (Mu'addab al-Dawla), a French-trained physician and member of Riza Khan's inner circle, was named minister of health, followed by Hassan Luqman Adham (Hakim al-Dawla), who assumed the position after the dissolution of the first postcoup cabinet three months later.[147] During Adham's tenure, the government incorporated the Sanitary Council, including its European members, into the new Ministry of Hygiene and Welfare, with jurisdiction over public health and epidemic disease prevention.[148] However, the council's multinational makeup ran against the new regime's policy of ethnic nationalism and hostility toward the historically meddlesome Western powers. Officials were intent on distancing themselves from the failed 1919 agreement. Even former clients of the European imperial powers in the government opposed, at an almost xenophobic level, foreign participation in any aspect of Iranian governance.[149] As a result, the Sanitary Council was prohibited from communicating directly with Iran's Cabinet of Ministers and foreign legations in Tehran, as had been the norm. Relegated back to an advisory role, its recommendations had to be transmitted through the minister of hygiene, and it would no longer receive an independent budget. When the Sanitary Council defiantly passed a resolution to ignore the government's new regulations, the Ministry of Hygiene suspended the organization. Amir Khan Amir A'lam, saved the moribund institution by convincing Riza Khan, personally, to incorporate the Sanitary Council into the newly formed Public Health Department, which worked independently of Iran's ministries and was led by Amir A'lam himself.[150]

Although the Sanitary Council continued to meet regularly, it effectively ceased to exist as an international organization in 1926 when the government mandated that it conduct its proceedings in Persian: "During this year the language of the council has been changed from French to Persian. This is a notable event as it practically drives the European members out of the discussions. During the whole year no European member has attempted to give his view on any subject unless expressly asked to do so when the language of the meeting was changed to enable him to reply. This means that the non-Persian members are reduced to mere puppets, who can only reply when spoken to."[151] The British saw this as an extension of the prevailing Iranian policy to "drive out the foreigner and to prevent him taking any part in Persian affairs." The growing nationalism even affected European medical practitioners, who faced heightened "tension and almost bitterness" from their Iranian counterparts.[152] Tehran began enforcing the 1911 medical licensing law on foreigners, dismissing two British physicians in 1922 and a French surgeon in 1926 from the government hospital, replacing them with licensed native doctors.[153] The marginalization of European influence in the Sanitary Council allowed Iranians to challenge established Great Power economic interests that had hindered past disease prevention efforts. In 1926 the council banned the importation of secondhand clothes from India, which were believed to be an indirect route for the transmission cholera and plague into Iran. Despite British protests and offers to put the imports through a rigorous disinfection process, the Sanitary Council refused to modify its decision.[154]

British management of Iran's quarantines in the Persian Gulf also began caving to the political currents of nationalism and administrative centralization. The Iranian government fired the first salvo in 1923 when it stopped paying its share of the operating costs of the quarantines. After London protested the arrears owed to its medical officers, the Iranian Foreign Ministry, which had never contracted for their services, demanded written evidence of their employment.[155] The Iranians eventually remunerated the British, despite refusing to formalize the position of their quarantine staff. Iran reasserted its rights during the 1926 International Sanitary Conference in Paris by declining to ratify the meeting's final convention that upheld Britain's privilege to establish and operate quarantines in Iran.[156] The Iranian delegation, galvanized by the dissolution of the Qajar dynasty and the coronation of Riza Khan in the months preceding the meeting, challenged the treaty on legal and technical grounds, arguing that Iran could manage its own quarantines. However, the Iranians were unable to obtain the unanimous votes needed to overturn the convention due to fears, perpetuated by the British, that Iran lacked the necessary medical expertise to run its quarantines without Western assistance.[157]

This setback did not stop domestic Iranian efforts to end Britain's control. By 1927, Husayn Bahrami, Amir A'lam's successor at the helm of the Public Health Department, convinced the British to relent by promising formal multiyear contracts and regular salaries for their physicians.[158] He reassured London that the Iranian Public Health Department, like the Sanitary Council before it, would assume only a supervisory role over the quarantines in the Persian Gulf and the day-to-day operations would remain in the hands of its physicians. The British government was inclined to accept the proposed takeover, seeing as the Soviet Union had similarly agreed to relinquish its control of quarantines in northern Iran.[159] Bahrami, however, did not intend to keep his side of the bargain. He replaced the British chief quarantine officer with an Iranian government physician as soon as the Public Health Department formally assumed control of the quarantines and forbade the remaining British from taking any action unless ordered to do so by their newly appointed chief.[160] The Iranian government, therefore, not only had an oversight role but also determined how the quarantines ran on a day-to-day basis. This was part of a grand plan to eject the British once and for all. London predictably protested the turn of events. However, it was unable to challenge the Iranian government's new policy against employing citizens of a bordering country, such as physicians from British-India, which made up the majority of their medical staff in the Persian Gulf. Even so, the British knew that their physicians' contracts were ultimately contingent on ratification by the National Assembly, which had not spoken on the issue.[161]

In 1928 the Iranian government did not include the salaries of the British medical officers in its annual budget. The National Assembly, therefore, could not ratify the multiyear contracts that Bahrami had promised. The Iranians formally dismissed the British medical staff in March of that year, ending more than half a century of British dominion over its quarantines.[162] Tehran rapidly appropriated funds to employ French and German physicians to replace the British. In this way, it forestalled international criticism, of the type Iran faced two years earlier at the Paris Conference—that it lacked the technical work force to run its quarantines. The British, in contrast to past claims, were unable to fight Iran's takeover due to their economic difficulties in the period immediately after the war. Their fiscal problems increased with wage and price inflation and the return to the gold standard in 1925.[163] This made them dependent on Iran's financial participation to continue their control of the quarantines in the Persian Gulf. Fundamentally, however, Iran built its success on a rapid turnaround of its central government's administrative organization and financial health. Improved domestic security under Riza Khan returned trade and agrarian life to its prewar levels. The new regime's success in modernizing its treasury and revenue collection allowed it to balance its budget by 1925 and end "the vicious circle of deficit and borrowing."[164] For the first time in decades, Iran had a fiscal surplus when it took over the quarantines from the British in 1928.[165]

Improved finances also allowed the government to begin developing the infrastructure of sanitation and public health in urban areas as recommended by the 1925 League of Nations report.[166] Thus began the gradual extinction of devastating large-scale cholera epidemics in Iran. When word of a cholera epidemic in Mesopotamia reached Tehran in 1927, Iran's Public Health Department immediately closed its frontier with Iraq, established new quarantine stations, and barred travel south of Shiraz, including pilgrimage to the Shi'ite shrines of Karbala and Najaf. The government also purchased one hundred thousand additional doses of cholera vaccines from Germany, which it rapidly flew by airplane to susceptible populations in the South.[167] Although the cholera epidemic did eventually spread into Iran, its impact was limited to 829 registered cases with fewer than 700 deaths.[168] The Public Health Department's successful containment of the outbreak signaled a new era of effective self-reliance on the part of Iran's sanitarians, nominally independent of foreign physicians and Western intervention.

The eradication of epidemics became a centerpiece of the Iranian government's public-sector development plans by the third decade of the twentieth century. This led the Iranian Ministry of Finance to create the Razi State Serum and Vaccine Institute, the country's first domestic pharmaceutical company in 1931. The institute initially focused on controlling contagions that decimated Iran's agricultural sector but, after several years, began manufacturing serum against human infectious diseases, multiplying Iran's domestic microbiological research and vaccine production capabilities.[169] When cholera returned in 1931 and again in 1939, it no longer had the devastating impact of the epidemics of the nineteenth and early twentieth centuries. The outbreaks caused only several hundred fatalities throughout the country and were largely contained in frontier districts by the Iranian government.[170] These outbreaks marked the beginning of the end for Iran's age of great cholera epidemics.

Conclusion

Despite cholera's attenuation, contagious diseases overall did not significantly decline in Iran until the advent of antibiotics and universal vaccination decades later, along with improvements in its water supply and sewage systems. These advances would have been impossible without the growth of Iranian nationalism, which held the central government responsible for the country's susceptibility to cholera and other outbreaks in the years leading up to the Constitutional Revolution. As a result, the nascent National Assembly took a more assertive role to improve public health in Iran. It not only provided the Sanitary Council with the financial and political backing to meet regularly and to carry out interventions against outbreaks but also ensconced public health in the Iranian government's policy-making process. The administration of public health throughout the country was increasingly centralized in Tehran. The Sanitary Council and its patrons in the Iranian government avoided the pitfalls of the Belgian-run Customs Department by making reforms that did not antagonize Russia or Britain. They enhanced medical education and enacted laws to regulate health care in Iran. These measures were critical to the formation and expansion of the vaccination service by the Sanitary Council in the second decade of the twentieth century.

The sanitary momentum created in the first decade after the Constitutional Revolution continued despite the chaos, famines, and epidemics of the First World War. Iranians assumed leadership of the Sanitary Council and expanded the vaccination service. The establishment of the Pasteur Institute of Iran after the war reflected the growing importance of public health in the Iranian state. Most importantly, World War I weakened the imperial powers, which had historically impeded the independent application of public health policy by the government in Tehran. This opened the door for the new post-Qajar regime to consolidate and centralize the administration of public health and break free from imperial sanitary constraints. The new Pahlavi dynasty capitalized on the administrative and intellectual developments in public health established in the postconstitutional period and instituted the framework that brought Iran's age of cholera epidemics to a close.

Epilogue

A pandemic wave of cholera from the east reinvaded Iran in 1965 for the first time in more than two generations.[1] The new outbreak began in Indonesia's Sulawesi (Celebes) Island, where it had incubated for several decades until political unrest and the growing reach of modern transportation helped it spread to neighboring Southeast Asian countries in 1961. Four years later, the pandemic had overrun Bangladesh (Eastern Pakistan), India, Pakistan, and Afghanistan. When news of the contagion's ingress into Iran's Khorasan Province reached Tehran in July of that year, the government immediately airlifted bacteriologists and epidemiologists from other parts of the country to the site of the outbreak in Torbat-i Heydarieh, near the Afghan frontier, to assess the situation. In less than four days, they identified the El-Tor strain of the *Vibrio cholerae* bacterium as the culprit.[2] The El-Tor biotype, named after a quarantine station on the Sinai Peninsula where it had been discovered in 1905, was known for its resilience. It could survive significantly longer in food, water, and the human body than classic cholera strains could. This allowed the microbe to lay dormant in various reservoirs for years until circumstances were favorable for its dissemination. The El-Tor strain also caused higher ratios of mild and asymptomatic infections, despite being indistinguishable from other virulent cholera outbreaks in the majority of cases. These "silent" carriers of the disease often evaded detection and treatment, increasing the microbe's range and dissemination. Even individuals with prominent symptoms had the potential of becoming long-term hosts, with documented cases of people shedding the El-Tor bacterium years after their active illness had resolved.[3]

Iran's ministry of public health conveyed the bacteriological profile of the Iranian outbreak to the World Health Organization (WHO), which immediately sent an expert to help establish a central reference laboratory in Tehran to type and confirm El-Tor cases as they emerged in the country. The Iranian government also convened a "national anti-cholera council," which included representatives from relevant government ministries, benevolent organizations, and the armed forces, to lead a unified response to the emergency.[4] A scientific working group in the ministry of public health ensured that the council's decisions were informed by the latest clinical standards in cholera prevention and treatment. Within a week of the outbreak, Tehran ordered the cessation of all communication with the country's East, mobilized and transported 1,428 medical personnel (Iran's total public health workforce) to the affected areas, and requisitioned hospitals in its eastern provinces to treat cholera cases. Local public health departments closed schools, banned public gatherings, and educated the population on sanitary precautions in cholera stricken districts.[5]

Iranian authorities rapidly established quarantine stations on the country's principal border crossings with Russia, Afghanistan, and Pakistan and detained local and world travelers for three days to ensure they were symptom free. Unlike the anti-cholera restrictions of the past, most of the backlash against the mandated detentions came from Western travelers on the overland hippie trail through Iran and Afghanistan to India. Influenced by the prevailing antiestablishment atmosphere of the 1960s, they typically suspected that the quarantines were politically motivated and lacked any preventive value. This viewpoint likely led to the shooting death of a West German traveler by Iranian gendarmes as he attempted to escape from an observation camp on the Afghan border.[6] The subsequent diplomatic row between Tehran and Bonn did not diminish Iran's rigorous efforts to stop the pandemic. In fact, the economic harm resulting from international travel bans and Russia's injunction on Iranian citrus fruits, chromate ore, and other important export commodities made Tehran especially eager to stop the outbreak as rapidly and decisively as possible.[7]

To keep the epidemic in check, the Iranian government initiated a prophylactic campaign in cholera-free parts of the country and among Shi'ite pilgrims using chloramphenicol, a powerful broad-range antibiotic known for its potentially lethal side effects, including bone marrow suppression and leukemia.[8] The United States Food and Drug Administration had almost banned chloramphenicol in the American market four years earlier, agreeing to its continued distribution only under the strictest labeling and prescription standards.[9] This did not stop Iranian physicians, under instructions from the government, from liberally dispensing the "magic pill," as chloramphenicol was called, both to prevent new infections and to treat active cases.[10] Alarming side effects notwithstanding, this first attempt at mass chemotherapy to halt the spread of a cholera epidemic showed some promising results in a subsequent appraisal of the intervention.[11]

The Iranians bolstered their cholera control strategy by sealing their borders to travelers and goods from India and Pakistan to avoid a reintroduction of the contagion. They also began immunizing the population around affected areas in their eastern provinces to contain the epidemic.[12] The vaccination program initially relied on the country's reserves and emergency contributions from Pakistan, Israel, and the Philippines. But these initial stocks rapidly dwindled as people throughout Iran, including Tehran's three million inhabitants, clamored to be vaccinated against the dreaded disease.[13] Their demand to be vaccinated was a far cry from the Qajar era when Iranians often evaded inoculation efforts and a testament to the population's evolving faith in the power of preventative immunization. Because the world's largest vaccine producers in the West lacked the necessary stockpiles, the government had to mobilize its own biomedical resources to rapidly manufacture the needed cholera sera.

In a matter of weeks, the Iranian Pasteur and Razi institutes successfully produced almost forty-four liters of the El-Tor specific vaccine, enough to immunize the whole country and export a million doses to Syria (fig. E.1).[14] The government began a compulsory national cholera vaccination program, enforced at checkpoints on the country's main roads where members of the newly established Iranian Health Corps, supported by armed military conscripts, screened and obliged anyone without an official proof of immunization to be vaccinated. Influential members of society and powerful political figures were not spared. Multiple injections were drawn from the same large serum supply, typically held in a receptacle that looked like a one pint Mason jar, because the government could not rapidly import or manufacture small, single-dose glass containers. The cloudy appearance of the serum itself often shocked the uninitiated, particularly Westerners, who were habituated to limpid vaccines in disposable glass ampules. The serum often contained visible 'floaters' that were remnants of monkey intestinal tissue used in the vaccine production process, prompting at least one American expatriate to exclaim that "nobody was going to inject crap like that into his children."[15] Its unsightliness aside, the Iranian cholera vaccine turned out to be two times more potent than its American-produced counterpart in a study conducted by the U.S. Public Health Service.[16]

Although the El-Tor pandemic eventually reached Tehran and a few cities in central Iran, the country's overall casualty rates did not surpass several

Fig. E.1. Laboratory technician inoculates eggs in the process of producing rabies vaccine
1967 [Razi Institute]. United States Agency for International Development (USAID),
Historical Archive, Washington, DC. Photo ID: 2117480.

hundred. These deaths occurred mainly in disadvantaged districts where the
severely ill could not access emergency care in a timely manner. The government's interventions largely limited the cholera outbreak to Iran's eastern
half and, by November 1965, extinguished the epidemic altogether.[17] Iran's
northern industrialized neighbor, the Soviet Union, did not fare as well. The
pandemic had entered through its southern frontier with Afghanistan around
the same time as the Iranian outbreak despite the communist state's iron-
clad border controls, draconian quarantines, and sophisticated biomedical
capabilities.[18] Moscow's political and administrative shortfalls, including its
unwillingness to be forthcoming about the epidemic's progress, hampered
prevention and treatment efforts. Cholera, along with plague, suicide, and

homicide did not even appear in the Soviet Union's reported cause-of-death statistics at the high point of the outbreak.[19] The Soviets also abandoned what they perceived as expensive low-efficacy preventive measures such as bacteriological surveillance and vaccination in favor of "low cost" quarantines. Once implemented, the quarantines paradoxically worsened Moscow's finances by restricting the country's commercial activities and failed to halt the contagion, which slowly crawled westward through the Russian heartland in the ensuing five years.[20]

Iran's relative success against the El-Tor pandemic was built on the foundations of modern biomedicine and administration that were established in the Qajar period. The Pasteur and Razi Institutes owed their existence to the emergence of public health as a professional discipline and the triumph of the germ theory of disease in Iran at the turn of the nineteenth century. The country's modern medical workforce and laboratory infrastructure, also established in that period, allowed the Iranian government to rapidly detect and indigenously manufacture an effective vaccine to contain the cholera outbreak. This would not have been possible without Tehran's constructive technical collaboration with the WHO, built on a century of international sanitary engagements that ranged from Iran's participation in the first international sanitary conferences against cholera to its membership in the International Office of Public Health, a forerunner to the WHO. The vaccine's delivery on a national scale relied on an organized network that had its origins in Iran's successful countrywide smallpox immunization efforts in the years leading up to the First World War.

The Iranian Ministry of Public Health, which directed the government's campaign against the pandemic, traced its institutional roots to the Sanitary Council that first convened almost a hundred years earlier to stop the spread of Asiatic cholera.[21] Like the specialized commissions of its predecessor, the ministry's central branch had three offices with distinct mandates including a technical office that managed disease surveillance, urban hygiene, national vaccination programs, and government laboratories (including the Pasteur and Razi Institutes); an administrative office that handled the ministry's supply chain; and an inspection office that oversaw the ministry's public health programs. The provincial branch of the ministry managed Iran's district hospitals and traveling clinics that delivered free health care, medications, and vaccinations to the country's rural population. It also directed the work of the provincial sanitary councils, which monitored and responded to local health matters and led the regional effort against the pandemic.[22] Officials from the provincial sanitary councils quarantined suspected cases of cholera in their homes for at least two days, where they were obliged to receive treatment from visiting teams of doctors and nurses until a laboratory could definitively verify their infection status.[23] The administrative sophistication of the Iranian Ministry of Public Health allowed Tehran to successfully implement a multilayered public health campaign, ranging from large-scale antibiotic prophylaxis and vaccination operations to public education efforts and targeted quarantines within affected areas of the country.

The Iranian government's ability to halt the El-Tor pandemic in its territory ran against the notion that the development of water and sanitation systems was an indispensable requirement for the eradication of cholera. At the time of the outbreak, Iran continued to have major sanitary shortcomings, with a large proportion of its population still lacking access to safe drinking water. Although Tehran received the country's first citywide piped underground water system in 1955, only a portion of the country's urban households had piped water by the time of the outbreak a decade later. The remaining eighty percent of the total population received its drinking water through open ditches and channels that were susceptible to contamination, as they had been during the Qajar era.[24] Rural communities were particularly vulnerable. They relied mainly on surplus irrigation for drinking, bathing, and washing clothes. Others obtained drinking water from uncased, open wells that were usually hand-dug next to the family outhouse.[25] Wastewater treatment and piped sewage disposal were virtually nonexistent. Even in Tehran, sewage typically flowed into hidden reservoirs under homes where it accumulated before diffusing into the surrounding soil.[26] The importance of modernizing Iran's public works had not been lost on the country's leaders after the Qajar era. But the government had put off large-scale developments due to its limited resources and recurrent economic crises in the 1940s and 1950s; even Tehran's modern waterworks was largely developed with American financial assistance.[27]

Iran did not significantly upgrade its sanitary infrastructure until 1966, over a year after the cessation of the El-Tor outbreak, when rising oil revenues allowed the government to increase its investments in the country's public works.[28] This meant that other undertakings, particularly the government's vaccination campaign, played an oversized role in stopping the pandemic. Cholera inoculations had been administered in Iran since the first years of the twentieth century despite questions about their safety and efficacy. Even after receiving international endorsement at the 1926 Sanitary Conference in Paris, the vaccine remained controversial due to the multiple boosters that it required to confer immunity and the high proportion of nonresponders.[29] But the practice never lost its luster in Iran, where the medical elite and policy makers, influenced by the "Pasteurian" approach to public health, considered even vaccines with limited immunogenicity as the preferred method to prevent and control infectious diseases.[30] Iran had developed one of the region's best laboratory operations with the necessary workforce, research, and manufacturing capacity to meet its diagnostic and immunization needs. This expertise allowed Iranian bacteriologists to produce a serum with multiple serotypes specific to the El-Tor antigen that could deliver a broader spectrum of protection against the cholera microbe than its predecessors. The Iranians also made the vaccine easier to administer by reducing the number of boosters necessary to develop and maintain immunity to the disease.[31] These improvements ensured that a sufficiently high proportion of the population was successfully vaccinated against cholera for "herd immunity" to occur, thereby protecting individuals who continued to be susceptible to the disease.

The Iranian government received international accolades from the WHO for its effective interventions against the 1965 El-Tor pandemic.[32] In the ensuing decades, other cholera outbreaks were similarly confronted and extinguished before they could extract a significant toll.[33] One of the worst spates occurred less than six months after the founding of the Islamic Republic in 1979, amid large-scale revolutionary purges in the government. The vacuum of leadership in Iran's municipal and public health administration likely precipitated the outbreak, which infected at least twelve thousand individuals and killed more than a dozen.[34] But even with the postrevolutionary chaos and brain drain, the Iranian government stopped the contagion in a matter of weeks. A century of intellectual secularization in hygiene, biomedicine, and associated public health practices had effectively ended the age of major contagions in Iran; however, the country retained elements of its Qajar era sociocultural vulnerabilities to cholera. Religious barriers to prevention and treatment continued to affect the epidemiology of diseases with more complex social and political determinants, such the growing scourge of addiction to narcotics. Before the 1979 revolution, Iran's drug policy had slowly moved away from a failed strategy of supply reduction and criminalization to an increasingly successful treatment-oriented medical paradigm.[35] But the country's new clerical leaders, who saw addiction not as a medical or public health matter but as a moral shortcoming, almost immediately reversed this scientific approach. They implemented strict zero-tolerance narcotics laws that characterized drug users as social and religious deviants deserving the worse punishment sanctioned in the Quran. Penalties included fining, imprisonment, and lashings; drug dealers and smugglers were often considered to be "at war with God" and executed. These draconian measures against drug users and dealers were matched with similarly aggressive operations to prevent the flow of opiates across the border from Afghanistan.

A decade of punitive policies had little effect on narcotics trafficking in Iran. Around fifty percent of the world's elicit opiate production still passed through its territory, flooding its markets with opium, heroin, and morphine. The Islamic Republic's burden of addicts also increased, forcing the government to send thousands to prisons and camps to detoxify and atone for their sins through forced labor. Ironically, the prisons and camps where addicts were expected to kick their habits became epicenters of drug use, in which people learned how to inject heroin and share primitive infection-prone needles. The rise in malignant drug use brought with it more deaths, more cases of addiction, and, most embarrassingly for Iran's leaders, a full-blown HIV/AIDS epidemic. After years of blaming the West's moral turpitude and decadence for the virus, Iran's leadership had to face an outbreak at home, fueled by its own failed antinarcotic policy. By the late 1990s, in some provinces, double-digit percentages of heroin users were succumbing to the disease.[36]

Iran's growing opiate addiction and accompanying HIV/AIDS epidemic forced the country's religious leadership to return to a "secular" approach to fighting narcotics. Instead of focusing on punishing addicts and trying to stop the drug supply, they decided to try to reduce the harm of narcotics and the demand for them. By 2002, more than fifty percent of the country's drug-control budget was dedicated to preventive public health campaigns, such as advertisement and education. Iran's conservative and previously intransigent leadership opened narcotics outpatient treatment centers and abstinence-based residential centers in Tehran and the provinces. The government began to implicitly support needle-exchange programs, going so far as to encourage the distribution of clean needles in Iran's prison system. Gradually, the road was paved for methadone maintenance treatment centers and clinics that dispensed locally produced opium pills, in a bid to turn injection drug users into medicated patients.[37]

But the damage was done. The remedicalization of opiate abuse did little more than contain the steady growth of addiction in the country. Following the election of the conservative president Mahmoud Ahmadinejad in 2005, the Iranian government shifted back toward a religiously inspired policy of interdiction and criminalization, which worsened the country's drug problem.[38] Iran today has one of the highest addiction rates in the world, with an estimated 2.2 million of its 80 million population hooked on drugs.[39] Female addicts are the fastest-growing segment of this population due to gender-based religious and cultural barriers to care and rehabilitation; women who use drugs are more likely to be seen as immoral wrongdoers than their male counterparts and therefore unworthy of treatment and more deserving of punishment.[40] Iran's seesawing drug policy reflects the continued unresolved tensions between tradition and modernity in the country's health sector. Not only has this cost the country dearly in lives lost, but it has also deprived Iran of much-needed international assistance to stem the westward flow of narcotics from Afghanistan.[41]

Iran's addiction epidemic appears to share underlying features with the outbreaks of Asiatic cholera in the Qajar period. The country's growing trade links with Afghanistan, the world's principal producer of illicit opiates, has increased the deluge of narcotics just as Iran's commercial globalization in the nineteenth century facilitated the ingress of cholera from the disease's endemic repositories in India. Just as imperial antagonisms worsened Iran's vulnerability to cholera, Tehran's decades-long hostility toward the United States, the dominant power in Afghanistan, continues to exacerbate Iran's drug problem by depriving Tehran of critical intelligence on traffickers and opportunities for coordinate cross-border police action.[42] Iran's counternarcotic efforts are also paralyzed by the same type of administrative disorganization, corruption, rigid ideologies, and turf wars in the government and security apparatus that hindered Tehran's efforts against cholera in the Qajar era.[43]

In the same vein, Iran's successful multifaceted struggle against Asiatic cholera probably holds the key to turning the tide against the scourge of addiction. This means that the country's policy makers will not only have to change traditional ideologies on drug treatment and remission, but they will also have to address broader social and political drivers of addiction in Iranian society, including the lack of jobs, entertainment, and social outlets, that have pushed a restless and hopeless Iranian youth to look to drugs for solace and escape from the realities of a restrictive life. Iran began to turn the tide against cholera in tandem with reforms in the country's social and economic policies that began after the Constitutional Revolution and matured under Pahlavi rule. Ensuing changes in the country's institutional, religious, and social attitudes, driven by the Qajar elite, transformed broader societal perspectives on medicine, disease, and public health, which allowed Iranians to intervene effectively to stop the ingress of cholera and other infectious diseases after the First World War. Unless Iran's rulers similarly change their illiberal approach to the current social, economic, and political determinants of addiction in the country, the epidemic of illicit drug use and accompanying HIV/AIDS will likely continue to grow in Iran.

Nasir al-Din Shah's 1879 Decree
on the Hygiene of Tehran

[The] directions of Amin Huzur [Aqa ʿAli Ashtiyani] for the cleaning of the streets of the city of Tehran and its surrounding districts that His Excellency [Mirza Yusuf Ashtiani Mustufi al-Mamalik] has mentioned should be given in decree to Amin Huzur:

Firstly, the city and the city surroundings and the districts inside and outside should be totally cleaned, and it is forbidden by any way or means for garbage or filth to be seen in streets, thoroughfares, or alleys.

Secondly, [regarding] the city and the areas outside the private quarters [andarun] including houses, shops, baths, properties, caravanserais, and open spaces, and roundabout, [if] any of these sites have an owner, each of these owners or proprietors is obliged and compelled to sweep and clean and beautify and get rid of garbage in front of and around their house and property, and if these houses or properties are owned by the Crown, Amin Huzur should, on behalf of the government, get rid of the garbage.

Thirdly, appoint a guard to control the rainwater that flows through streets so as not to soil the streets [and cover it with] algae and to prevent stench/miasma [ufunat] from appearing in the atmosphere; in addition, the storage water cisterns of the city should be changed with new water twice a month.

Fourthly, wherever the ceiling is too short, the walls are damaged, the ceiling is destroyed or broken, the street is in ruins, or the localities [in the city] are in dilapidated condition, [you] must pay attention to demolish the whole thing from top to bottom and make the owner rebuild the whole thing by the most rigorous standards, and if it is Crown lands, the Government should rebuild it.

Fifthly, if it is needed to repair a wall somewhere, designate an attendant to prevent such scenes to be viewed and to tell the chief architect [miʿmar bashi] to build as Amin Huzur orders and not to put it off.

Two-page edict with Nasir al-Din Shah's seal issued in 1879 [1296 AH] located in the Archives of the Institute for Iranian Contemporary Studies. See Jalal Farahmand, "*Tanzif-i shahr-i Tehran* [Hygiene in the city of Tehran]," *Mahnamih-i iliktruniki-yi baharistan* [Baharistan online monthly], 1, no. 5, accessed March 10, 2017, http://www.iichs.org/magazine/sub.asp?id=85&theme=Orange&magNumber=1&magID=6&magIMG=coverpage1.jpg.

Sixthly, camels and donkeys and other load-carrying and non-load-carrying animals that pass through streets and roundabouts and bring and take loads should not be allowed to be idle in the streets without aim or purpose or to sleep in the middle of streets or thoroughfares; to wit, the sooner the animals leave the streets the better.

Seventhly, wherever there is a lot or a mound of rubbish and waste in the private sanctuaries [*harim*], in the house, in the store, or properties of whomever that may be, [you] should command them to take [the rubbish] and make it disappear from sight, and if this [garbage] is in the Crown's private sanctuary, then draft animals of the government [*bark-ishha-yi divan*] must remove them.

Eighthly, based on the information [supplied] by Chief Physician Tholozan, it has been decided that laundry houses in the city and surroundings, outhouses (septic pits), and graveyards will be built outside the city surrounded by walls from now on [and] the cost and plans and locations of these things must be determined [by and] with the knowledge of the chief architect by the spring; Amin Huzur himself must be mindful of it being built and brought to completion.

Ninthly, the totality of the government's sheep sellers, draft mules, and laborers should be under the responsibility of Amin Huzur and are obligated to work, and if draft mules or laborers are wanting, the government will supply however many are needed.

Muzaffar al-Din Shah's 1897 Decree
Regulating Sanitary Stations and Quarantines
on Iran's Eastern Frontiers

His Majesty Muzaffar al-Din Shah king of Persia, in his foresight and magnanimity for the Persian people, decided on the 12th day of the month of Ramezan of the year 1314 Hijra (14th February 1897) that amongst other sanitary measures already ordered in the Persian Gulf against arrivals from India who could carry plague to Persia, there is need to establish Sanitary Posts along the frontier of Afghanistan and Baluchistan.

I. For the Afghan frontier, His Majesty the Shah has decreed that all the passes and all the roads shall be closed by troops with the exception of the grand route from Herat to Mashhad to the north and the large highway from Kandahar to Birjand and from there to Kerman, Yazd, and the South.

II. In as what concerns the Baluchistan frontier, the surveillance will be very difficult; but, on the other hand, the passages and traffic are much less.

A sanitary fort/station shall be established on the road that comes from Baluchistan to Bampour and Bam, and frequent and numerous patrols will guard that no contraband crosses the frontier beyond this pass.

At the entrance of these three routes on Persian territory will be installed sanitary stations, supplied with troops, physicians, medicines, and necessary materials to perform the requisite inspection, observation, and, if need be, quarantine of materials and peoples passing through the stations.

Sanitary Posts

A. Installations: The sanitary station shall be installed in the proximity of potable water, either in gardens, or under tents, on each side of the road, at least one thousand meters from inhabited areas.

Each sanitary station will include as personnel: 1. a chief with numerous officers and soldiers necessary to enforce the respect of laws and compliance with regulations, 2.

Translated from original French *farman* that was printed in Tehran by Imp. A. Beknazar in FO 60/586 Commercial Dispatch (n. 4) from Charles Hardinge to Marquis of Salisbury. Tehran, March 3, 1897. [Enclosure n. 1] Règlement des Postes Sanitaires des Frontières.

two physicians: one of whom will be charged with inspection and the other with the treatment of plague victims, if there are any, and 3. nurses to help the physicians carry out their duty.

The chief of the sanitary station shall in addition to the physicians be personally responsible for the strict application of the regulation. He should ensure that the physicians' decisions are applied in a rigorous way as it concerns the observation or quarantine imposed on merchandise as well as the isolation of the different category of travelers, based on their date of arrival and their suspected or contested contamination.

The chief of the station and the chief physician shall each week individually produce a detailed report as to the operations of the post as well as the number and status of the sick who are under observation or in quarantine. They will address the report to the governor, who will then forward it to the concerned minister.

The materials at the sanitary station will include:

1. A special tent for each one of the chiefs as well as other tents that will be necessary for the troops and subaltern personnel.
2. A tent for the medical and disinfectant supplies.
3. A certain number of tents divided in three principal groups (personnel, observation, isolation). This number shall be different depending on the roads and their traffic.
4. One of those tents to be used as a disinfection tent.
5. A number of objects needed for alimentation and hospital functioning.
6. A sufficient quantity of medication and disinfectants, mentioned below.
7. A number of utensils, such as buckets, shovels, and earthen pots, that are indispensable for disinfection.

B. Functioning: The most senior physician at the station will be charged with inspecting patients and goods.

The second physician will treat the patients who are recognized to have the plague and will direct the disinfecting operation. This physician and his aides will live completely apart, that it to say in the lazaretto proper, that is to say in the tents that are reserved for plague victims and are completely isolated from the rest of the group at a distance of at least three hundred meters. The personnel of the lazaretto should in no way communicate with the sanitary station. The materials of the lazaretto will be completely distinct from the rest of the station.

No frontier crossings will be allowed at night. The station chief will place sentries and the necessary cordon to stop travelers until daytime, and by consequence the inspection will be conducted on the same day.

Upon their arrival, the travelers, after an evaluation of their destination, will report to the chief physician, who will inspect one by one each traveler and will immediately separate the sick who will be sent, with their baggage, to the Lazarettos.

The other travelers, found to be well, will then be retained for fifteen days for observation. They will be divided into five groups, separated as much as possible and containing one hundred persons so that the disinfections can be done based on the time of their arrival. If during these fifteen days one of the isolated travelers comes down with the plague, he will be immediately sent to the lazaretto with his baggage. His tent will be evacuated, disinfected, and transported to another location; his companions must restart a period of fifteen days observation from this date. The location of the infected tent will be rigorously disinfected. In the lazaretto the second physician and his aides will give the necessary treatments to the sick. In case of deaths, the cadaver will be buried at a depth of one meter and surrounded by lime if possible.

The effects of the deceased, cloths, bedding, etc. shall be immediately incinerated.

In case of cure, the convalescing will be placed in quarantine for a month following his cure, and his effects, clothes, bedding will be incinerated. Following his exit from the lazaretto, the recovered patient will be given new clothing from the sanitary post. His quarantine will take place apart from the rest of the sick.

The location of the plague tents will be frequently changed and disinfected.

Latrines will be installed at the proximity of the tents but as much as possible in a direction opposite to the reigning winds. The latrines will be made up of narrow trenches, sufficiently deep, placed far from potable water and; they will be filled with earth everyday after being watered with a solution of lime We will also pour in the trenches the vomitus of the sick, the waters used for washings, and the detritus that cannot be destroyed with fire.

Alimentation

The governors of the provinces will give the necessary orders so that bazaars will be installed near each sanitary station. The governors will also ensure a sufficient supply of wheat, rice vegetables, and in a word all the necessities of alimentation for the personnel of the post and the travelers undergoing observation or quarantine.

We should also look to the feeding of the animals.

A special bazaar will be dedicated to the lazaretto, and like the lazaretto, it will be excluded from all communication with the outside.

A special reduced tariff of prices shall be posted at the bazaar by the station chief so that there will be no abuse by the vendors.

Alimentation shall be supplied free of charge to the indigent patient.

The source of drinking water shall be the object of particular scrutiny. It will be protected against all pollution, and notably it will be forbidden to wash cadavers, laundry, or clothes, or to bathe and water animals, to allow excrement to contaminate it, or to throw garbage or butchers' leftovers into the water sources, streams, or basins of drinking water.

Very severe penalties should be decreed against all delinquents.

The best approach would be to create designated washing places and watering holes for the animals and to watch them and to frequently clean and change the water in them.

Disinfection

1. The effects, clothes, and bedding of plague victims will be immediately burned.

2. The clothes and baggage of healthy travelers should be washed in disinfectant prior to the completion of the quarantine imposed on them.

A. Modes of disinfection:
   1. Incineration.
   2. Soaking in water for one hour.
   3. The aqueous solution of carbolic acid a 5 parts per 100 (or 1 misqal of carbolic acid for 20 misqal of water) [1 misqal was approximately 4–6 ounces or 4.6 grams].
   4. The aqueous solution of mercuric chloride (Hgcl) at 1 part per 1,000 (or 1 misqal of mercuric chloride per 1,000 misqal of water).
   5. The aqueous solution of cupric sulfate at 2 parts per 100 (or 1 misqal of cupric sulfate per 50 misqal of water).
   6. The "milk" or solution of lime at 20 parts per 100 (or 2 parts of water and 1 of lime).
   7. Sulfur (or 1 batman for an ordinary chamber) [a batman was equivalent to 300 Tabrizi misqal].

B. Preparation of disinfectants:

Incineration can be done directly in a fireplace in the open air.

The ablution and washings will require large boilers.

The chemical solutions can be prepared in any kind of container, with the exception of the mercuric chloride, which is corrosive to metals and should be prepared in an earthen glazed vessel, taking care of adding a misqal of ordinary salt per batman of water.

To prepare the solution of lime, we dilute the lime in a pot with double its volume of water.

To disinfect with the use of sulfur, we spread the belongings of travelers on cords in the room designated for disinfection, we pour the powdered sulfur on several well-lit braziers [*manqal*]. We shut all openings from the room to the outside and seal all locations with humid earth. We do not withdraw the fumigated belongings until twenty-four hours have passed.

C. Mode of application to various objects:

We have to employ:

For belongings made of canvass or cotton (shirts, underwear, sox, handkerchiefs, belts, towels, etc.) mercuric chloride, carbolic acid, or cupric sulfate or soaking the objects in water for one hour.

For belongings made of wool (tunics, overcoats, trousers, sox, vests, beddings, covers, etc.) the same solutions as above or fumigation with sulfur.

For objects made of oil-cloth, skin, wood, or pasted, same solutions or sulfur fumigation but no soaking in water.

Fur clothing and those made of animal skins will be refused entry at the frontier (in case of absolute necessity, fur clothing will be allowed only after having spent forty-eight hours in a solution of mercuric chloride).

The belongings, clothing, bedding, and any object without value (straw, hay, chiffon, old papers, bandages, and animal or vegetable garbage) of those stricken with the plague should be burned on a hearth. To help the fire burn the objects they can be covered in petrol.

The post, the letters, and important documents will be disinfected through fumigation with sulfur.

The tents reserved for those stricken with plague shall be burned after the quarantine has been removed.

The other tents will be washed with a solution of mercuric chloride.

The emplacement of the tents shall be disinfected by covering the locations with dry wood and petrol and lighting it on fire.

The excrements of the sick together with the water used for washings shall be disposed of immediately in the latrines.

The latrines will be disinfected on a daily basis with a solution of iron sulfate, mercury, or lime before being covered with earth.

The receptacles for excrements and the urinals will be made of metal and will be washed every time they are used with a solution of carbolic acid.

The physicians, their aides, and the companions of those struck by the plague as well as those convalescing should before anything wash their whole body several times with warm water and soap, followed by rinsing with a solution of mercuric chloride.

As far as beards and hair are concerned, they should shave it as much as possible or at least washed energetically with a solution of mercuric chloride.

Note 1a: All travelers who refuse to submit to these methods of prophylaxis and disinfection shall be pushed back across the frontier without mercy.

Note 2a: The physicians and their aides should take all necessary precautions to disinfect themselves carefully, but they should in no way fear contagion from which they will surely be safeguarded, if they are clean, sober, and do not engage in excesses. At Hong Kong and in India, the plague did not strike the medical personnel, with only one exception.

They should see to it that poisoning does not occur with the disinfectants being used (mercuric chloride, carbolic acid, etc.).

Medical Practice Act of 1911

Art. 1. The practice of medicine or its different branches in any part of Persia is forbidden unless a permit be obtained from the Ministry of Education and registered at the Home Office.

Art. 2. In future the documents which are recognized, or to whose possessors a permit will be given, are as follows:

(a) Diplomas which are given in Persia by the Medical Schools of the Government;

(b) Diplomas of the Government schools of foreign countries.

Art. 3. Persons who do not possess any of the certificates mentioned in Art. 2 of this law, and who, at the time of the promulgation of this law, have practiced medicine less than five years in Tehran, will not be given a permit, and they will be prevented from continuing to practice.

Art. 4. Persons who do not possess the certificates mentioned in Art. 2 of this law, and who, at the time of the promulgation of this law, have practiced medicine less than five years in Tehran, will not be given a permit, and they will be prevented from continuing to practice.

Art. 5. Persons who do not possess any of the diplomas mentioned in Art. 2 of this law, and who have not practiced medicine for more than five years and less than ten years, may obtain a permit, on the condition that, from the time of the promulgation of this law up to three years, they appear before a special committee organized according to Art. 8 of this law, and pass an examination; or on the condition that they present to the Ministry of Education, within four months, one of the following certificates:

(a) A certificate of study in the clinic of one of the well-known physicians, which certificate must certify that the bearer has been continually for five years one of the physician's assistants or pupils;

John Gilmour, *Report on an Investigation into the Sanitary Conditions in Persia: Undertaken on behalf of the Health Committee of the League of Nations at the Request of the Persian Government* (Geneva: League of Nations, 1925), 32–34.

(b) A certificate from the physician of the Government Hospital or American Hospitals. (The last-named certificates are valid up to four months after the promulgation of the law; after this date the documents mentioned in Art. 2 are required.)

Art. 6. Persons who have continually practiced medicine for more than ten years shall have the right to be given a permit.

Art. 7. For persons mentioned in Art. 5 and Art. 6 of this law who can obtain permits, from the promulgation of this law up to four months, there shall be a committee in the Ministry of Education consisting of the following:

One of the professors of the School of Medicine.
One of the doctors of the Government Hospital.
One of the well-known Persian physicians.
One of the well-known doctors of the American Hospital.
Two representatives of the Ministry of Education, one of whom shall be President of the Committee.

Art. 8. From the date of the promulgation of this law up to three years, there shall be twice a year, in spring and autumn, a committee in the Ministry of Education to examine those who, according to Art. 5, have practiced medicine for more than five and less than ten years, have no certificates, and have appeared before the committee mentioned in the preceding article. This committee will consist of:

Professors of the School of Medicine.
Four of the well-known physicians.
A representative of the Ministry of Education, who shall be President.

Art. 9. Persons who have been practicing medicine in the provinces, or dentistry in Tehran or the provinces, at the time of the promulgation of this law, shall appear before the Ministry of Education, from the promulgation of this law up to one year, and shall obtain a permit. After one year, in no part of Persia shall any permit be given, except in accordance with Art. 2.

Art. 10. Whoever practices medicine contrary to this law shall first be warned, the second time he shall be imprisoned for four months, the third time he shall be imprisoned for one year.

Art. 11. After the promulgation of this law no physician shall be taken into the Government service unless he is in possession of a permit delivered according to Art. 2.

Art. 12. Cipher prescriptions [prescription where the ingredient is not mentioned] are forbidden, and the doctor who gives one shall be punished by imprisonment for a four-month period.

1914 Sanitation Ordinances for Tehran

I
Decree on Removal of Sweepings and Domestic Garbage

Art. 1- The deposition on public throughways of sweepings and domestic garbage is rigorously forbidden at all hours of the day and night.

Art. 2- The removal of the domestic garbage will occur in the morning, by way of a cart of the type adopted by the municipality; and in noncarriagable roads, by hand driven carts.

Art. 3- Proprietors of apartments must provide a communal receptacle where residents will deposit their domestic garbage from ten o'clock at night.

Art. 4- The receptacles must be of uniform model for all apartments and should be supplied with a cover conforming to the model adopted. Gasoline barrels, measuring 10, 20, or 30 batman, in relation to the importance of apartments, supplied with a cover constitute the type that has been adopted.

Art. 5- In the morning, beginning at six o'clock, the receptacles should be placed in proximity of the agents of the roads.

Art. 6- Materials coming from the sweepage of merchandise, paper, straw, debris of wood, etc. will be swept and immediately transported to the interior of the home, to be placed in the garbage cart at the time of its passage.

Art. 7- The removal of the domestic garbage should begin at sunrise (six o'clock in the morning) and should finish at eight o'clock in the summer and nine o'clock in the winter.

Art. 8- It is forbidden to throw refuse or trash onto the streets.

A violation of any of the articles in this decree will result in a fine of . . .

NLM, Procés-Verbal 129eme Séance ordinaire du Conseil Sanitaire de l'Empire de Perse, 6 Janvier 1914, 209–212.

II
Decree on the Hygiene of Bazaars and Markets

Art. 1- The merchants should keep their stands and surroundings of their display at a constant state of cleanliness. It is forbidden to allow residue of any kind to linger on the floor of their stand; peelings, debris of meat, meat-stuff, and fowl-stuff, animals, paper, straw, etc.

Art. 2- It is forbidden to throw in the passages, reserved for circulation, papers or debris of any kind, or to block the passages with displays, or objects of any sorts.

Art. 3- The merchants should procure themselves a garbage bin, supplied with a cover, of the same type that is used in cities, and in which they will be held responsible to place any residues originating from their displays. These garbage bins will be transported to designated points in the bazaars, wherein the mornings carts shall come and collect them [the garbage].

Art. 4- It is forbidden to spit on the floor in the interior of markets or public edifices.

Art. 5- In the stands of butchers, pork butchers, merchants of salted produce, grease, and butter, all parts of the material finding itself in contact with the merchandise, or serving as their cutlery, and in their preparation shall be scrubbed and washed at night, after the closure of the markets, and more frequently if necessary.

Art. 6- Merchandise that is destined to be consumed raw, such as butter, cheese, ham, etc., as well as pastries, shall be protected from dust and flies and, as a result, will be placed under glass coverings or displays, or covered with nets, in a perfect state of cleanliness.

Art. 7- The bakers established in the markets or on the streets will be required to place their bread in (glass) displays or [in some other way] protected from dust.

Art. 8- All alimentary products sold in the markets shall be submitted to the surveillance of the police, who are permitted to require the destruction or forbid the sale of a rotten product.

Art. 9- The veterinary inspectors should exercise a surveillance of the fresh meat or preserves sold in the bazaar, or in the city. They must inspect the carcass, fish, etc., and they have the right to confiscate if [the produce] is of bad quality.

Art. 10- The inspector-veterinarians will visit daily all animals sacrificed at the abattoirs, as well as the merchandise meat arriving into the city. They will be vigilant to everything that relates to the salubriousness of the abattoirs.

Art. 11- In the bazaars, all proprietors, or owners of boutiques are held to assure the destruction of rats, mice, insects, etc.

A violation of any of the articles in this decree will result in a fine of . . .

III
Decree on the General Measures of Salubriousness

Art. 1- The public thoroughfares should be kept in a constant state of cleanliness by the [municipal] Sweeping Service.

Art. 2- It is forbidden to deposit any garbage in the public thoroughfares, to push garbage, or products of sweeping of houses onto them, or to throw paper, garbage, or any kind of detritus there.

The domestic waters, the salted waters, or used waters cannot be thrown into the streets or gutters.

Art. 3- The ground-level proprietors and the proprietors on the banks of passages, roads, embankments, or other private passages open to the public should keep them in good upkeep and propriety.

Art. 4- Rugs or any objects cannot be dusted from windows after eight o'clock in the morning. Objects from a sick-room should never be dusted from windows.

Art. 5- Dusting of rugs cannot be performed in the city at less than fifty meters of all habitation and public thoroughfares.

Art. 6- It is forbidden to spit on the floors of public buildings, schools, ministries, mosques, etc.

Art. 7- The public-service cars should be carefully cleaned every day. The hygiene service could require the disinfection or cleaning of these cars (e.g., times of epidemics, after the transport of sick or cadavers).

Art. 8- It is forbidden to wash in the public washhouses, or to give to washers rags and clothes that have served a contagiously sick person, until it has been disinfected beforehand. It is forbidden to sell to the public the clothes, covers, or clothing that have served a contagiously ill person without having been disinfected beforehand. The disinfection should be done in the domicile of the sick person by immersion lasting six hours in an aqueous solution of mercuric chloride [corrosive antiseptic] at one part per hundred.

Art. 9- The transport of butchered meat can happen only in closed cars or by not revealing any part of their load; the transport can be done on the back of animals on the condition that the meats are wrapped hermetically with cloth in a perfect state of cleanliness.

Art. 10- It is forbidden to throw refuse or fecal matter in ditches or streams. It is also forbidden to wash clothing, animals, or any kind of utensils in running water that serve to supply basins used to store drinking water.

Art. 11- Basins that are not covered should contain fish, perchs, carps and cyprinids that destroy the eggs of insects.

Art. 12- If the basins and reservoirs do not contain fish, one should pour 3 meskals (15 cc) of petroleum by meter square on the surface to destroy the larvae of mosquitoes.

Art. 13- The deposits of dung, refuse, or other dirtied materials in the proximity of qanats or flowing drinking water is formally forbidden.

Introduction

1. Maryam Sinaiee, "Iran Battles to Contain Cholera Outbreak," *National*, August 27, 2008, http://www.thenational.ae/news/world/middle-east/iran-battles-to-contain-cholera-outbreak.

2. Sinaiee, "Iran Battles to Contain Cholera Outbreak."

3. Kamran B. Lankarani and Seyed Moayed Alavian, "Lessons Learned from Past Cholera Epidemics, Interventions Which Are Needed Today," *Journal of Research in Medical Sciences: The Official Journal of Isfahan University of Medical Sciences* 18 (2013): 630–631.

4. Taj al-Saltana, *Crowning Anguish: Memoirs of a Persian Princess from the Harem to Modernity, 1884–1914*, ed. Abbas Amanat and trans. Anna Vanzan and Amin Neshati (Washington, DC: Mage, 1993), 282.

5. Abbas Amanat characterizes modern Iranian historiography as failing to "provide any meaningful interpretation of the past" due to the moral and ideological distortions of the historical realities. See Abbas Amanat, "The Study of History in Post-Revolutionary Iran: Nostalgia, or Historical Awareness?," *Iranian Studies* 22, no. 4 (1989): 5.

6. William Floor, *Public Health in Qajar Iran* (Washington, DC: Mage, 2004); Hormoz Ebrahimnejad, *Medicine, Public Health, and the Qajar State: Patterns of Medical Modernization in Nineteenth-Century Iran* (Leiden: E. J. Brill, 2004); Cyrus Schayegh, *Who Is Knowledgeable Is Strong: Science, Class, and the Formation of Modern Iranian Society, 1900–1950* (Berkeley: University of California Press, 2009); Firoozeh Kashani-Sabet, *Conceiving Citizens: Women and the Politics of Motherhood in Iran* (New York: Oxford University Press, 2011); Hormoz Ebrahimnejad, *Medicine in Iran: Profession, Practice, and Politics, 1800–1925* (New York: Palgrave Macmillan, 2014).

7. Early on, the cholera bacterium likely thrived in the brackish, alkaline, and organically rich waters of the estuaries without human or animal hosts and only later adapted to the growing human presence in its ecosystem by acquiring mutations that allowed it to traverse the digestive tract, multiply in the small intestine, and more easily disseminate among people. The long-standing presence of cholera in India is bolstered by millennia-old Sanskrit texts and sixteenth- and seventeenth-century European accounts that describe lethal disease outbreaks with cholera-like features. See Rita R. Colwell, "Infectious Disease and Environment: Cholera as a Paradigm for Waterborne Disease," *International Microbiology* 7 (2004): 285–289; Andrew David Cliff, Matthew Smallman-Raynor, Peter Haggett, Donna F. Stroup, and Stephen B. Thacker, *Infectious Diseases: Emergence and Re-Emergence; A Geographical Analysis* (Oxford: Oxford University Press, 2009), 81–82; Rita R. Colwell, "Global Climate and Infectious Disease: The Cholera Paradigm," *Science* 274, no. 5295 (1996): 2025–2031; Myron Echenberg, *Africa in the Time of Cholera: A History of Pandemics from 1817 to the Present* (Cambridge: Cambridge University Press, 2011), 17.

8. I often challenged Serge Lang, a mathematical theorist and human immunodeficiency virus (HIV) skeptic, on his stance against the virus's causative role in acquired immunodeficiency syndrome (AIDS). Not unlike the postcolonial critics of today, he justified his skepticisms by arguing that "the power structure does what they want, when they want; then they try to find reasons to justify it." See Kenneth Chang and Warren Leary, "Serge Lang, 78, a Gadfly and Mathematical Theorist, Dies," *New York Times*, September 25, 2005; "Politicians, Muslim Scholars Join Vaccination Effort as Violence Hinders Pakistan Polio Drive," *Washington Post*, January 11, 2014.

9. Charles E. Rosenberg, *The Cholera Years: The United States in 1832, 1849, and 1866* (Chicago: University of Chicago Press, 1962), 2.

10. Dilip Mahalanabis, A. M. Molla, and David Sack, "Clinical Management of Cholera," in *Cholera*, ed. Dhiman Barua and William B. Greenough III (New York: Plenum, 1992), 270.

11. James E. Baker (M.R.C.S., English Medical Superintendent, Her Britannic Majesty's Telegraph Staff in Persia), *A few Remarks on the most prevalent Diseases and the Climate of the North of Persia, appendix to Herbert, Report on the present State of Persia and her Mineral Resources*, House of Commons, Parliamentary Papers, Accounts and Papers 67 (1886), 325.

12. Gertrude Bell, *Persian Pictures* (1894; New York: Boni and Liveright, 1926), 67–68.

13. Mary Leonora Woulfe Sheil, *Glimpses of life and manners in Persia: With Notes on Russia, Koords, Toorkomans, Nestorians, Khiva, and Persia* (London: John Murray, 1856), 197–198.

14. Najaf, the burial place of Imam 'Ali ibn Abi Talib; Karbala, the burial place of Imam Husayn ibn 'Ali; Kadhimiya, the burial place of the fifth and ninth Imams; Mashhad, the burial place of Imam 'Ali ibn Musa al-Ridha; and Samarra, the burial place of the tenth and eleventh Imams and where the twelfth imam went into occultation.

15. G. Bell, *Persian Pictures*, 68.

16. Arthur S. Hardy, "Persia. Sanitary Report from Tehran," *Public Health Reports* 14, no. 10 (March 10, 1899): 331.

17. John Gilmour, *Report on an Investigation into the Sanitary Conditions in Persia* (Geneva: League of Nations, 1925), 50–57; Baker, *A few Remarks on the most prevalent Diseases and the Climate of the North of Persia*," 324–325.

18. Iran lacks an accurate population census from the nineteenth century, but an estimate puts the total population in 1867 at 4 million, of which only 850,000 inhabited the country's principal cities. See Charles Issawi, ed., *The Economic History of Iran, 1800–1940* (Chicago: University of Chicago Press, 1971), 28–29.

19. Frank G. Clemow, "A Contribution to the Epidemiology of Cholera in Russia," *Transactions of the Epidemiological Society of London* 13 (1893–1894): 60–82; Amir A. Afkhami, "Disease and Water Supply: The Case of Cholera in Nineteenth-Century Iran," in *Transformations of Middle Eastern Natural Environments*, ed. Jeff Albert, Magnus Bernhardsson, and Roger Kenna (New Haven: Yale University Press, 1998): 212.

Chapter 1 · *Cholera and the Globalization of Health in Iran, 1821–1889*

1. Michael Axworthy, *History of Iran: Empire of the Mind* (New York: Basic Books, 2008), 167.

2. Scott C. Levi, "India," *Encyclopedia Iranica* 8, fasc. 1 (2004): 44–47, updated online version accessed January 6, 2016, http://www.iranicaonline.org/articles/india-xiii-indo-iranian -commercial-relations.

3. Claude Mathieu de Gardane, *Mission du général Gardane en Perse sous le premier empire*. Documents historiques publiés par son fils, le comte Alfred de Gardane (Paris: Librairie de AD. Lainé, 1865), 102 (published documents by General Gardane's son).

4. Andrew Jukes, a surgeon with the East India Company, arrived in Iran in 1808 as a member of Sir Harford Jones Brydges's diplomatic mission. He lived in Iran until his death,

probably due to Asiatic cholera in 1821, and remains buried in the Vank Cathedral's cemetery in Isfahan. For a more detailed account of Anglo-Persian diplomacy and the medical developments in this period, see Denis Wright, *The English amongst the Persians* (London: Heinemann, 1977), 122–123; Cyril Elgood, *A Medical History of Persia and the Eastern Caliphate from the Earliest Times until the Year A.D. 1932* (Cambridge: Cambridge University Press, 1951), 443–448; John Malcolm, *The History of Persia from the Most Early Period to the Present Times* (London: John Murray, 1815), 2:383; Mahmud Nadjmabadi, "Les relation médicales entre la Grande Bretagne et l'Iran et les médecins anglais serviteurs de la médecine contemporaine de l'Iran," in *Proceedings of the XXIII International Congress of the History of Medicine* (London: Wellcome Institute of the History of Medicine, 1974), 704; James Morier, *A Second Journey Through Persia, Armenia, and Asia Minor, to Constantinople, Between the Years 1810 and 1816* (London: Longman, 1818), 191; Abbas Iqbal, "*Abila kubi*" [Smallpox inoculation], *Yadigar* 4, no. 3 (1326 AS [1947]): 69.

5. John Cormick came to Iran as an assistant surgeon with John Malcolm's third mission to Fath 'Ali Shah in 1810 and became 'Abbas Mirza's personal physician. During Iran's wars with Russia, he was attached to 'Abbas Mirza's army as a surgeon and died of typhus during one of 'Abbas Mirza's last campaigns in 1833. See Wright, *The English*, 122–123; Elgood, *A Medical History*, 445–446, 465–466; Kamran Ekbal and Lutz Richter-Bernburg, "Cormick, John," *Encyclopedia Iranica* 6, fasc. 3 (2011): 274–275, updated online version accessed January 5, 2016, http://www.iranicaonline.org/articles/cormick-john.

6. William Floor, "Cap," *Encyclopedia Iranica* 4, fasc. 7 (1990): 760–764, updated online version accessed January 5, 2016, http://www.iranicaonline.org/articles/cap-print-printing-a -persian-word-probably-derived-from-hindi-chapna-to-print-see-turner-no.

7. Cyril Elgood, *Safavid Medical Practice; or, The Practice of Medicine, Surgery and Gynaecology in Persia between 1500 A.D. and 1750 A.D.* (London: Lucaz, 1970); Cyril Elgood, *Safavid Surgery* (Oxford: Pergamon Press, 1966).

8. Abu 'Ali al-Husayn Ibn 'Abdullah Ibn Sina, *The Canon of Medicine of Avicenna [al-Qanun fi al-tibb]*, ed. and trans. Oskar Cameron Gruner (New York: AMS Press, 1973).

9. Cupping, leeching, and bloodletting were staples of heroic interventions in humoral Iranian medicine and commonly practiced by barbers in Iranian bathhouses well into the second half of the twentieth century. See Amir Arsalan Afkhami, "Humoralism," *Encyclopedia Iranica* 12, fasc. 6 (2004): 566–570, updated online version accessed January 5, 2016, http://www .iranicaonline.org/articles/humoralism-1; Seyyed Hossein Nasr, *Science and Civilization in Islam*, 2nd ed. (Cambridge: Islamic Texts Society, 1987), 219–220.

10. Henri Massé, *Croyances et coutumes persanes: Suivi de contes et chansons populaires* [Persian beliefs and customs] (Paris: Librairie Orientale et Américaine, 1938), 1:331.

11. After beginning his studies in 1811, Mirza Hajji Baba Afshar finished his education at Oxford in 1234 AH [1818–1819] and entered 'Abbas Mirza's service under the supervision of John Cormick. After the crown prince's death, he became Muhammad Shah's chief physician, a post he occupied until 1254 AH [1838–1839]. See Elgood, *A Medical History of Persia*, 465, 474, 482; Mahmud Nadjmabadi, "Tarikh-i tibb va danishkada-yi pizishki-yi Iran" [History of medicine and the Medical College of Iran], *Jahan-i pizishki* 1, no. 3 (1947): 43–46; Husayn Mahboubi Ardakani, "Afsar, Hajji Baba," *Encyclopedia Iranica* 1, fasc. 6 (1984): 586, updated online version accessed January 10, 2016, http://www.iranicaonline.org/articles/afsar-hajji-baba-court-physician -under-mohammad-shah-qaar.

12. Alexandre Moreau de Jonnès, *Rapport au Conseil supérieur de santé sur le choléra-morbus pestilentiel: les caractères pathologiques de cette maladie, les moyens curatifs et hygiéniques qu'on lui oppose, sa mortalité, son mode de propagation et ses irruptions dans l'Indoustan, l'Asie orientale, l'archipel indien, l'Arabie, la Syrie, la Perse, l'empire russe et la Pologne* (Paris: Imprimerie de Cosson, 1831), 258–265; Graham Williamson, "The Turko-Persian War of 1821–1823:

Winning the War but Losing the Peace," in *War and Peace in Qajar Persia: Implications Past and Present*, ed. Roxane Farmanfarmaian (London: Routledge, 2008), 95–101; George C. Kohn ed., *Encyclopedia of Plague and Pestilence* (New York: Facts on File, 1995), 237; Great Britain, Public Record Office, Foreign Office (FO) 60/20 [pp. 108–109] Dispatch (n. 22) from Henry Willock to the Marquess of Londonderry. Tehran, October 8, 1821; FO 60/20 [pp. 113–115] Dispatch (n. 21) from Henry Willock to the Marquess of Londonderry. Tehran, October 15, 1821; FO 60/20 [pp. 117–118] Dispatch (n. 22) from Henry Willock to the Marquess of Londonderry. Tehran, December 16, 1821. Transcripts of unpublished Crown Copyright material in the Public Record Office appear by permission of Her Majesty's Stationary Office.

13. James Baillie Fraser, *Narrative of a Journey into Khorasan, in the Years 1821 and 1822: Including Some Account of the Countries to the North-East of Persia; with remarks upon the national character, government, and resources of that kingdom* (London: Longman, Hurst, Rees, Orme, Brown, and Green, 1825), 57–85; FO 60/20 [pp. 169–173] Dispatch (n. 3) from George Willock to Joseph Planta Esq. (His Maj's Under Secretary of State). Tabriz, August 6, 1822; FO 60/20 [pp. 214–216] Dispatch (n. 3) from George Willock to Lord Viscount Strangford. Tabriz, October 19, 1822.

14. The English medical director of the Indo-European Telegraph Department experienced this firsthand during an outbreak in Shiraz almost half a century later, when a high ranking Shi'ite cleric advised him to swallow a strip of paper containing written "charms against the cholera" at first sign of illness. See Charles James Wills, *In the Land of the Lion and Sun, Or Modern Persia: Being Experiences of Life in Persia from 1866 to 1881* (London: Macmillan, 1883), 290–291.

15. The outbreak killed Fath 'Ali Shah's eldest son and one of his ablest field commanders, Muhammad 'Ali Mirza, and the crown prince's most capable administrator, Mirza Isa Buzurg Qa'im Maqam Farahani.

16. Elgood, *A Medical History of Persia*, 462; Ahmad Seyf, "Iran and the Great Plague," *Studia Islamica* 69 (1989): 155.

17. Christoph Werner, *An Iranian Town in Transition: A Social and Economic History of the Elites of Tabriz, 1747–1848* (Wiesbaden: Harrassowitz Verlag, 2000), 91.

18. John Maynard Woodworth, *The Cholera Epidemic of 1873 in the United States* (Washington, DC: Government Printing Office, 1875), 595.

19. The French government sent Louis-André-Ernest Cloquet (1787–1855), known for his work on the anatomy of the nervous system, in 1846. He treated the shah's family during the cholera epidemic and cared for the monarch on his deathbed. He maintained his standing, following Muhammad Shah's death, as Nasir al-Din Shah's physician and successfully extracted a bullet from the young shah after an assassination attempt in 1852. His career was cut short in 1855 after he drank a fatal amount of tincture of cantharides (Spanish fly), which he had mistaken for brandy. For a more complete biography, see Lutz Richter-Bernburg, "Cloquet, Louis-André-Ernest," *Encyclopedia Iranica* 5, fasc. 7 (1992): 718, updated online version accessed January 17, 2016, http://www.iranicaonline.org/articles/cloquet-louis-andr-ernest-b.

20. FO 60/127 Letter from Charles W. Bell M.D. to H. U. Addington Esq. (Her Majesty's Under Secretary of State for Foreign Affairs). Manchester, June 1, 1846.

21. Qajar, Muhammad Husayn va Mons. Jibril, *Risala dar vaba*. Melli Library, Teheran, MS no. 735F, 1262 AH [1845–1846]; Jakob Eduard Polak, *Safarnama-yi Polak: Iran va Iranian*, trans. KeyKavoos Jahandari (Tehran: Khawrazmi, 1361 AS [1982]), 5.

22. FO 60/123 Dispatch (n. 78) from Martin Sheil to the Earl of Aberdeen. Camp near Tehran, July 25, 1846.

23. FO 60/126 Dispatch (n. 23) from Keith Abbot to Viscount Palmerston. Tabriz, December 6, 1846; FO 60/174 Dispatch (n. 25) from Consul Richard W. Stevens to the Earl of Malmesbury. Camp near Tabriz, July 28, 1852; FO 60/126 Dispatch (n. 22) Keith Abbot to Viscount

Palmerston. Tabriz, November 19, 1846. The short-term entombments were in fact a liability because the bodies were potentially infectious, even after death, and the Shi'ite tradition of transporting cadavers for permanent burial in holy sanctuaries could disseminate cholera to other parts of the country. See Amir A. Afkhami, "Disease and Water Supply," 206–220.

24. "Cholera and Locusts in Persia," *Christian Observer*, December 10, 1847.

25. FO 60/141 Dispatch (n. 13) from Keith Abbot to Viscount Palmerston. Goolahek, August 26, 1848; FO 60/124 Dispatch (n. 92) from Martin Sheil to Viscount Palmerson. Camp near Tehran, September 1, 1846.

26. Jean François Dequevauviller, *Notices sur le Docteur Ernest Cloquet, Médecin et Conseiller du Schah de Perse, Membre Correspondant de l'Académie Impériale de Médecine, Officier de la Légion D'honneur, Etc.* (Paris: Imprimerie de L. Martinet, 1856), 10–11; Kohn, *Encyclopedia of Plague and Pestilence*, 237.

27. FO 60/124 Dispatch (n. 88) from Martin Sheil to the Earl of Aberdeen. Camp Lavasaun, August 10, 1846.

28. Abbas Amanat, *Resurrection and Renewal: The Making of the Babi Movement in Iran, 1844–1850* (Ithaca: Cornell University Press, 1989), 97–98. This messianic expectation also extended to Iranian Jews. For more on this, see Mehrdad Amanat, "Judeo-Persian Communities of Iran," *Encyclopedia Iranica* 15, fasc. 2 (2009): 117–124, updated online version accessed January 17, 2016 http://www.iranicaonline.org/articles/judeo-5-2-qajar-conversion-of-jews.

29. Edward Granville Browne, *The Press and Poetry of Modern Persia: Partly Based on the Manuscript Work of Mirza Muhammad 'Ali Khan "Tarbiyat" of Tabriz* (Cambridge: Cambridge University Press, 1914), 10.

30. *Ruznama-yi vaqayi'-i ittifaqiya*, no. 3, 19 Rabi' al-Thani 1267 AH [January 22, 1851], cited in M. Hassan Beygi, *Tehran-i qadim* [Old Tehran] (Tehran: Qafnus, 1366 AH [1946–1947]), 199.

31. Maryam Dorreh Ekhtiar, "Nasir al-Din Shah and the Dar al-Funun: The Evolution of an Institution," *Iranian Studies* 34 (2001): 153–154.

32. Feza Gunergun, "Science in the Ottoman world," in *Imperialism and Science: Social Impact and Interaction*, ed. George N. Vlahakis, Isabel Maria Malaquias, Nathan M. Brooks, François Regourd, Feza Gunergun, and David Wright (Santa Barbara: ABC-CLIO, 2006), 96–97; Ekhtiar, "Nasir al-Din Shah and the Dar al-Funun," 153–154; John Gurney and Negin Nabavi, "Dar al-Fonun," *Encyclopedia Iranica* 6, fasc. 6 (1993): 662–668, updated online version accessed January 17, 2016, http://www.iranicaonline.org/articles/dar-al-fonun-lit.

33. Hamid Algar, "Amir Kabir, Mirza Taqi Khan," *Encyclopedia Iranica* 1, fasc. 9 (1989): 959–963, updated online version accessed July 15, 2016, http://www.iranicaonline.org/articles/amir-e-kabir-mirza-taqi-khan.

34. Johannes Lodewick Schlimmer (1818–1876) was born in Rotterdam and studied medicine in his native city and Leiden for four years before traveling to Aleppo in 1839 to investigate the etiology and treatment of the Aleppo sore (cutaneous leishmaniasis). He settled in northern Iran five years later, where he claimed to have successfully used strychnine against cholera while working as a private practitioner during the 1846 outbreak in Tabriz and Hamedan. Schlimmer likely came to the attention of Mirza Taqi Khan during his sojourn in Tabriz. Johanne L. Schlimmer, *Du Presage et de L'Avortement de L'Imminence Cholerique* (Rotterdam: Nijgh & Van Ditmar, 1874), 101; *Ruznama-yi vaqayi'-i ittifaqiya*, no. 7, 12 Jumada al-Awwal 1267 AH [March 15, 1851]; Christoph Werner, "Polak, Jakob Eduard," *Encyclopedia Iranica*, online edition, accessed August 7, 2016, http://www.iranicaonline.org/articles/polak-jakob-eduard; Stephanie Cronin, "Army iv a. Qajar Period," *Encyclopedia Iranica*, online edition accessed August 7, 2016, http://www.iranicaonline.org/articles/army-vii-qajar.

35. William Farr, *Report on the Mortality of Cholera in England, 1848–49* (London: W. Clowes, 1852); John M. Eyler, *Victorian Social Medicine: The Ideas and Methods of William Farr*

(Baltimore: Johns Hopkins University Press, 1979); John M. Eyler, "William Farr on the Cholera: The Sanitarian's Disease Theory and the Statistician's Method," *Journal of the History of Medicine* 28 (1973): 79–100.

36. John Snow, "On the pathology and mode of communication of the cholera," *London Medical Gazette* 44 (1849): 730–732, 745–752, 923–929.

37. John Duffy, *The Sanitarians: A History of American Public Health* (Urbana: University of Illinois Press, 1990), 79–92.

38. FO 60/174 Dispatch (n. 40) from Consul Richard W. Stevens to the Earl of Malmesbury. Oroomeeya, November 15, 1852; FO 60/174 Dispatch (n. 92) from George Alexander Stevens (for Consul Richard W. Stevens absent from town) to Lieut. Col. Sheil. Oroomeeya, December 26, 1852.

39. Muzaffar al-Din's symptoms were more consistent with malaria, prevalent in Tehran during that period, than cholera. Charles W. Bell, "Report on the Epidemic of Ague or 'Fainting Fever' of Persia, A Species of Cholera, occurring in Tehran in the Autumn of the year 1842," *British and Foreign Medical Review, or Quarterly Journal of Practical Medicine and Surgery* 16 (1843): 558–566.

40. FO 60/179 [pp. 212–213] Dispatch (n. 32) from Taylor Thompson to the Earl of Clarendon. Camp Near Tehran, May 12, 1853.

41. Louis André Ernest Cloquet, "Sur le Choléra en Perse," *Bulletin de l'Académie Nationale de Médecine* 18 (1852–53): 1190–1192.

42. Government of Iran, *Ruznama-yi vaqayi'-i ittifaqiya*, no. 139, 25 Dhu al-Hijja 1269 AH [September 28, 1853].

43. FO 60/174 Dispatch (n. 39) from George Alexander Stevens (for Consul Richard W. Stevens absent from town) to the Earl of Malmesbury. Tabreez, October 20, 1852.

44. FO 60/181 [pp. 112–115] Dispatch (n. 89) from Taylor Thompson to the Earl of Clarendon. Camp near Tehran. August 21, 1853.

45. FO 60/174 Dispatch (n. 42) from George Alexander Stevens (for Consul Richard W. Stevens absent from town) to Lieut. Col. Sheil. Oroomeeya, December 27, 1852; FO 60/179 [pp. 257–260] Dispatch (n. 38) from Taylor Thompson to the Earl of Clarendon. Tehran, May 21, 1853.

46. Jakob Eduard Polak, "La médicine militaire en Perse.—Par le docteur J.-E. Polak, ancien médecin particulier du schah de Perse," *Revue scientifique et administrative des médecins des armées de terre et de mer* 7 (1865): 651.

47. Jakob Eduard Polak (1818–1891) was a Jewish physician from Bohemia (Habsburg Empire) who trained in medicine, surgery, and obstetrics in Prague and Vienna. He worked primarily at Vienna's General Hospital (*Allgemeines Krankenhaus*) until 1851, when he agreed to teach medicine and pharmacology at the Polytechnic College in Tehran. Polak quickly mastered Persian, which he used in lectures and publications within a year of his arrival. He became the Iranian army's chief medical inspector after the death of Fortunato Casolani in 1852 and Nasir al-Din Shah's personal physician after the death of Louis-André-Ernest Cloquet in 1855. Polak introduced modern surgical techniques and the use of anesthesia to Iran. He was an expert in renal stone removal, having operated on 158 individuals during his stay. Polak returned to his practice in Vienna in 1862 and became a recognized expert on Iran. The Iranian government frequently turned to him as an adviser in Europe, and the Austrian Foreign Ministry sought his help with cultural and political matters relating to Iran and the Islamic world. Polak played an important role in bringing the plight of his coreligionists to the attention of the European Jewish community and gave a balanced and sympathetic perspective of Iranian culture in his seminal memoir. For a concise biography of Polak, see Werner, "Polak, Jakob Eduard," *Encyclopedia Iranica*; Habib Levy, *Comprehensive History of the Jews of Iran: The Outset of the Diaspora* (Costa Mesa: Mazda, 1999), 452–453.

48. Polak, "La médicine militaire en Perse," 648–651.

49. Jakob Eduard Polak, *Bimari-yi vaba* [The illness of cholera], Melli Library, Tehran, MS no. 349F, 1269 AH [1852–1853].

50. Schlimmer, *Du Présage et de L'Avortement*, 116–118.

51. Polak translated and published a book on anatomy (1853–1854), a two-volume work on surgery (1856–1857), and a work on ophthalmology (1853–1854). He also wrote a number of minor treatises on therapeutics and general internal medicine. Schlimmer's books include a monograph on basic anatomy (1859–1860), a 160-page text on elementary chemical physiology (1862–1863), and other notable works on the diseases of the eye and skin.

52. Schlimmer's medical glossary, written with input from Polak, was his final publication in Iran. The book translated Persian medical terms into English, French, and German and included observations on epidemics, disease prevention, and hygiene in Iran. This was the first medical reference book of its kind in Iran and articulated the definition of public health in the Persian scientific language by formally defining "*Préservation*" (prevention) as "*Héfze Sihhte: Jelow-e Maraz Gereftan*" (hygiene / public health: to prevent disease). Johannes L. Schlimmer, *Terminologie médico-pharmaceutique et anthropologique français-persane, avec des traductions anglais et allemand des termes français* (Tehran: Lithographie d'Ali Gouli Khan, 1874), 476.

53. Jakob Eduard Polak, *Safarnama-yi Polak: Iran va Iranian*, trans. KeyKavoos Jahandari (Tehran: Khawrazmi, 1361 AH [1982]), 212–214.

54. Schlimmer, *Du Presage et de L'Avortement*, 43.

55. Polak, *Safarnama-yi Polak*, 210–211.

56. *Ruznama-yi vaqayi'-i ittifaqiya*, no. 153, 5 Rabi' al-Thani 1270 AH [December 6, 1853].

57. Polak, *Safarnama-yi Polak*, 210–211.

58. David Menashri, *Education and the Making of Modern Iran* (Ithaca: Cornell University Press, 1992), 50.

59. *Ruznama-yi vaqayi'-i ittifaqiya*, no. 2, 11 Rabi' al-Thani 1267 AH [January 14, 1851]; no. 6, 10 Jumada al-Awwal 1267 AH [March 13, 1851]; no. 15, 14 Rajab 1267 AH [May 15, 1851]; no. 40, 11 Muharram 1268 AH [November 5, 1851].

60. *Ruznama-yi vaqayi'-i ittifaqiya*, no. 140, 2 Muharram 1270 AH [October 5, 1853];no. 6, 10 Jumada al-Awwal 1267 AH [March 13, 1851];no. 35, 6 Muharram 1267 AH [November 11, 1850]

61. *Ruznama-yi vaqayi'-i ittifaqiya*, no. 156, 26 Rabi' al-Thani 1270 AH [January 26, 1854].

62. *Ruznama-yi vaqayi'-i ittifaqiya*, no. 145, 8 Safar 1270 AH [November 9, 1853].

63. *Ruznama-yi Vaqayi'-i Ittifaqiya*, no. 196, 8 Safar 1271 AH [October 30, 1854].

64. *Ruznama-yi vaqayi'-i ittifaqiya*, no. 150, 13 Rabi' al-Awwal 1270 AH [December 14, 1853].

65. *Ruznama-yi vaqayi'-i ittifaqiya*, no. 158, 11 Jumada al-Awwal 1270 AH [February 9, 1854]; no. 161, 2 Jumada al-Thani 1270 AH [March 2, 1854].

66. Great Britain General Board of Health, *Report of the General Board of Health on the Epidemic Cholera of 1848 & 1849* (London: Her Majesty's Stationery Office, W. Clowes, 1850).

67. Hakim Kazulani [Fortunato Casolani],'*Arizi-yi majlis-i mamuran bi padishah-i Inglis* [Petition of the assembled commission to the King of England], Melli Library, Tehran, MS no. 313F, n.d. Polak was appointed to this largely ceremonial position after Casolani's death in 1852.

68. Kazulani, '*Arizi-yi majlis-i mamuran*, 7–9.

69. Kazulani, '*Arizi-yi majlis-i mamuran*, 10–13; John Sutherland, *Appendix (A) to the Report of the General Board of Health on the Epidemic Cholera of 1848 & 1849* (London: Her Majesty's Stationery Office, W. Clowes, 1850), 18–21.

70. Hormoz Ebrahimnejad, "Un traité d'epidémiologie de la médecine traditionnelle persane: Mofarraq ol-Heyze Va'l-Vaba de Mirza Mohammad-Taqi Shirazi (ca. 1800–1873)," *Studia Iranica* 27 (1998), 93; Muhammad Taqi Mir, *Pizishkan-i nam-i Pars* [Persian physicians], 2nd ed.

(Tehran, 1363 AS [1984–1985]), 50; "Bayan-i kiyfiyat-i tap-i vaba-yi va ta'un ki dar tibb-i akbar muzu' ast nivisht-i mishavad; Or History of the Pestilential Fever and Plague from the Tib-e-Akhbar," in *A Practical Treatise on Epidemic Cholera, Ague, and Dysentery; Illustrating the Principals of Treatment By Their Anatomical Physiology Pointing Out Their Contagion and Westering Inclination; to Which Is Added A Persian Treatise on Plague and Cholera*, by W. G. Maxwell (Calcutta: T. Ostell and Co. British Library, 1838), 1–23.

71. Sayyid 'Ali 'ibn-i Muhammad Tabrizi Mar'ashi, *Qanun al-'alaj* [The cannon of treatments], Kitabkhana-yi Markazi Danishgah-i Tehran [Central Library, University of Tehran] (Danishgah). MS no. 4115, 1269 AH [1852–1853].

72. Sheil appears to have confused the quality of the remedy with the quality of the disease—hence, hot peppers used for cold diseases. See Mary Leonora Woulfe Sheil, *Glimpses of life and manners in Persia: With Notes on Russia, Koords, Toorkomans, Nestorians, Khiva, and Persia* (London: John Murray, 1856), 213.

73. Polak, *Bimari-yi vaba*.

74. Abbas Amanat, *Pivot of the Universe: Nasir al-Din Shah and the Iranian Monarchy, 1831–1896* (Berkeley: University of California Press, 1997), 351–405.

75. Iranians experienced limited outbreaks of cholera in 1854 (Azerbaijan), 1856 (Gilan, Azerbaijan), 1857 (Azerbaijan, Tehran, Yazd, Khorasan, Qom, Kashan, Fars), 1858, 1860 (Azerbaijan, Tehran), 1861, 1863 (Gorgan), and 1863. For more on the outbreaks of cholera in this period, see Schlimmer, *Du Presage et de L'Avortement*, 43; Ahmad Seyf, "Iran and Cholera in the Nineteenth Century," *Middle Eastern Studies* 38, no. 1 (2002): 172.

76. In 1861 cholera in Tabriz killed about 2,500 individuals in a one-month period. Robberies and brigandage in the city and surroundings were rampant during the outbreak. See FO 60/259 Dispatch (n. 37) from Keith Abbot to Lord John Russel. Camp near Tabreez, October 6, 1861; FO 60/259 Dispatch (n. 44) from Keith Abbot to Lord John Russel. Tabreez, December 31, 1861.

77. Amanat, *Pivot of the Universe*, 383.

78. For more on the proliferation of broadsheets in this period, see Negin Nabavi, "Journalism," *Encyclopedia Iranica* 11, fasc. 1 (2009): 46–54, updated online version accessed August 10, 2015, http://www.iranicaonline.org/articles/journalism-i-qajar-period; Sayyid Farid Qasemi, *Sarguzasht-i matbu'at-i Iran* [The history of press in Iran], vol. 2 (Tehran: Sazman-i Chap va Intisharat-i Vizarat-i Farhang va Irshad-i Islami, 1380 AS [2001–2002]), 1302–1313.

79. This included Mirza Husayn Duktur, who wrote a thesis on the use of therapeutic arsenic in treating fevers (based on his experience at Tehran's Imperial Hospital); Mirza Riza bin Muqim, who wrote a thesis on polyuria (diabetes); and Mirza 'Ali Naqi, who wrote a thesis on the diagnosis and treatment of hydropsy (congestive heart failure). Mirza Husayn was the son of Mirza Tabib Afshar. He was among the first students of the Polytechnic College who wrote a book on the treatment of cholera, which he dedicated to the shah. He left for further medical studies in Paris and returned to teach at the Dar al-funun as early as 1861. *Ruznama-yi vaqayi'-i ittifaqiya*, no. 469, 6 Muharram 1277 AH [July 25, 1860]; Maryam Dorreh Ekhtiar, "The Dar al-Funun: Educational Reform and Cultural Development in Iran (PhD diss., New York University, 1994), 136–137; Husayn Mahbubi Ardakani, *Tarikh-i mu'assasat-i tamadduni-yi jadid dar Iran* [The history of the institutions of modern civilization in Iran], 2nd ed. (Tehran: Anjuman-i Danishjuyan-i Danishgah-i Tehran, 1370 AS [1992]), 1:287–288.

80. Werner, "Polak, Jakob Eduard," *Encyclopedia Iranica*.

81. Joseph Désiré Tholozan (1820–1897), a French military physician and a veteran of the Crimean War, became Nasir al-Din Shah's personal doctor in 1858 and remained in Iran until his death almost four decades later. In the 1850s he carried out a series of experiments in collaboration with famed neurologist, Charles-Edouard Brown-Séquard, which led to the description of the vasomotor reflex. He published more than fifty articles and books between 1847 and 1892 mainly on infectious pathology and epidemiology. His mouthpiece was the *Gazette*

Medical de Paris, which he edited from 1850 to 1856. He was also a corresponding member of the Epidemiological Society of London, the National Academy of Medicine in France [Académie Nationale de Médecine], and the French Academy of Sciences [Académie des Sciences]. Ornithodoros Tholozani, a tick vector for a recurring fever infection, which he described in 1882, bears his name. During his stay in Iran, Tholozan wrote several books in Persian, including two in the field of internal medicine (*Asrar al-atibba* and *Risala dar tibb*) and the first Iranian treatise on the benefits of quinine. Tholozan was made commander of the Legion of Honor by the French government and appointed to the Order of St. Michael and St. George by Queen Victoria of England. For more on Tholozan, see Jean Theodorides, "Tholozan et la Perse [Tholozan and Persia]," *Histoire des Sciences Médicales* 32, no. 3 (1998): 287–296; Jean Theodorides, "Un grand épidémiologiste franco-mauricien: Joseph Désiré Tholozan (1820–1897) [A great Franco-Mauritian epidemiologist: Joseph Désiré Tholozan (1820–1897)]," *Bulletin de la Société de Pathologie Exotique* 91, no. 1 (1998): 104–108; Louis-Cyril Celestin, *Charles-Edouard Brown-Séquard: The Biography of a Tormented Genius* (London: Springer, 2014), 104–105; FO 60/508 Chancellor's draft letter to the Marquis of Salisbury (Foreign Office). London, August 13, 1889; Muhammad Hassan Khan I'timad al-Saltana, ed., "Jinab-i Duktur Tholozan [His Excellency Doctor Tholozan]," *Sharaf*, no. 35, 1303 AH [1885–1886].

82. Joseph Désiré Tholozan, *Rapport à Sa Majesté le Chah sur l'état actuel de l'hygiène en Perse; progrès à réaliser; moyens de les effectuer; résultats obtenus depuis un an* [Report to His Majesty the Shah on the current sanitary state of Persia, progress to be realized, modes of bringing about progress, and results obtained since a year ago], 1869, Bibliothèque de l'Académie National de Médecine, Paris, Archives et manuscrits de la bibliothèque, cote 50.426, 36–38.

83. Charles Ambrose Storey, *Persian Literature: A Bio-bibliographical Survey*, vol. 2, pt. 2: *E. Medicine* (London: Luzac [for] the Royal Asiatic Society of Great Britain and Ireland, 1971), 299.

84. "Médecine et Chirurgie: Prix Bréant," *Comptes rendus hebdomadaires des séances de l'Académie des sciences* 75 (1872): 1362–1369.

85. Mirza Sayyid Raz Hakim-bashi, Mirza Muhammad Hakim-bashi-yi Tupkhana, Mirza Sayyid 'Ali Nayib Hakim-bashi, Mirza Riza Quli Jarah-bashi were some of the first notable medical graduates who served in the military. For more on this, see Muhammad Hasan Khan I'timad al-Saltana, *Tarikh-i muntazam-i Nasiri* [History of Nasir al-Din Shah's reign], ed. Muhammad Isma'il Rizvani (Tehran: Dunya-yi kitab, 1367 AS [1988]), 3:2070–2071, 2097.

86. Soli Shahvar, "Telegraph i. First Telegraph Lines in Persia," *Encyclopedia Iranica*, online edition accessed August 10, 2016, http://www.iranicaonline.org/articles/telegraph-i-first -telegraph-lines-in-persia; Patrick Clawson and Michael Rubin, *Eternal Iran: Continuity and Chaos* (New York: Palgrave Macmillan, 2005), 39; Amanat, *Pivot of the Universe*, 404.

87. *Ruznama-yi dawlat-i 'alliya-yi Iran*, no. 532, 13 Jumada al-Awwal, 1279 AH [October 26, 1862], 7; no. 604, 10 Safar, 1284 AH [June 13, 1867], 7.

88. Persian transcript of Tholozan's response to Dr. Proust, Yale University Library, Manuscripts and Archives, New Haven, Ghassem Ghani Collection (MS 235), ser. 5, box 2, folder 18.

89. Antoine Sulpice Fauvel, *Le Choléra: Étiologie et Prophylaxie* (Paris: J.-B. Ballière et Fils, 1868), 91. Mirza Malkam Khan (1833–1908) at the time was a consular aide-de-camp at the Iranian legation in Istanbul. For more on Malkam Khan, see Hamid Algar, *Mirza Malkum Khan* (Berkeley: University of California Press, 1973).

90. Britain had not always opposed quarantines. In 1710 it passed the first Quarantine Act, followed by successively stricter laws. By the first quarter of the nineteenth century, new considerations brought the British quarantine system into question. From one side, the rising tide of economic liberalism called for free and unhampered trade among countries. From the other, the absence of plague from the British Isles for over a century perpetuated a liberal interpretation of contagions and reduced the need for maritime quarantines. The advent of cholera in 1830s increased this anticontagionist intellectual trend in Britain. Cholera's fecal-oral

transmission and its preponderance among the poorer segments of the population challenged the doctrine of direct contagion, inspiring the work of German scientist Max Joseph von Pettenkofer (1818–1901), who postulated that factors such as a particular climate, elevation, and individual predisposition were necessary conditions for the spread of the new disease. For more on this, see J. C. McDonald, "The History of Quarantine in Britain during the 19th Century," *Bulletin of the History of Medicine* 25, no. 1 (1951): 22–27; Max Joseph von Pettenkofer, "Cholera," *Lancet* 2 (1884): 769–771, 816–819, 861–864, 904–905, 1042–1043, 1086–1088; Max von Pettenkofer, *Relations of The Air to The Cloths we Wear, the House we Live in, and The Soil We Dwell On: Three Popular Lectures Delivered Before The Albert Society at Dresden*, trans. August Hess (London: N. Trubner, 1873); Hermann Weber, "On Professor Pettenkofer's Theory of the Mode of Propagation Cholera," *Transactions of the Epidemiological Society of London* 2 (1867): 404–413.

91. Filippo Pacini's (1812–1883) research indicated the existence of disease-causing microbes in the intestines of cholera victims. His work epitomized the empirical Italian approach to cholera and anticipated Robert Koch's discoveries by thirty years. Filippo Pacini, "Osservazioni microscopiche e deduzioni patologiche sul cholera asiatico," *Gazzetta Medica Italiana: Federativa Toscana*, 2, ser. 4 (1854): 397–401, 405–412.

92. Fauvel, *Le Choléra*, 645.

93. Fauvel, *Le Choléra*, 484–490, 648–652.

94. Conférence Sanitaire International, *Rapport sur Les Mesures à Prendre en Orient Pour Prévenir de Nouvelles Invasions du Choléra en Europe* [Août 1866] (Constantinople: Imprimerie du Levant Herald, 1866), 52.

95. Conférence Sanitaire International, *Rapport sur Les Mesures à Prendre*, 52–55.

96. Fauvel, *Le Choléra*, 652; Norman Howard-Jones, "The Scientific Background of the International Sanitary Conferences, 1851–1938. 2," *WHO Chronicle* 28, no. 5 (1974): 236.

97. Howard-Jones, "The Scientific Background. 2," 236.

98. Howard-Jones, "The Scientific Background. 2," 239.

99. FO 60/300 Dispatch (n. 13) from Keith Abbot to the Earl of Clarendon. Tabreez, June 10, 1866.

100. The Iranian government reported 20,000 to 30,000 deaths out of the total population of 130,000 though the British thought this estimate exaggerated and placed the mortality in the five percent range. See FO 60/300 Dispatch (n. 34) from Keith Abbot to the Lord Stanley. Tabreez, December 28, 1866; FO 248/232 Dispatch (n. 46) from K. E. Abbot to Mr. Allison. Zenjenab near Tabreez, September 1, 1866.

101. FO 60/306 Dispatch (n. 117) from Allison to Lord Stanley. Tehran, November 14, 1867. [Enclosure n. 1] Astrabad Agent to Mr. Allison October 23, 1867.

102. E. D. Dickson, "On Cholera in Persia, 1866–1868," *Transactions of the Epidemiological Society of London* 3 (1866–1876): 259, 264; FO 60/305 Dispatch (n. 89) from Allison to Lord Stanley. Gulahek, August 28, 1867.

103. Tholozan, *Rapport à Sa Majesté*, 4.

104. They included Mirza Sayyid 'Ali in Bushehr, Monsieur Faquergeen in Shiraz, Mirza Zayn al-'Abidin in Kermanshah, Mirza Habib Allah in Hamedan, Mirza Abul Qasim in Isfahan, Mirza 'Abd al-Rahim in Kerman, Monsieur Stagno in Tabriz, Mirza Mustafa in Mashhad, Mirza Ahmid in Semnan, Mirza Hassan in Saveh, Mirza Sayyid 'Abd al-Karim in Kashan, and Mirza Riza in Zanjan. Tholozan, *Rapport à Sa Majesté*, 58.

105. FO 60/313 Dispatch (n. 109) from Allison to Lord Stanley. Gulahek, August 15, 1868.

106. FO 60/320 Dispatch (n. 73) from Ronald Thomson to the Earl of Clarendon. [Enclosure n. 3] Translation of Extract of Report by the Asterabad Agent. Asterabad, September 1869; FO 60/320 Dispatch (n. 86) from Ronald Thomson to the Earl of Clarendon. Tehran, November 17, 1869. [Enclosure n. 2] Translation of Report by Asterabad Agent. Asterabad, November 4, 1869.

107. FO 60/313 Report from Dr. Dickson to Charles Allison. Gulhek, August 25, 1868.

108. Dickson, "On Cholera in Persia," 261–263.

109. Tholozan, *Rapport à Sa Majesté*, 4–7. Tholozan believed that the cost of public health in Iran would amount to no more than one thousand tuman (less than two thousand dollars in the 1892 tuman to dollar exchange rate).

110. Tholozan, *Rapport à Sa Majesté*, 9–10.

111. Tholozan, *Rapport à Sa Majesté*, 10; FO 60/323 Draft of Letter (n. 50) from Earl of Clarendon to Council Office. Tehran, November 13,1869; Joseph Désiré Tholozan, "Rapport à S.M. le Shah sur l'état actuel de l'hygiène en Perse, progrès à réaliser, moyens de les effectuer, résultats obtenus [Texte présenté à la séance du 11 octobre par H. Larrey]," *Comptes rendus hebdomadaires des séances de l'Académie des sciences* 69 (1869): 838–840.

112. Tholozan, *Rapport à Sa Majesté*, 12.

113. Shoko Okazaki, "The Great Persian Famine of 1870–71. I," *Bulletin of the School of Oriental and African Studies* 49 (1986): 183–192; Charles Melville, "The Persian Famine of 1870–1872: Prices and Politics," *Disasters* 12, no. 4 (1988): 309–325.

114. FO 60/334 Dispatch from Allison to Viceroy of India. Tehran, May 4, 1871. [Enclosure n. 1] Dispatch from British Agent in Mashed to Mr. Allison. Shiraz, Mashed, April 9, 1871.

115. FO 60/336 Dispatch (n. 162) from Allison to Earl of Granville. Tehran, November 7, 1871.

116. FO 248/271 Dispatch (n. 399/19) from Lieutenant Colonel Lewis Pelly to Charles Alison. Bushire, April 17, 1871. [Enclosure n. 2] Copy of Telegram from Dr. S. M. D. Cummings to Col. Pelly. Shiraz, April 18, 1871.

117. FO 60/342 Dispatch (n. 2) from Allison to Earl of Granville. Gulhek, January 1, 1872.

118. Mirza Husayn Khan Mushir al-Dawla (1828–1881) began his career in the foreign service arm of the government; serving first in India and Tbilisi and eventually as the Iranian envoy to the Ottoman Empire, a position he held for eleven years. He was appointed minister of justice by the shah in December 1870 and prime minister in November 1871; see Shaul Bakhash, *Iran: Monarchy, Bureaucracy and Reform under the Qajars 1858–1896* (London: Ithaca Press, 1978), 44, 79.

119. Guity Nashat, *The Origins of Modern Reform in Iran, 1870–80* (Urbana: University of Illinois Press, 1982), 157–158.

120. British Library, London. Oriental and India Office Collections, India Office (IO) R/15/1/182 Pelly to Mirza Muhammad Khan, no. 390, Bushire, July 21, 1871. Transcripts of unpublished Crown Copyright material in the India Office Records appear by permission of Her Majesty's Stationary Office; IO/R/15/1/182 Pelly to Mirza Muhammad Khan, no. 399, Bushire, July 24, 1871.

121. FO 60/334 Dispatch (n. 84) from Allison to Earl of Granville. Gulhek, July 5, 1871. [Enclosure n. 2] Translation of a letter from Mirza Saeed Khan, Minister of Foreign Affairs, to Meerza Mahomed, Foreign Office Agent at Bushire. Tehran, June 22, 1871.

122. Hasan 'Ali Khan Garusi served as Iran's envoy to France between 1858 and 1865. Guity Nashat, *The Origins of Modern Reform in Iran*, 159; Abbas Amanat, "Amir Nezam Garrusi," *Encyclopedia Iranica* 1, fasc. 9 (1989): 966–969, updated online version accessed January 10, 2016, http://www.iranicaonline.org/articles/amir-e-nezam-garrusi.

123. Johanne L. Schlimmer, *Du Présage et de L'Avortement*, 5.

124. FO 60/345 Letter (n. 24484) from Arthur Helps, Privy Council Office, Whitehall, to Under Secretary of State, Foreign Office. London, July 12, 1872.

125. Henry Creswicke Rawlinson, *England and Russia in the East: A Series of Papers on the Political and Geographical Condition of Central Asia* (London: John Murray, 1875), 393.

126. Amanat, "Amir Nezam Garrusi," *Encyclopedia Iranica*.

127. Jalal Farahmand, "Tanzif-i shahr-i Tehran [Hygiene in the city of Tehran]," *Mahnama-yi iliktruniki-yi baharistan* [Baharistan online monthly], 1: 5, accessed March 10, 2017, http://www

.iichs.org/magazine/sub.asp?id=85&theme=Orange&magNumber=1&magID=6&magIMG
=coverpage1.jpg.

128. Abdollah Mostofi, *The Administrative and Social History of the Qajar Period* (Costa
Mesa: Mazda, 1997), 3:891; Hosayn Farhudi, "City Councils," *Encyclopedia Iranica* 5, fasc. 6
(1991): 646–648, updated online version accessed August 10, 2015, http://www.iranicaonline.org
/articles/city-councils-anjoman-e-sahr-in-persia.

129. Afshin Marashi, *Nationalizing Iran: Culture, Power, and the State, 1870–1940* (Seattle:
University of Washington Press, 2008), 15–48.

130. Nobuaki Kondo, *Islamic Law and Society in Iran: A Social History of Qajar Tehran*
(New York: Routledge, 2017), 117; "Teheran Gets Pure Piped Water but Mourns Quaint Old Cus-
toms: New System, Result of Efforts Begun in 1880's, Will Eliminate Horse-Drawn Carts and
Open-Street Conduits," *New York Times*, November 9, 1955.

131. John Netten Radcliffe, assistant medical officer to the Local Government Board, au-
thored the report. John Netten Radcliffe, *Recent diffusion of cholera in Europe* (London: Local
Government Board [printed by T. Harrison], 1872); FO 60/345 from Radcliffe to Simon, London, May 7, 1872. [Enclosure n. 1] Simon to Granville (n. 417) London, June 8, 1872.

132. FO 60/343 Dispatch (n. 91) from Ronald Thompson to Earl of Granville. Tehran,
December 3, 1872.

133. FO 60/343 Dispatch (n. 91) from Ronald Thompson to Earl of Granville. Tehran, December 3, 1872. [Enclosure n. 2] Copy and translation of Letter from His Excellency Mirza Said
Khan, Minister for Foreign Affairs, to Ronald Thompson Esq. Tehran, November 24, 1872.

134. FO 60/286 Dispatch (n. 54) from William Abbot (Acting Consul General) to Allison.
Tabreez, December 26, 1863.

135. FO 60/286 Dispatch (n. 51) from William Abbot (Acting Consul General) to Allison.
Tabreez, December 11, 1863; Charles Issawi, "The Tabriz-Trabzon Trade, 1830–1900: Rise and
Decline of a Route," *International Journal of Middle East Studies* 1 (1970): 18–27.

136. FO 60/345 from Radcliffe to Simon. London, May 7, 1872. [Enclosure n. 1] from Simon
to Granville, no. 417. London, June 8, 1872.

137. FO 60/345 from Radcliffe to Simon. London, May 7, 1872. [Enclosure n. 1] from Simon
to Granville, no. 417. London, June 8, 1872.

138. FO 248/297 from Derby to Thomson, circular, London, December 22, 1874.

139. Edward Cator Seaton, "A brief account of the proceedings of the international sanitary
conference held at Vienna in 1874," *Transaction of the Epidemiological Society of London* 3
(1874): 556–570.

140. Seaton, "A brief account"; and Frank G. Clemow, "The Constantinople Board of
Health," *Lancet* 204 (1923): 1074–1076, 1126–1127, 1170–1171, 1180–1181.

141. FO 248/297 Circular Commercial Dispatch from the Earl of Derby to Taylor Thomson
Esq. Foreign Office. London, December 22, 1874.

142. Joseph Désiré Tholozan, *La Peste en Turquie Dans Les Temps Modernes : Sa Prophylaxie Défectueuse, Sa Limitation Spontanée* (Paris: G. Masson, 1880); Joseph Désiré Tholozan,
*Histoire de La Peste Bubonique en Perse ou Détermination de Son Origine, de sa Marche, du Cycle
de Ses Apparitions et de la cause de sa Prompte Extinction* (Paris: G. Masson, 1874).

143. Hossein Modarressi Tabataba'i, *Ashna-yi ba chand nuskhaha-yi khatti* [introduction to
several written manuscripts] (Qom: Chapkhana-yi mihr, 1355 AS [1976]), 250–251.

144. 'Aliquli Mirza I'tizad al-Saltana (b. 1238 AH [1822]-1298 AH [1880]) was the shah's first
minister of public instruction during the Qajar period. I'tizad al-Saltana's responsibilities included the supervision of the expanding Iranian telegraph lines and the management of the
Dar al-funun, including the supervision of its hospital. I'tizad al-Saltana's contribution to science and medicine in Iran included sending forty-two students, mostly Dar al-funun graduates, to further their scientific education in France in 1858/1859. Following I'tizad al-Saltana's

death, many of his responsibilities, including the direction of the Imperial Hospital, were taken over by his successor 'Aliquli Khan Mukhbir al-Dawla. See Abbas Amanat, "E'tezad al-Saltana, Aliqoli Mirza," *Encyclopedia Iranica* 8, fasc. 6: 669–672, updated online version accessed November 2, 2015, http://www.iranicaonline.org/articles/etezad-al-saltana; *Ruznama-yi dawlat 'alliya-yi Iran*, no. 604, 10 Safar 1284 AH [June 13, 1867], 3.

145. FO 60/382 Dispatch letter (n. 45) Taylor Thompson to Earl of Derby. Tehran, April 20, 1876. [Enclosure n. 1] Copy of Memorandum from Joseph Dickson to Taylor Thomson. Tehran, April 14, 1876.

146. FO 60/382 Dispatch letter (n. 45) Taylor Thompson to Earl of Derby. Tehran, April 20, 1876. [Enclosure n. 2] Copy of Memorandum from Joseph Dickson to Taylor Thomson. Tehran, April 17, 1876.

147. FO 881/3332 Confidential Dispatch (n. 1) from Count Beust to Earl of Derby. Belgrave Square, June 27, 1877; FO 881/3600 Confidential Dispatch (n. 1) from Count Beust to Earl of Derby. Belgrave Square, October 31, 1877. [Enclosure n. 3] from Houssein Khan to M. von Kuczynski, 10 Shaban 1294 [August 20, 1877].

148. FO 60/382 Dispatch letter (n. 58) Taylor Thompson to Earl of Derby. Tehran, May 16, 1876. [Enclosure n. 1] Copy of Memorandum from Sir Joseph Dickson to Taylor Thomson. Tehran, May 16, 1876.

149. FO 60/382 Dispatch letter (n. 67) Taylor Thompson to Earl of Derby. Tehran, June 5, 1876. [Enclosure n. 4] Translation of Letter from the I'tizad al-Saltana to the Persian Minister for Foreign Affairs. Tehran, May, 1876.

150. The plague killed over twenty percent of the population in the southwestern town of Shushtar in 1876. See Ahmad Seyf, "The Plague of 1877 and the Economy of Gilan," *Iran* 27 (1989): 81; FO 60/382 Dispatch letter (n. 63) Taylor Thompson to Earl of Derby. Tehran, May 23, 1876. [Enclosure n. 1] Copy of Memorandum from Sir Joseph Dickson to Taylor Thomson. Tehran, May 21, 1876.

151. FO 881/3186 Confidential Letter (n. 2) from Taylor Thompson to Earl of Derby. Tehran, February 21, 1877. [Enclosure n. 2] Copy of Memorandum from Sir Joseph Dickson to Taylor Thomson. Tehran, January 14, 1877.

152. FO 881/3332 Confidential Dispatch (n. 1) from Count Beust to Earl of Derby. Belgrave Square, June 27, 1877.

153. Edward G. Browne, *A Year amongst the Persians: Impressions to the Life, Character, and Thought of the People of Persia Received during Twelve Months' Residence in That Country in the Year 1887–1888* (Cambridge: Cambridge University Press, 1927), 107.

154. Browne, *A Year amongst the Persians*, 107; FO 881/3186 Confidential Letter (n. 2) from Taylor Thompson to Earl of Derby. Tehran, February 21, 1877. [Enclosure n. 2] Copy of Memorandum from Sir Joseph Dickson to Taylor Thomson. Tehran, January 14, 1877.

155. Browne, *A Year amongst the Persians*, 108.

156. 'Abd al-Rasul Husayni Tabib, *Mémoire sur l'épidémie de choléra qui éclata à Téhéran en 1892* [An account of the cholera epidemic that broke out in Tehran in 1892] (handwritten Persian manuscript). Bibliothèque Nationale de France, Richelieu, Paris, Département des Manuscrits, cote: Supplément Persan 1290, 15–16.

Chapter 2 · The 1889–1893 Cholera Epidemics

1. Shabaz Shahnavaz, "Karun River iii. The Opening of the Karun," *Encyclopedia Iranica* 15, fasc. 6 (2011): 633–640, updated online version accessed January 6, 2017, http://www.iranicaonline.org/articles/karun_3.

2. Shabaz Shahnavaz, *Britain and South-West Persia, 1880–1914: A Study in Imperialism and Economic Dependence* (New York: Routledge Curzon, 2005), 162–173; Charles Issawi, ed., *The Economic History of Iran, 1800–1940* (Chicago: University of Chicago Press, 1971), 166–177;

George Nathaniel Curzon, "The Karun River and the Commercial Geography of Southwest Persia," *Proceedings of the Royal Geographical Society* 9 (1890): 527.

3. Issawi, *The Economic History*, 161–162.

4. Alexander Morrison, *Russian Rule in Samarkand, 1868–1910: A Comparison with British India* (Oxford: Oxford University Press, 2008), 63–65.

5. E. D. Dickson, "The Outbreak of Cholera in Mesopotamia and Syria in 1889, 1890, and 1891," *Transactions of the Epidemiological Society of London* 13 (1893–1894): 152.

6. Dickson, "The Outbreak of Cholera," 150; John Gordon Lorimer, *Gazetteer of the Persian Gulf, Oman and Central Arabia* (Calcutta: India Superintendent Government Printing, 1915), 1:2524.

7. National Archives, Washington, DC (NA), Diplomatic Series (Dip. Ser.), State Department M223, enclosure 1, no. 390, W. W. Torrence, MD, to E. Spencer Pratt, United States Envoy to Persia (Torrence to Pratt), Tehran, September 4,1889.

8. Dickson, "The Outbreak of Cholera," 144–145.

9. NA, Dip. Ser., State Department M223, enclosure 1, no. 390, Torrence to Pratt, September 4, 1889.

10. Dickson, "The Outbreak of Cholera," 151; NA, Dip. Ser., State Department M223, enclosure 1, no. 390, Torrence to Pratt, September 4, 1889.

11. "Aid for the Tehran Hospital: Woman's Presbyterian Board of Missions of the Northwest Decides to Give help," *Chicago Daily Tribune*, August 30, 1892; NA, Dip. Ser., State Department M223, no. 383, E. Spencer Pratt, United States Envoy to Persia, to James G. Blainie, Secretary of State, Tehran, August 7, 1889 (Pratt to Blainie); FO 60/505 Telegram (n. 130) from Mr. R. T. Kennedy to the Marquis of Salisbury. Tehran, August 26, 1889; FO 60/505 Telegram (n. 131) from Mr. R. T. Kennedy to the Marquis of Salisbury. Tehran, September 1, 1889.

12. NA, Dip. Ser., State Department M223, no. 389, Pratt to Blainie, August 31, 1889.

13. Dickson, "The Outbreak of Cholera," 146.

14. Dickson, "The Outbreak of Cholera," 150.

15. Lorimer, *Gazetteer of the Persian Gulf*, 1:2524; Dickson, "The Outbreak of Cholera," 146, 151; and NA, Dip. Ser., State Department M223, enclosure 1, no. 390, Torrence to Pratt, September 4, 1889.

16. The meeting was attended by Dr. Albu, Dr. Camposampiero of the Ottoman Empire, Russia's Dr. Danilov, and Dr. T. F. Odling, who was the acting medical superintendent of the Indian Government Telegraph Department in Iran.

17. FO 60/505 Dispatch letter (n. 165) Mr. R. T. Kennedy to the Marquis of Salisbury. Gulahek, September 5, 1889. [Enclosure n. 2] Dr. Casson, physician to H.M.'s Legation in Persia, to R. T. Kennedy. Tehran, August 27, 1889.

18. The twenty-two-year-old Isma'il Amin al-Mulk (1284–1316 AH [1867–1898]) was the brother of premier, 'Ali-Asghar Khan Amin al-Sultan. When Amin al-Sultan became prime minister in 1306 [1888–1889], Amin al-Mulk became the head of the Treasury and chief deputy to his brother at the Ministry of Finance and royal court. Amin al-Mulk served as regent while Nasir al-Din Shah and his brother were out of the country on a European tour. See Abbas Amanat, "Amin al-Molk, Mirza Esmail," *Encyclopedia Iranica* 1, fasc. 9 (1989): 948–949, updated online version accessed March 6, 2015, http://www.iranicaonline.org/articles/amin-al-molk -mirza-esmail-1284-1316-1867-98-a-high-ranking-official-towards-the-end-of-naser-al-din -shahs-reign.

19. FO 60/505 Dispatch letter (n. 165) Mr. R. T. Kennedy to the Marquis of Salisbury. Gulahek, September 5, 1889. [Enclosure n. 1] Dr. Casson to R. T. Kennedy "Proceedings Sanitary Council Meeting," Tehran, September 4, 1889. Ja'far Quli Khan Nayyir al-Mulk (known as Lala Bashi) was born in 1247 AH. [1831]. His father was the principal of the Polytechnic College and his brother was Mukhbir al-Dawla, the minister of education, who had succeeded

I'tizad al-Saltana. Nayyir al-Mulk enrolled at the Polytechnic College completing his education in 1276 A H [1859] with a degree in engineering. Like his father, Nayyir al-Mulk became the principal of his alma mater in 1277 A H [1860], a position that he held until 1312A H [1894], and by 1313 A H [1895] he succeeded his brother to the post of minister of education. He served his ministerial post for nine years and died at the age of eighty-four in Tehran in 1293 A S [1915]. See Mahdi Bamdad, *Sharh-i hal-i rijal-i Iran dar qarn-i 12, 13, 14 hijri* [The biography of Iranian personalities during the 12, 13, 14 A H centuries] vol. 5, 4th ed. (Tehran: Intisharat-i zavar 1371 A S [1992]), 50–51.

20. NA, Dip. Ser., State Department M223, no. 390, Torrence to Pratt, September 4, 1889; FO 60/505 Dispatch letter (n. 165) Mr. R. T. Kennedy to the Marquis of Salisbury. Gulahek, September 5, 1889. [Enclosure n. 1] Dr. Casson to R. T. Kennedy "Proceedings Sanitary Council Meeting," Tehran, September 4, 1889.

21. FO 60/505 Copy Dispatch Commercial letter (n. 9) Mr. R. T. Kennedy to the Marquis of Salisbury. Gulahek, September 28, 1889.

22. FO 60/505 Copy Dispatch Commercial letter (n. 9) Mr. R. T. Kennedy to the Marquis of Salisbury. Gulahek, September 28, 1889. [Enclosure n. 3] Copy of dispatch letter (n. 71) Colonel Ross to R. T. Kennedy Esq. Bushire, September 13, 1889.

23. FO 60/505 Copy Dispatch Commercial letter (n. 10) Mr. R. T. Kennedy to the Marquis of Salisbury. Gulahek, October 8, 1889. [Enclosure n. 2] Copy of Report by Tom Odling to Mr. R. T. Kennedy Esq. Tehran, October 5, 1889.

24. Joseph Albu was a physician of German-Jewish origin who came to Iran from Berlin with 'Ali Quli Khan Mukhbir al-Dawla in 1882 as an instructor of medicine at the Polytechnic College and remained in that position for the better part of the decade. See Edward G. Browne, *The Press and Poetry of Modern Persia: Partly Based on the Manuscript Work of Mirza Muhammad 'Ali Khan "Tarbiyat" of Tabriz* (Cambridge: Cambridge University Press, 1914), 154–155; Cyril Elgood, *A Medical History of Persia and the Eastern Caliphate from the Earliest Times until the Year A.D. 1932* (Cambridge: Cambridge University Press, 1951), 502, 512; FO 60/505 Copy Dispatch Commercial letter (n. 10) Mr. R. T. Kennedy to the Marquis of Salisbury. Gulahek, October 8, 1889. [Enclosure n. 1] Copy of Report of Tom Odling to Mr. R.T. Kennedy Esq. Tehran, October 1.

25. FO 60/505 Copy Dispatch Commercial letter (n. 10) Mr. R. T. Kennedy to the Marquis of Salisbury. Gulahek, October 8, 1889. [Enclosure n. 2] Copy of Report by Tom Odling to Mr. R. T. Kennedy Esq. Tehran, October 5.

26. Roger T. Olsen, "Persian Gulf Trade and the Agricultural Economy of Southern Iran in the Nineteenth Century," in *Modern Iran: The Dialectics of Continuity and Change*, ed. Michael E. Bonine and Nikki R. Keddie (Albany: State University of New York Press, 1981), 176–177.

27. Shaul Bakhash, *Iran: Monarch, Bureaucracy and Reform under the Qajars, 1858–1896* (London: Ithaca Press, 1978), 264–265.

28. Mirza Mohammad Hosayn Farahani, *A Shi'ite Pilgrimage to Mecca, 1885–1886: The Safarnameh of Mirza Mohammad Husayn Farahani*, ed. and trans. Hafez Farmayan and Elton L. Daniel (Austin: University of Texas Press, 1990), 289–290.

29. FO 60/505 Copy Dispatch Commercial letter (n. 10) Mr. R. T. Kennedy to the Marquis of Salisbury. Gulahek, October 8, 1889. [Enclosure n. 2] Copy of Report of Tom Odling to Mr. R. T. Kennedy Esq. Tehran, October 5.

30. FO 60/505 Copy Dispatch Commercial letter (n. 13) Mr. R. T. Kennedy to the Marquis of Salisbury. Gulahek, October 22, 1889. [Enclosure n. 1] Copy of Report of Dr. Casson to Mr. R. T. Kennedy Esq. Tehran, October 19.

31. Dickson, "The Outbreak of Cholera," 146.

32. NA, Dip. Ser., State Department M223, no. 394, Torrence to Pratt, September 16, 1889.

33. NA, Dip. Ser., State Department M223, no. 397, Torrence to Pratt, September 17, 1889.

34. NA, Dip. Ser., State Department M223, no. 400, Torrence to Pratt, September 17, 1889.

35. The National Archives of the Presbyterian Church in the USA, Presbyterian Historical Society, Philadelphia (PHS), Board of Foreign Missions Correspondence and Reports (Foreign Missions), 1833–1911, Persia, Letters 1889–1890, vol. 6, no. 45, Dr. John Gillespie to Rev. Wear (Gillespie to Wear), Hamadan, Persia, November 26, 1889.

36. PHS, Foreign Missions, vol. 6, no. 45, Gillespie to Wear, November 26, 1889; NA, Dip. Ser., State Department M223, no. 452, E. W. Alexander to E. Spencer Pratt (Alexander to Pratt), United States Envoy to Persia, Hamadan, April 19, 1890.

37. NA, Dip. Ser., State Department M223, no. 452, Alexander to Pratt, April 19, 1890.

38. In Hamadan, about a hundred Jewish students were enrolled at the American Presbyterian Mission School; see George Nathaniel Curzon, *Persia and The Persian Question* (New York: Longmans, Green, 1892), 1:510.

39. PHS, Foreign Missions, vol. 6, no. 45, Gillespie to Wear, November 26, 1889.

40. NA, Dip. Ser., State Department M223, no. 452, Alexander to Pratt, April 19, 1890.

41. PHS, Foreign Missions, vol. 6, no. 45, Gillespie to Wear, November 26, 1889.

42. NA, Dip. Ser., State Department M223, no. 418, Pratt to Blainie, November 4, 1889.

43. NA, Dip. Ser., State Department M223, no. 408, Pratt to Blainie, October 14, 1889. By 1904, as many as sixty thousand Iranian migrant workers traveled to the Russian-Azerbaijan border for jobs. See Hassan Hakimian. "Wage Labor and Migration: Persian Workers in Southern Russia, 1880–1914," *International Journal of Middle East Studies* 17 (1985): 443–462.

44. Jean-Baptiste Feuvrier (1842–1926) was a French military physician and a veteran of the Franco-Prussian War. He replaced Tholozan as the royal physician when the latter fell ill in Paris during the shah's visit there in August 1889. Feuvrier remained in Nasir al-Din Shah's service through 1892. See Jean Calmard, "Feuvrier, Jean-Baptiste," *Encyclopedia Iranica* 9, fasc. 6 (1999): 569–571, updated online version accessed November 2, 2015, http://www.iranicaonline.org/articles/feuvrier.

45. "Iran," *Akhtar*, no. 23, 21 Jamadi al-Akhar 1307 AH [February 3, 1890], 188; *Times* (London), April 19, 1890, 7.

46. Andreas David Mordtmann, "Statistique de L'Épidémie Cholérique en Perse en 1889 et 1892," *Gazette Médicale D'Orient* 35 (1892–1893): 297; Adrien Achille Proust, "Le choléra de Mésopotamie, de Perse et de Syrie, en 1889 et 1890," *Bulletin de l'Académie de Médecine* 26 (1891): 150.

47. "Iran," *Akhtar*, no. 7, 18 Safar 1307 AH [October 14, 1889].

48. "Iran," *Akhtar*, no. 7, 18 Safar 1307 AH [October 14, 1889].

49. *Akhtar* was a weekly Persian newspaper published in Istanbul beginning on January 13, 1876, until 1895–1896. Covering primarily politics and social conditions, the paper counted among its famed contributors Mirza Aqa Khan Kirmani, Shaykh Ahmad Ruhi, and Mirza Mahdi Tabrizi. The famed British orientalist E. G. Browne, writing in 1888, described the paper as "the only newspaper worth reading," and although the paper was moderate in tone, its criticisms was a source of annoyance for the Iranian government, which attacked the paper via state-sponsored newspapers. See L. P. Elwell-Sutton, "Aktar newspaper," *Encyclopedia Iranica* 1, fasc. 7 (1984): 730, updated online version accessed January 2, 2015, http://www.iranicaonline.org/articles/aktar-2-persian-newspaper; Browne, *Press and Poetry*, 17–18, 36–37; H. L. Rabino (di Borgomale), "La presse persane depuis ses origines jusqu'à nos jours," *Revue du monde musulman* 22 (1913): 287–315; Muhammad Sadr Hashemi, *Tarikh-i jarayid va majallat-i Iran* [The history of newspapers and magazines in Iran] (Isfahan: Intisharat-i kamal, 1332 AS [1953–1954]), 1:63–65.

50. "Iran," *Akhtar*, no. 9, 3 Rabi' al-Awwal 1307 AH [October 28, 1889].

51. "Iran," *Akhtar*, no. 9, 3 Rabi' al-Awwal 1307 AH [October 28, 1889].

52. Richard J. Evans, "Epidemics and Revolutions: Cholera in Nineteenth Century Europe," *Past and Present* 120 (1988): 123–146.

53. In contrast to Iran's experience with Shi'ite Muslim clerical militancy, most of the upheavals in response to cholera outbreaks in the West were secular in nature. The church had a stabilizing influence on Western society by discouraging rebellious tendencies and supporting restrictive sanitary measures. See Frank M. Snowden, "Cholera in Barletta, 1910," *Past and Present* 132 (1991): 98–99.

54. The succession to the Prophet Muhammad is central to the schism separating the Shi'ite branch of Islam from the Sunni majority. Shi'ites believe that had the prophet been alive to select his successor, his cousin and son-in-law 'Ali b. Abu Talib would have occupied the temporal and spiritual leadership of the faithful. This position would subsequently be passed on to his descendants. See Moojan Momen, *An Introduction to Shi'i Islam: The History and Doctrines of Twelver Shi'ism* (New Haven: Yale University Press, 1985).

55. Abbas Amanat, "Between the *Madrasa* and the Marketplace: The Designation of Clerical Leadership in Modern Shi'ism," in *Authority and Political Culture in Shi'ism*, ed. Said Amir Arjomand (Albany: State University of New York Press, 1988), 98–132.

56. Ahmad Ashraf and H. Hekmat, "Merchants and Artisans and the Development Processes of Nineteenth-Century Iran," in *The Islamic Middle East, 700–1900*, ed. A. L. Udovitch (Princeton: Darwin Press, 1981), 731–733.

57. Mansoureh Ettehadieh [Nezam Mafi] and William Floor, "Concessions," *Encyclopedia Iranica* 6, fasc. 2 (1992): 119–122, updated online version accessed November 2, 2015, http://www .iranicaonline.org/articles/concessions.

58. William Floor, "The Economic Role of the Ulama in Qajar Persia" in *The Most Learned of the Shi'a*, ed. Linda S. Walbridge (Oxford: Oxford University Press, 2001), 71.

59. Shaykh Muhammad Taqi Aqa Najafi (1846–1914) was a leading cleric and nominal rival of Zill al-Sultan, eldest son of Nasir al-Din Shah and the prince governor of Isfahan. The violence that he instigated against "apostates" was a recurrent problem for the Iranian government, causing him to be summoned to Tehran for reprimand in 1889 and 1903. See Muhammad Hasan Khan I'timad al-Saltana, *Chehel sal tarikh-i Iran dar durih-yi padishahi-i Nasir al-Din Shah (al-ma'sir va al-athar)* [Forty year history of Iran during Nasir al-Din Shah's Reign], ed. Iraj Afshar (Tehran: Intishart-i asatir, 1368 AS [1989–1990]), 2:773; Ibrahim Safa'i, *Rahbaran-i mashruta* [The leaders of the constitutional revolution], 2nd ed. (Tehran: Intishart-i javidan, 1364 AS [1985]), 226; Abbas Amanat, *Resurrection and Renewal: The Making of the Babi Movement in Iran, 1844–1850* (Ithaca: Cornell University Press, 1989), 353.

60. Amanat, *Resurrection*, 353.

61. Gertrude Bell to Frank Bell, September 12, 1892, in *The Earlier Letters of Gertrude Bell*, ed. Elsa Richmond (London: Ernest Benn, 1937), 336–337.

62. Curzon, *Persia and The Persian Question*, 1:511.

63. Nikki R. Keddie, *Religion and Rebellion in Iran: The Tobacco Protest of 1891–1892* (London: Frank Cass, 1966), 44.

64. *Times* (London), February 1, 1890, 5.

65. Keddie, *Religion and Rebellion*, 65–66.

66. Keddie, *Religion and Rebellion*, 44.

67. Zayn al-'Abidin Maragha'i, *Siyahatnama-yi Ibrahim Big* [The travels of Ibrahim Beg] (Tehran: Kitabha-yi sadaf, 1344 AS [1965]), 178.

68. "Nakhoshi-yi anfluanza," *Akhtar*, no. 22, 5 Jamadi al-Akhar 1307 AH [January 27, 1890], 174.

69. Mateo Mohammad Farzaneh, *The Iranian Constitutional Revolution and the Clerical Leadership of Khurasani* (Syracuse: Syracuse University Press, 2015), 54.

70. Joseph Désiré Tholozan, "La grippe en Perse en 1889–1890," *Bulletin de l'Académie de Médecine*, 26 (1891): 251–262.

71. "Akhtar," *Akhtar*, no. 62, 4 Rajab 1307 AH [February 24, 1890], 210.

72. "Maktub az Tehran," *Akhtar*, no. 27, 11 Rajab 1307 AH [March 3, 1890], 221–222; Tholozan, "La grippe," 251; Abdollah Mostofi, *The Administrative and Social History of the Qajar Period* (Costa Mesa: Mazda, 1997), 1:271.

73. *Akhtar*, no. 27.

74. "Nakhoshi-yi anfluanza," *Akhtar*, no. 24, 19 Jamadi al-Akhar 1307 AH [February 10, 1890], 198–199.

75. *Akhtar*, no. 62, 210; "Tasfiya-yi hava," *Ittila'*, no. 354, Saneh 1312 AH [September 11, 1894].

76. FO 60/510 Dispatch (n. 36) from Sir H. Drummond Wolff to the Marquis of Salisbury. Tehran, February 5, 1890.

77. Even when Pasteur's revolutionary treatment for rabies was introduced to the readers of *Akhtar* in November 1889, very little was said of the microbial etiology of rabies or of Pasteur's germ theory of disease in general. Pasteur's monetary reward was of greater concern than the science that led to the cure. See "Maratib-i qadr shinasi-i Ingilisan darbaray-i danishmandan va sahiban-i 'ilm az har tabaqa" [The stages of gratitude of the English in regards to scholars and scientists of every level], *Akhtar*, no. 12, 24 Rabi' al-Awwal 1307 AH [November 18, 1889], 103.

78. "Hava va vaba," [Air and cholera] *Akhtar*, no. 15, 15 Rabi' al-Akhar 1307 AH [December 9, 1889], 123.

79. "Jinn va vaba," *Ittila'*, no. 354, Saneh 1312 AH [September 11, 1894]. *Ittila'* was the semiofficial organ of Iranian government founded in 1881, under guidance of the Ministry of Information. Civil employees, landlords, and members of the government were obliged to buy it, and the subscription fee was deducted from their salary. It focused on foreign affairs, with articles on politics in Europe and the United States, which included aspects of Iran's foreign relations. The newspaper was published until 1909. See Sayyid Farid Qasemi, *Sarguzasht-i matbu'at-i Iran* [The history of the press in Iran] (Tehran: Sazman-i Chap va Intisharat-i Vizarat-i Farhang va Irshad-i Islami, 1380 AS [2001–2002]), 2:1639–1703.

80. *New York Times*, July 22, 1891, 3.

81. See both the *Times* of London and the *New York Times* throughout 1890 and 1891; *New York Times*, July 29, 1891, 4.

82. *New York Times*, July 19, 1891, 1.

83. Dr. Camposampiero, "On The Recent Outbreak of Cholera in Persia," *Transactions of the Epidemiological Society of London* 13 (1893–1894): 154.

84. Camposampiero, "On The Recent Outbreak of Cholera in Persia," 155.

85. Joannes Feuvrier, *Trois Ans à la Cour de Perse* [Three years at the Persian court] (Paris: F. Juven, 1894), 356; Camposampiero, "On The Recent Outbreak of Cholera in Persia," 154–157.

86. 'Ali Asghar Khan Amin al-Sultan Atabak-i 'Azam (1274–1325 AH /1858–1907), Nasir al-Din Shah's last prime minister, was instrumental in securing concessions for British subjects in Iran, including the controversial tobacco and banking monopolies. Following the shah's assassination, he continued to be a powerful figure in Iranian politics and regained the premiership after the Constitutional Revolution. He was assassinated in front of the Iranian Parliament on August 31, 1907.

87. Camposampiero, "On The Recent Outbreak of Cholera in Persia," 155.

88. Gertrude Bell, *Persian Pictures* (1894; New York: Boni and Liveright, 1926), 59.

89. C. E. Yate, *Khurasan and Sistan* (Edinburgh: William Blackwood and Sons, 1900), 348.

90. Curzon, *Persia and The Persian Question*, 1:152.

91. Curzon, *Persia and The Persian Question*, 1:153. The necropolis next to Fatima's shrine in the city of Qom rivaled that of Mashhad in the size and the number of graves.

92. The epidemic killed five thousand in Mashhad by June 20, 1892. See FO 60/533 Paraphrase of Telegram (n. 77) from Sir F. Lascelles to the Marquis of Salisbury. Tehran, May 27, 1892; Bell, *Persian Pictures*, 59; Camposampiero, "On The Recent Outbreak of Cholera in Persia," 155.

93. FO 248/545 Telegram (not numbered) from Ney Elias, Officer at Mashhad, to Col. Wells, Physician Indo-European Telegraph Department. Mashhad, May 29, 1892.

94. Camposampiero, "On The Recent Outbreak of Cholera in Persia," 155–156.

95. Ashgabat is located some 250 miles to the west of Merv. The first deaths due to cholera in Turkmenbashi (Uzun-Ada) occurred on May 20 1892. For more on the 1892 outbreak in Russia's Central Asian territories, see Frank G. Clemow, *Cholera epidemic of 1892 in the Russian Empire: with notes upon treatment and methods of disinfection in cholera, and a short account of the conference on cholera held in St. Petersburg in December 1892* (London: Longmans, Green, 1893), 1–9.

96. Clemow, *Cholera epidemic of 1892*, 25.

97. J. H. Shedd, "Persia: The Cholera. Persecution. Mar Shimon," *The Independent. Devoted to the consideration of politics, of social and economic tendencies, of history, literature, and the arts* 44, no. 2294 (November 17, 1892): 17; Camposampiero, "On The Recent Outbreak of Cholera in Persia," 158–159.

98. Thomas Edward Gordon, *Persia Revisited* (London: E. Arnold, 1896), 8–10; Camposampiero, "On The Recent Outbreak of Cholera in Persia," 158.

99. FO 60/546 Consular Sanitary Dispatch (n. 1) from Harry L. Churchill, H.M.'s Consul at Tabriz, to Earl of Rosebery. Resht, January 22, 1893.

100. Camposampiero, "On The Recent Outbreak of Cholera in Persia," 159.

101. FO 248/547 Dispatch (n. 32) from Mr. Robert M. Paton, Acting British Consul General at Tabriz, to Sir F. Lascelles. Tabriz, July 13, 1892.

102. FO 248/547 Dispatch (n. 34) from Mr. Robert M. Paton, Acting British Consul General at Tabriz, to Sir F. Lascelles. Astari, August 5, 1892.

103. The cholera outbreak in Tabriz officially began on July 26; see FO 248/547 Dispatch (n. 34) from Paton to Lascelles. Astari, August 5, 1892.

104. Shedd, "Persia," 17; Henry Heylyn Hayter, *Victorian Yearbook, 1892* (Melbourne: Printed for the Government Printer by Sands & McDougall Limited, 1893), 1:134.

105. FO 248/547 Dispatch (n. 38) from Mr. Robert M. Paton, Acting British Consul General at Tabriz, to Sir Frank Cavendish Lascelles. Tabriz, October 7, 1892.

106. "Saved by an American Woman; Many Persons at Tauris Owe Their Lives to Miss Bradford; A Missionary," *Chicago Daily Tribune*, September 24, 1892.

107. FO 248/547 Dispatch (n. 38) from Mr. Robert M. Paton, Acting British Consul General at Tabriz, to Sir Frank Cavendish Lascelles. Tabriz, October 7, 1892.

108. FO 60/534 Letter (n. 107) Sir F. Lascelles to the Marquis of Salisbury. Gulhek, June 10, 1892.

109. Camposampiero, "On The Recent Outbreak of Cholera in Persia," 156.

110. FO 60/534 Letter (n. 107) Sir F. Lascelles to the Marquis of Salisbury. Gulhek, June 10, 1892. [Enclosure n. 1] Copy of Telegram from Sir F. Lascelles to Amin-es-Sultan. Tehran, June 1, 1892; G. Bell, *Persian Pictures*, 59.

111. Curzon, *Persia and The Persian Question*, 1:306.

112. Joseph Désiré Tholozan, *Prophylaxie du choléra en Orient: l'hygiène et la réforme sanitaire en Perse* (Paris: Victor Masson et Fils, 1869), 23.

113. Arthur S. Hardy, "Persia. Sanitary Report from Tehran," *Public Health Reports* 14, no. 10 (March 10, 1899): 330.

114. "Vaqaya-yi mukhtalif-i mahalat-i dar al-khalafih, mahala-yi Dawla" [Police reports of Tehran, Dawla District], 4 Zu'l Qa'deh 1303 AH [August 4, 1886], in *Guzarishha-yi nazmiya az*

mahalat-i Tehran: rapurt-i vaqa'i mukhtalif-i mahalat-i dar al-khalafih (1303 A.H.–1305 A.H.) [Police reports on the various happenings in the various districts of Tehran, 1886–1888], ed. Ensiya Shaykh Rezaei and Shahla Azari (Tehran: Intisharat-i sazman-i asnad-i milli-yi Iran, 1377 AS [1999]), 1:13.

115. G. Bell, *Persian Pictures*, 60.

116. Camposampiero, "On The Recent Outbreak of Cholera in Persia," 156.

117. Camposampiero probably made an error in his report by referring to the Sarcheshmeh quarter as "Serghendesh." Parts of Nasiriya Road survive today as Naser Khosrow Street in the Pamenar district of Tehran. See Camposampiero, "On The Recent Outbreak of Cholera in Persia," 156, and Mirza Muhammad Tabib (Mushir al-Attiba), *Risala-yi vaba'ih* [Treatise on cholera] MS: Tehran, 1310 [1892], cited in Homa Nategh. *Musibat-i vaba va bala-yi hukumat* [The calamity of cholera and the pain of government] (Tehran: Nashr-i gustarish, 1358 AS [1979]), 19.

118. Sirus Sadvandian and Mansoureh Ettehadieh, eds., *Amar-i dar al-khalafih-yi Tehran: asnadi az tarikh-i ijtimai-yi Tehran dar asr-i Qajar* [Statistics of Tehran: Manuscript pertaining to the social history of Tehran in the Qajar era] (Tehran: Nashr-i tarikh-i Iran, 1368 AS [1989]), 366.

119. Navid Jamali and Mina Fatemi "A Visit to Oudlajan, Unremembered Remnant of Old Tehran," *Tavoos Art Magazine*, accessed June 20, 2017, http://www.tavoosonline.com/Articles /ArticleDetailEn.aspx?src=214&Page=1.

120. Aqa Khan Nuri, Nasir al-Din Shah's premier after Amir Kabir, built the Sadr Abad and Kazim qanats that supplied drinking water to the Udlajan and Bazaar quarters of Tehran in the 1850s. See Nabuaki Kondo, *Islamic Law and Society in Iran: A Social History of Qajar Tehran* (New York: Routledge, 2017), 107.

121. John Gilmour, *Report on an Investigation into the Sanitary Conditions in Persia* (Geneva: League of Nations, 1925), 50.

122. Feuvrier, *Trois Ans*, 203.

123. Samuel Graham Wilson, *Persian Life and Customs: with scenes and incidents of residence and travel in the land of the lion and the sun* (New York: F. H. Revell, 1899), 70.

124. Rezaei and Azari, *Guzarishha-yi nazmiya*, 1:13.

125. Jakob Eduard Polak, *Safarnama-yi Polak: Iran va Iranian*, trans. KeyKavoos Jahandari (Tehran: Khawrazmi, 1361 AS [1982]), 50–51.

126. John G. Wishard, *Twenty Years in Persia: A Narrative of Life under the Last Three Shahs* (New York: Flemming H. Revell, 1908), 219.

127. Joseph Désiré Tholozan, *Rapport à Sa Majesté le Chah sur l'état actuel de l'hygiène en Perse; progrès à réaliser; moyens de les effectuer; résultats obtenus depuis un an* [Report to His Majesty the Shah on the current sanitary state of Persia, progress to be realized, modes of bringing about progress, and results obtained since a year ago], 1869, Bibliothèque de l'Académie National de Médicine, Paris, Archives et manuscrits de la bibliothèque, cote 50.426, 49–50.

128. G. Bell, *Persian Pictures*, 67–68; Gilmour, *Report on an Investigation*, 51.

129. Muhammad Isma'il al-Bukhari, *Jami' Hadith* 23: 9, cited in Muhammad Ali Maulana, *A Manual of Hadith* (Lahore: Ahmadiyya Anjuman, n.d.), 191.

130. FO 248/545 Telegram (n. 27) from Ney Elias, Officer at Mashhad, to Sir F. Lascelles. Mashhad, June 15, 1892.

131. G. Bell, *Persian Pictures*, 61.

132. FO 60/532 Letter (n. 139) Sir F. Lascelles to the Earl of Rosebery. Gulahek, September 3, 1892.

133. Mostofi, *The Administrative*, 1:287.

134. Gertrude Bell to Frank Bell, August 29, 1892, in Richmond, *The Earlier Letters*, 330–331.

135. 'Abd al-Rasul Husayni Tabib, *Mémoire sur l'épidémie de choléra qui éclata à Téhéran en 1892* [An account of the cholera epidemic that broke out in Tehran in 1892] (handwritten Persian manuscript). Bibliothèque Nationale de France, *Richelieu, Paris*, Département des Manuscrits, cote: Supplément Persan 1290, 4–5.

136. Tabib, *Mémoire sur l'épidémie*, 4–5; Mostofi, *The Administrative*, 1:269–270.

137. "Horrors of the Plague: A Letter Giving a Graphic Picture of the Cholera Near Tehran, Persia," *Washington Post*, October 17, 1892, 4.

138. Tabib, *Mémoire sur l'épidémie*, 4–5; Feuvrier, *Trois Ans*, 411.

139. Archives du Ministère des Affaires étrangères, Paris Quai d'Orsay (AMAE), Correspondance Politique, 1871–1896 (CP), Perse, vol. 44. Dépêche (direction politique n. 51), extrait n. 1, Mr. de Balloy, ministre de France à Téhéran à Son Excellence le ministre des affaires étrangères à Paris. Téhéran, 27 Août, 1892.

140. Hajji Muhammad Hasan Amin al-Zarb to Sayyid 'Abd al-Rahim Amin al-Tujar, Tehran to Kerman, 6 Safar 1310, *Dafatar-i kupi-yi*, cited in Nategh, *Musibat-i vaba*, 19.

141. Curzon, *Persia and The Persian Question*, 1:332; "Horrors of the Plague," 4.

142. Amin al-Zarb to Amin al-Tujar, 19 Muharram 1310, cited in Nategh, *Musibat-i vaba*, 19.

143. "Horrors of the Plague," 4.

144. AMAE, Correspondance Politique, 1871–1896 (CP), Perse, vol. 44. Dépêche (direction politique n. 51), extrait n. 1, Mr. de Balloy, ministre de France à Téhéran à Son Excellence le ministre des affaires étrangères à Paris. Téhéran, 27 Août, 1892.

145. AMAE, Correspondance Politique, 1871–1896 (CP), Perse, vol. 44. Dépêche (direction politique n. 54), extrait n. 1, Mr. de Balloy, ministre de France à Téhéran à Son Excellence le ministre des affaires étrangères à Paris. Téhéran, 17 Septembre, 1892; Feuvrier, *Trois Ans*, 413; Muhammad Hasan Khan I'timad al-Saltana, *Ruznama-yi khatirat-i I'timad al-Saltana* [The memoirs of I'timad al-Saltana], ed. Iraj Afshar (Tehran: Amir Kabir, 1345 AS [1976]), 830.

146. Feuvrier's house was uninhabitable, forcing the doctor to live with the French legation for the rest of his stay in Iran. This unpleasant experience was the "last straw" that convinced him at the end of the month to return to France. See Feuvrier, *Trois Ans*, 411–412.

147. Sadid al-Saltana Kababi, *Khatirat-i sal-i 1310* [Memoirs of the year 1892] (lithographed manuscript, Tehran University), cited in Nategh, *Musibat-i vaba*, 19.

148. Nategh, *Musibat-i vaba*, 19.

149. Gerald L. Mandell, *Principles and Practice of Infectious Diseases*, 5th ed. (London: Churchill Livingstone, 2000), 2266–2271.

150. Frank M. Snowden, *Naples in the Time of Cholera, 1884–1911* (Cambridge: Cambridge University Press, 1995), 149–154; Peter Baldwin, *Contagion and the State in Europe, 1830–1831* (Cambridge: Cambridge University Press, 1999), 63–65.

151. Wishard, *Twenty Years*, 220.

152. Feuvrier, *Trois Ans*, 356, 357.

153. Feuvrier, *Trois Ans*, 400–406.

154. Feuvrier, *Trois Ans*, 407; G. Bell, *Persian Pictures*, 65.

155. Feuvrier, *Trois Ans*, 407.

156. By August, the epidemic claimed more than 5,000 souls in Tehran. It also killed 1,500 in nearby Shah Abdol Azim, and 1,200 in the highland villages of Shemiran; see AMAE, Correspondance Politique, 1871–1896 (CP), Perse, vol. 44. Annexe n. 2: Irfan Bey, Le Chargé d'Affaire de Turquie à Mr. De Balloy. Téhéran, 31 Août 1892; Camposampiero, "On The Recent Outbreak of Cholera in Persia," 158.

157. Cholera in Tehran killed 10,000 out of a total population of approximately 120,000 by November 23 when the city was officially declared disease free. See Feuvrier, *Trois Ans*, 411; Habibola Zandjani, "Téhéran et sa population: Deux siècles d'histoire," in *Téhéran: Capitale*

bicentenaire, ed. Chahryar Adle and Bernard Hourcade (Paris: Institut Français de Recherche en Iran, 1992), 252–253.

158. Camposampiero, "On The Recent Outbreak of Cholera in Persia," 157.

159. FO 248/548 Dispatch (n. 54) from J. R. Preece, Consul at Isphahan, to Sir Frank Lascelles. Isfahan, August 24, 1892. [Enclosure n. 1] Report received from Yezd, August 14, 1892.

160. FO 248/548 Dispatch (n. 54) Preece to Lascelles. Isfahan, August 24, 1892.

161. Camposampiero, "On The Recent Outbreak of Cholera in Persia," 158.

162. FO 248/547 Paton to Lascelles. Tabriz, July 13, 1892; FO 248/548 Dispatch (n. 68) from Mr. S. D. Aganoor, British Agent at Isphahan, to Sir Frank Lascelles, Julfa-Isfahan, November 5, 1892.

163. Camposampiero, "On The Recent Outbreak of Cholera in Persia," 158.

164. Camposampiero, "On The Recent Outbreak of Cholera in Persia," 156.

165. FO 248/548 Preece to Lascelles. Isfahan, August 24, 1892; Mordtmann, "Statistique de L'Épidémie Cholérique," 298–299.

166. FO 60/535 Confidential Dispatch (n. 357) from Mr. Walter Townley to the Foreign Office. Fulbourn Manor, Cambridge, April 22, 1892. [Enclosure n. 1] Report of a Journey from Tehran to Alexandretta, by way of Isfahan, Shiraz, Bushier, Basra, Baghdad, Mosul, Diarbekr, and Aleppo; by Walter Townely.

167. I'timad al-Saltana, *Ruznama-yi Khatirat*, 830.

168. G. Bell, *Persian Pictures*, 65.

169. I'timad al-Saltana, *Ruznama-yi Khatirat*, 830.

170. FO 539/6411, Confidential Memorandum, Mr. Bertie on the Affairs of Persia, August 1892 to October 1893, November 1893; AMAE, Correspondance Politique, 1871–1896 (CP), Perse, vol. 44. Dépeche (direction politique n. 51) Mr. de Balloy, ministre de France à Téhéran à Son Excellence le ministre des affaires étrangères à Paris. Téhéran, 27 Août, 1892.

171. FO 539/6411, Confidential Memorandum, Mr. Bertie on the Affairs of Persia, August 1892 to October 1893, November 1893.

172. FO 60/542 Dispatch (n. 9) from Sir F. Lascelles to the Earl of Rosebery. Tehran, January 16, 1893. [Enclosure 1] Dispatch from J. R Preece to Sir F. Lascelles. Isfahan, January 7, 1893.

173. FO 539/6411, Confidential Memorandum, Mr. Bertie on the Affairs of Persia, August 1892 to October 1893, November 1893.

174. FO 539/6411, Confidential Memorandum; Mr. Bertie on the Affairs of Persia, August 1892 to October 1893, November 1893.

175. Sayyid Jamal al-Din Va'iz Isfahani, *Libas al-tiqwa* [Virtuous garb], ed. Huma Rizvani (Tehran: Nashr-i tarikh-i Iran, 1363 AS [1984]), 23–24; Sayyid Jamal al-Din Va'iz Isfahani and Malik al-Mutakallimin, *Ru'ya-yi sadiqah* [True dream], ed. Sadiq Sajjadi (Tehran: Nashr-i tarikh-i Iran, 1363 AS [1984]), 34–42.

176. G. Bell, *Persian Pictures*, 65.

177. AMAE, Correspondance Politique, 1871–1896 (CP), Perse, vol. 44. Dépeche (direction politique n. 54), extrait n. 1, Mr. de Balloy, ministre de France à Téhéran à Son Excellence le ministre des affaires étrangères à Paris. Téhéran, 17 Septembre, 1892.

178. Gertrude Bell to Frank Bell, September 12, 1892, in Richmond, *The Earlier Letters*, 336–337.

179. AMAE, Correspondance Politique, 1871–1896 (CP), Perse, vol. 44. Dépeche (direction politique n. 54), extrait n. 1, Mr. de Balloy, ministre de France à Téhéran à Son Excellence le ministre des affaires étrangères à Paris. Téhéran, 17 Septembre, 1892.

180. Iran's younger delegate at the conference, Khalil Khan A'lam al-Dawla Saqafi (1862–1943), was born into a family of physicians in Tehran in 1862. His father Hajj Mirza Abd-al Baqi Tabib Hakim Bashi I'tizad al-Attiba taught Galenic medicine at the Polytechnic College in Tehran,

where he graduated with distinction in 1880. Eight years later, he received an advanced diploma in medicine and became an assistant professor at the school's medical faculty. In 1893, Tholozan, who considered Khalil Khan one of his brightest students, convinced Nasir al-Din Shah to allow him to further his medical training in France. He returned in 1898, became one of Muzaffar al-Din Shah's chief physicians, and assumed a number of high administrative and diplomatic positions under the Constitutional Government and the Pahlavi regime. Mirza Mahmoud Khan Saghaphi, *In the Imperial Shadow: Page to the Shah* (New York: Doubleday, Doran, 1928); Mahmud Nadjmabadi, "Duktur Khalil Khan Saqafi (A'lam al-Dawla)" [Doctor Khalil Khan Saqafi (A'lam al-Dawla)], *Jahan-i pizishki* 11, no. 12 (1958): 447–456.

181. Norman Howard-Jones, "The Scientific Background of the International Sanitary Conferences, 1851–1938. 5. The Ninth Conference: Paris, 1894," *WHO Chronicle* 28, no. 10 (1974): 455–458.

182. Ministère des Affaires étrangères, *Conférence Sanitaire Internationale de Paris, Procès-Verbaux [7 Février–3 Avril 1894]* (Paris: Imprimerie National, 1894), 446.

183. Ministère des Affaires étrangères, *Conférence Sanitaire Internationale de Paris, Procès-Verbaux [7 Février–3 Avril 1894]*, 447.

184. Ministère des Affaires étrangères, *Conférence Sanitaire Internationale de Paris, Procès-Verbaux [7 Février–3 Avril 1894]*, 165.

185. Ministère des Affaires étrangères, *Conférence Sanitaire Internationale de Paris, Procès-Verbaux [7 Février–3 Avril 1894]*, 165.

186. Lorimer, *Gazetteer of the Persian Gulf*, 1:2527–2528.

187. FO 60/545 Commercial Dispatch (n. 3a) from Sir Frank Lascelles to the Earl of Rosebery. Tehran, March 26, 1893. [Enclosure n. 1] A. C. Talbot, Consul General at Bushire, to Frank C. Lascelles. Bushire, March 6, 1893.

188. FO 60/534 Dispatch (n. 138) from Sir F. Lascelles to the Earl of Rosebery. Tehran, September 2, 1892. [Enclosure 4] Memorandum from Capt. J. C. Trench, H.B.M.'s Consul at Bussorah, to Major R. H. Jennings, Assistant Political Agent and Consul at Bassorah. Bassorah, June 11, 1891.

189. Howard-Jones, "The Scientific Background. 5," 458; Lorimer, *Gazetteer of the Persian Gulf*, 1:2528.

190. Issawi, *The Economic History*, 250–251.

191. Ministère des Affaires étrangères, *Conférence Sanitaire Internationale de Paris, Procès-Verbaux [7 Février–3 Avril 1894]*, 447.

192. Suzanne Maloney, *Iran's Political Economy since the Revolution* (New York: Cambridge University Press, 2015), 23–24.

193. Edward Granville Browne, *The Persian Revolution of 1905–1909* (Cambridge: Cambridge University Press, 1910), 63–94.

Chapter 3 · Epidemics and Sanitary Imperialism, 1896–1904

1. Suzanne Maloney, *Iran's Political Economy since the Revolution* (New York: Cambridge University Press, 2015), 25.

2. Firuz Kazemzadeh, *Russia and Britain in Persia, 1864–1914* (New Haven: Yale University Press, 1968); Roger Platt Churchill, *The Anglo-Russian Convention of 1907* (Cedar Rapids, IA: Torch Press, 1939); Briton Cooper Busch, *Britain and the Persian Gulf, 1894–1914* (Berkeley: University of California Press, 1967).

3. Myron Echenberg, "Pestis Redux: The Initial Years of the Third Bubonic Plague Pandemic, 1894–1901," *Journal of World History* 13, no. 2 (2002): 431–438.

4. Qahraman Mirza 'Ain al-Saltana, *Ruznama-yi khatirat 'Ain al-Saltana* [The memoirs of 'Ain al-Saltana], ed. Massoud Salur and Iraj Afshar (Tehran: Intisharat-i asatir, 1376 AS

[1997–1998]), 2:1075; Robert Peckham, *Epidemics in Modern Asia* (Cambridge: Cambridge University Press, 2016), 137; Mark Harrison, *Public Health in British India: Anglo-Indian Preventive Medicine, 1859–1914* (Cambridge: Cambridge University Press, 1994), 140–146.

5. FO 60/586 Commercial Dispatch (n. 2) from Sir Mortimer M. Durand, British Minister in Tehran, to Marquis of Salisbury. Tehran, January 18, 1897.

6. IO L/P & S/20/C248B Jerome Antony Saldanha, *Précis on naval arrangements in the Persian Gulf, 1862–1905* (Simla: Government Central Press, 1906); James Onley, *The Arabian Frontier of the British Raj: Merchants, Rulers, and the British in the Nineteenth Century Gulf* (Oxford: Oxford University Press, 2007), 45.

7. John Gordon Lorimer, *Gazetteer of the Persian Gulf, Oman and Central Arabia* (Calcutta: India Superintendent Government Printing, 1915), 1:2547.

8. FO 60/586 Commercial Dispatch (n. 2) from Sir Mortimer M. Durand, British Minister in Tehran, to Marquis of Salisbury. Tehran, January 18, 1897.

9. FO 60/586 Commercial Dispatch (n. 2) from Sir Mortimer M. Durand to Marquis of Salisbury. Tehran, January 18, 1897.

10. FO 60/586 Commercial Dispatch (n. 2) from Sir Mortimer M. Durand to Marquis of Salisbury. Tehran, January 18, 1897.

11. The meeting was attended by an unprecedented number of ranking political notables including Foreign Minister Mirza Yahya Mushir al-Dawla, War Minister 'Abd al-Husayn Mirza Farman Farma, Justice Minister 'Abbas Mirza Mulk Ara, and Interior Minister 'Aliquli Khan Mukhbir al-Dawla; see FO 60/586 Commercial Dispatch (n. 3) from Charles Hardinge, Her Britannic Majesty's Chargé d'Affaires, to Marquis of Salisbury. Tehran, February 18, 1897. [Enclosure n. 1] Tom F. Odling, Physician to Charles Hardinge. Tehran, February 18, 1897.

12. FO 60/586 Commercial Dispatch (n. 3) from Hardinge to Marquis of Salisbury. Tehran, February 18, 1897; and Kazemzadeh, *Russia And Britain*, 304–305.

13. Jean-Etienne Justin Schneider, a French military doctor (Médecin Major de 1ère Classe) of Alsatian origin, came to Iran in 1893 to replace the aging Joseph Désiré Tholozan, who was in the process of retiring. He became the shah's physician-in-chief on March 21, 1894, after serving the monarch's son (and minister of war), Na'ib al-Saltana, for more than a year. When Muzaffar al-Din Shah ascended the throne, Schneider and Tholozan, both named chief physician to the new shah, joined Hugh Adcock and Mirza Mahmud Khan Hakim al-Mulk in this position. Schneider also served the French legation in Tehran as its official physician. AMAE, Correspondance Politique, 1871–1896: Correspondance Politique des ambassades et légations, vol. 45. Dépêche (direction politique n. 30) Mr. de Balloy, ministre de France à Téhéran à Son Excellence Monsieur Develle, le ministre des affaires étrangères à Paris. Téhéran, 12 Août 1893; AMAE, Affaires Divers Politiques, 1815–1896 IV (ADP), Perse, marge 4, numéro 93: Français au service de la Perse: Schneider, médecin du Shah, 1893–1895, "Contrat Passé entre General Nazere Aga et Docteur Schneider, Paris 15 Septembre 1893," Annexe à la lettre de la ministre de la guerre du 21 Septembre 1893; AMAE, Correspondance Politique et Commerciale (Nouvelle Série), 1897–1918 (CP), Perse, Français au Service de Perse: médecins dossier personnelles, 1896–1905, vol. 56, dossier 6A. Dépêche (direction politique n. 48) Dr. Schneider à Son Excellence le ministre des affaires étrangers à Paris. Paris, 27 Juillet, 1902.

14. AMAE, Correspondance Politique et Commerciale (Nouvelle Série), 1897–1918 (CP), Perse, Français au Service de Perse: médecins dossier personnelles, 1896–1905, vol. 56, dossier 6D. Lettre adressée par Mr. Le Docteur Tholozan à Son Excellence Mokber ed Dowlé, ministre de l'intérieur; et au Conseil des Ministres: "Au sujets des quarantaines dans la province du Fars et de la nécessité de les mettre en vigueur promptement dans la Golfe Persique," Annexe n. 2 dans 1897.

15. AMAE, Correspondance Politique et Commerciale (Nouvelle Série), 1897–1918 (CP), Lettre adressée par Mr. Le Docteur Tholozan à Son Excellence Mokber ed Dowlé.

16. Tholozan wrote his dissertation on the gross pathology of tumors, though most of his research as a faculty member at the Military Teaching Hospital of Val-de-Grâce in Paris was on the pathophysiology of cholera. See Joseph Désiré Tholozan, *Des Métastases Thèse Présentée et Soutenue à la Faculté de Médicine* [Metastases thesis presented and defended at the Faculty of Medicine] (Paris: Librairie de A. Delahaye et E. Chatel, 1857); Joseph Désiré Tholozan, "Recherche sur quelques points d'anatomie et de physiologie et de pathologiques du cholera," *Gazette Médical de Paris* 4, no. 3 (1849): 557–558.

17. AMAE, Correspondance Politique et Commerciale (Nouvelle Série), 1897–1918 (CP), Perse, Français au Service de Perse: médecins dossier personnelles, 1896–1905, vol. 56, dossier 6D. Lettre adressée par Mr. Le Docteur Tholozan à Son Excellence Mokber ed Dowlé, ministre de l'intérieur; et au Conseil des Ministères: "Au sujets des quarantaines dans la province du Fars et de la nécessité de les mettre en vigueur promptement dans la Golfe Persique," Annexe n. 2 dans 1897.

18. Kourosh Ahmadi, *Islands and International Politics in the Persian Gulf: The Abu Musa and the Tunbs in Strategic Perspective* (London: Routledge, 2008), 33; Sanchari Dutta, "Plague, Quarantine, and Empire: British-Indian Sanitary Strategies in Central Asia, 1897–1907," in *The Social History of Health and Medicine in Colonial India*, ed. Biswamoy Pati and Mark Harrison (London: Routledge, 2009), 79.

19. Norman Howard-Jones, "The Scientific Background. 5. The Ninth Conference: Paris, 1894," *WHO Chronicle* 28, no. 10 (1974): 458.

20. Myron Echenberg, *Plague Ports: The Global Urban Impact of Bubonic Plague, 1894–1901* (New York: New York University Press, 2007), 79.

21. The meeting was presided by Ja'far Quli Khan Hidayat Nayyir al-Mulk, minister of public instruction. Members of the Council included Tholozan, Basil Composanpiro, the Ottoman representative, Müller of the German legation, Schneider of the French legation, Tom F. Odling of the British legation, John G. Wishard, Presbyterian missionary and superintendent of the American Hospital in Tehran, the Russian legation's physician, and prominent Iranian physicians in Tehran. See AMAE, Correspondance Politique et Commerciale (Nouvelle Série), 1897–1918 (CP), Perse, Français au Service de Perse: médecins dossier personnelles, 1896–1905, vol. 56, Dossier 6D. Dépêche (direction politique n. 18) Mr. de Balloy, ministre de France à Téhéran à Son Excellence le ministre des affaires étrangères à Paris. Téhéran, Février 12, 1897: "Conseil de Santé: Procès-verbal de la séance du 12 Janvier 1897," Annexe n. 3 dans 1897.

22. AMAE, Correspondance Politique et Commerciale (Nouvelle Série), 1897–1918 (CP), Perse, Français au Service de Perse: médecins dossier personnelles, 1896–1905, vol. 56, dossier 6D. Dépêche (direction politique n. 18) Mr. de Balloy, ministre de France à Téhéran à Son Excellence le ministre des affaires étrangères à Paris. Téhéran, 12 Février 1897: Annexe n. 4: "Lettre adressé par le Dr. Tholozan au Dr. Schneider [Tehran January 20, 1897] au sujet du projet de procès-verbal de la séance du 12 Janvier 1897;" Musée du Service de Santé des Armées, au Val-de-Grâce, Cote C/1082, dossier 1541: Joseph Désiré Tholozan, Lettre à Son Excellence Mokber ed Dowlé, ministre de l'intérieur de Perse.

23. FO 60/586 Commercial Dispatch (n. 3) from Charles Hardinge to Marquis of Salisbury. Tehran, February 18, 1897. [Enclosure n. 1] Tom F. Odling, Physician to Charles Hardinge. Tehran, February 18, 1897.

24. FO 60/586 Commercial Dispatch (n. 3) from Charles Hardinge to Marquis of Salisbury. Tehran, February 18, 1897. [Enclosure n. 1] Tom F. Odling, Physician to Charles Hardinge. Tehran, February 18, 1897.

25. FO 60/586 Commercial Dispatch (n. 6) from Charles Hardinge to Marquis of Salisbury. Tehran, March 18, 1897. [Enclosure n. 1] Tom F. Odling, Physician to Charles Hardinge. Tehran, March 17, 1897; and Lorimer, *Gazetteer of the Persian Gulf*, 2547.

26. FO 60/586 Commercial Dispatch (n. 3) from Charles Hardinge to Marquis of Salisbury. Tehran, February 18, 1897. [Enclosure n. 1] Tom F. Odling, Physician to Charles Hardinge. Tehran, February 18, 1897.

27. FO 60/586 Commercial Dispatch (n. 3) from Charles Hardinge to Marquis of Salisbury. Tehran, February 18, 1897. [Enclosure n. 1] Tom F. Odling, Physician to Charles Hardinge. Tehran, February 18, 1897. The British legation physician in Tehran, Tom F. Odling, was a surgeon by training and came to Iran in 1872 as the Indo-European Telegraph Department's physician. He became the legation's doctor in 1891, replacing Joseph Dickson who also gave him his seat on the Sanitary Council. Odling's multiple responsibilities were typical for physicians attached to European embassies in Iran during the late Qajar era. By 1899 he ran a charitable dispensary at the British legation in addition to meeting the health care needs of the staff at the Telegraph Department and the embassy. He also used his position as physician to the prime minister (Amin al-Dawla) and other high officials to expedite British interests. Odling died in Tehran in 1905. See FO 60/608 Dispatch (n. 62) from Sir Mortimer M. Durand, British Minister in Tehran, to the Marquess of Salisbury. Gulahek, June 22, 1899; and Denis Wright, *The English amongst the Persians* (London: Heinemann, 1977), 125.

28. FO 60/586 Commercial Dispatch (n. 3) from Charles Hardinge to Marquis of Salisbury. Tehran, February 18, 1897. [Enclosure n. 1] Tom F. Odling, Physician to Charles Hardinge. Tehran, February 18, 1897.

29. FO 60/586 Commercial Dispatch (n. 6) from Charles Hardinge to Marquis of Salisbury. Tehran, March 18, 1897. [Enclosure n. 1] Tom F. Odling, Physician to Her Majesty's Legation to Charles Hardinge. Tehran, March 17, 1897.

30. FO 60/732 Copy of Foreign Office Memorandum: "Russian Sanitary Cordon at Turbat-i Haidari," July 22, 1903. [Copie.] "Lettre Adressée au Gouvernement de S.M. le Shah par le Conseil de Santé" [Letter addressed to the Government of H.M. the Shah by the Sanitary Council], February 24, 1897.

31. Panayote Bey, a delegate to the Sanitary Council in Istanbul, represented Iran at the conference. Howard-Jones, "The Scientific Background. 5," 461–463; Ministero degli affari esteri, *Conférence Sanitaire International de Venise, Procès-Verbaux [16 Février–19 Mars 1897]* (Rome: Forzani et Cie. Imprimeurs du Senat, 1897).

32. Ministero degli affari esteri, *Conférence Sanitaire International de Venise*, 111–115. It is now agreed that Alexandre Yersin (1863–1943) was the discoverer of the causative organism of bubonic plague named *Yersinia pestis* in 1894 in Hong Kong. Yersin went on to start the medical school of Hanoi, Pasteur Institute of Saigon, and revolutionized the rubber farming industry in French Indo-China. See Howard-Jones, "The Scientific Background. 5," 462–463; Christopher Wills, *Yellow Fever, Black Goddess* (Reading: Helix Books, 1996), 74–75; Elmer Bendiner, "Alexandre Yersin: Pursuer of Plague," *Hospital Practice* 24 (1989): 121–138; Henri H. Mollaret and Jacqueline Brossollet, *Yersin, un pasteurien en Indochine* (Paris: Belin, 1993).

33. Ministero degli affari esteri, *Conférence Sanitaire International de Venise*, 25.

34. Harrison, *Public Health in British India*, 137.

35. Adrien Achille Proust (1834–1903), father of writer Marcel Proust, was a practicing physician and a professor of hygiene at the faculty of medicine in Paris. He was a member of the Comité d'Hygiène publique de France and the Académie Nationale de Médecine, serving as its secretary from 1883 to 1888. His rivalry with Tholozan, particularly over the efficacy of quarantines against cholera, played out in the proceedings of the academy during these years. See Adrien Achille Proust, *Exposé Des Titres Et Travaux Scientifiques du Docteur A. Proust* (Paris: Imprimerie Emile Martinet, 1877); Persian transcript of Tholozan's response to Dr. Proust, in Yale University Library, Manuscripts and Archives, New Haven, Ghassem Ghani Collection (MS 235), ser. 5, box 2, folder 18; Howard-Jones, "The Scientific Background. 5," 465.

36. Ministero degli affari esteri, *Conférence Sanitaire International de Venise*, 109.

37. Adrien Achille Proust, *La défense de l'Europe contre le cholera* (Paris: G. Masson, 1892), 256–266.

38. Ministero degli affari esteri, *Conférence Sanitaire International de Venise*, 106–107.

39. Ministero degli affari esteri, *Conférence Sanitaire International de Venise*, 107.

40. Ministero degli affari esteri, *Conférence Sanitaire International de Venise*, 195.

41. Thus, Australia, a maritime British commonwealth, did not ratify the Venice convention on quarantines and epidemic diseases until 1902. It even refused to notify concerned powers of any plague outbreaks in its territories until that date. See Ministry of Foreign Affairs, The Center for Documents and the History, Tehran, doc. 90. Dispatch (no. 200) from The Royal Italian Legation to His Excellency Moshir al-Dawla, Iranian Minister of Foreign Affairs (on the 1897 public health agreement signed in Vienna to prevent the outbreak of plague). Tehran, 10 Ramazan 1320 AH [December 11, 1902]; Ministero degli affari esteri, *Conférence Sanitaire International de Venise*, 30–43.

42. Ministero degli affari esteri, 1 *Conférence Sanitaire International de Venise*, 97–198.

43. Ministero degli affari esteri, *Conférence Sanitaire International de Venise*, 214, 246–248.

44. FO 60/586 Commercial Dispatch (n. 8) from Charles Hardinge, Her Britannic Majesty's Chargé d'Affaires, to Marquis of Salisbury. Tehran, May 15, 1897. [Enclosure n. 1] Tom F. Odling, Physician to Her Majesty's Legation, to Charles Hardinge. Tehran, May 14, 1897; FO 60/586 Commercial Dispatch (n. 10) from Charles Hardinge, Her Britannic Majesty's Chargé d'Affaires, to Marquis of Salisbury. Tehran, July 1, 1897.

45. The Russians were using a recommendation by the Venice Conference that, in cases of deficient quarantine measures, "it is preferable to close frontiers rather than establishing quarantine stations, since quarantine stations only affect economic life, whereas the other does both economic and sanitary damage." See Ministero degli affari esteri, *Conférence Sanitaire International de Venise*, 151–152.

46. FO 60/586 Commercial Dispatch (n. 10) from Charles Hardinge to Marquis of Salisbury. Tehran, July 1, 1897.

47. FO 60/586 Commercial Dispatch (n. 13) from Charles Hardinge to Marquis of Salisbury. Tehran, July 28, 1897.

48. FO 60/586 Commercial Dispatch (n. 13) from Charles Hardinge to Marquis of Salisbury. Tehran, July 28, 1897.

49. FO 60/586 Commercial Dispatch (n. 15) from Charles Hardinge to Marquis of Salisbury. Tehran, August 20, 1897.

50. FO 60/586 Commercial Dispatch (n. 15) from Charles Hardinge to Marquis of Salisbury. Tehran, August 20, 1897; and FO 60/586 Commercial Dispatch (n. 8) from Charles Hardinge to Marquis of Salisbury. Tehran, May 15, 1897. [Enclosure n. 1] Tom F. Odling, Physician to Her Majesty's Legation, to Charles Hardinge. Tehran, May 14, 1897.

51. FO 60/586 Commercial Dispatch (copy) from Charles Hardinge, Her Britannic Majesty's Chargé d'Affaires, to Marquis of Salisbury. Gulahek, August 11, 1897.

52. FO 60/586 Commercial Dispatch (n. 118) from Charles Hardinge to Marquis of Salisbury. Tehran, September 13, 1897.

53. Echenberg, *Plague Ports*, 79.

54. Maryam Dorreh Ekhtiar, "Nasir al-Din Shah and the Dar al-Funun: The Evolution of an Institution," *Iranian Studies* 34 (2001): 158.

55. Ministère des Affaires étrangères, *Conférence Sanitaire Internationale de Paris, Procès-Verbaux [10 Octobre–3 Décembre 1903]* (Paris: Imprimerie National, 1904), 456.

56. Ardishir Khan, chief secretary to the Iranian legation in France, successfully completed his doctorate at the Medical Faculty in Paris at this time. See Ministère des Affaires étrangères, *Conférence Sanitaire Internationale de Paris, Procès-Verbaux [10 Octobre–3 Décembre 1903]*, 457.

57. 'Ain al-Saltana, *Ruznama-yi khatirat*, 2:1276–1277.

58. FO 60/608 Dispatch (n. 1) from Sir Mortimer M. Durand, British Minister in Tehran, to the Marquess of Salisbury. Gulahek, June 23, 1899.

59. Viedomosti of St. Petersburg [Sankt-Peterburgskie Vedomosti], October 20, 1899, cited in Cyril Elgood, *A Medical History of Persia and the Eastern Caliphate from the Earliest Times until the Year A.D. 1932* (Cambridge: Cambridge University Press, 1951), 523.

60. FO 60/608 Dispatch (no. 1) from Sir Mortimer M. Durand, British Minister in Tehran, to the Marquess of Salisbury. Gulahek, June 23, 1899. [Enclosure n. 1] "Memorandum respecting the Plague at Bushire."

61. Mirza 'Ali Khan Amin al-Dawla (1844–1904) became prime minister under Muzaffar al-Din Shah in March 1897 and, as his first priority, sought to improve Iran's financial and educational systems. However, a nagging loan scandal and the opposition of political rivals led to his fall from favor and dismissal by the shah on June 5, 1898. See Mirza 'Ali Khan Amin al-Dawala, *Khatirat-i siyasi-yi Mirza 'Ali Khan Amin al-Dawla* [The political memoirs of Mirza 'Ali Khan Amin al-Dawla], ed. Hafez Farmayan, 3rd ed. (Tehran: Amir Kabir, 1370 AS [1992]); Hafez Farmayan, "Amin-al-Dawla, Mirza 'Ali Khan" *Encyclopedia Iranica* 1, fasc. 9 (1989): 943–945, updated online version accessed November 3, 2017, http://www.iranicaonline.org/articles /amin-al-dawla-mirza-ali-khan; Hafez Farmayan, "Portrait of a Nineteenth-Century Iranian Statesman: The Life and Times of Grand Vizier Amin ud-Dawlah, 1844–1904," *International Journal of Middle East Studies* 15, no. 3 (1983): 337–351.

62. FO 60/608 Dispatch (no. 1) from Sir Mortimer M. Durand to the Marquess of Salisbury. Gulahek, June 23, 1899. [Enclosure n. 1] "Memorandum respecting the Plague at Bushire."

63. Lorimer, *Gazetteer of the Persian Gulf*, 2548.

64. FO 60/609 Dispatch (n. 91) from Sir Mortimer M. Durand to the Marquess of Salisbury, September 16, 1899. [Enclosure n. 3] Reports of Meeting of Sanitary Council, August 24, 1899.

65. FO 60/608 Dispatch (n. 1) from Sir Mortimer M. Durand to the Marquess of Salisbury. Gulahek, June 23, 1899. [Enclosure n. 1] "Memorandum respecting the Plague at Bushire."

66. Lorimer, *Gazetteer of the Persian Gulf*, 2548; FO 60/609 Dispatch (no. 79) from Sir Mortimer M. Durand, British Minister in Tehran, to the Marquess of Salisbury. Gulahek, July 27, 1899. [Enclosure n. 1] Monthly summery from June 29 to July 26, 1899 by Lt.-Col. H. Picot.

67. FO 60/609 Dispatch (n. 88) from Sir Mortimer M. Durand, British Minister in Tehran, to the Marquess of Salisbury. Gulahek, August 24, 1899. [Enclosure n. 1] Memorandum by Mr. Spring-Rice upon the outbreak of plague at Bushire from July 24 to August 24, 1899.

68. FO 60/609 Dispatch (n. 88) from Sir Mortimer M. Durand to the Marquess of Salisbury, August 24, 1899. [Enclosure n. 1] Memorandum by Mr. Spring-Rice.

69. FO 60/609 Dispatch (n. 88) from Sir Mortimer M. Durand to the Marquess of Salisbury, August 24, 1899. [Enclosure n. 1] Memorandum by Mr. Spring-Rice.

70. FO 60/609 Dispatch (n. 88) from Sir Mortimer M. Durand to the Marquess of Salisbury. Gulahek, August 24, 1899. [Enclosure n. 1] Memorandum by Mr. Spring-Rice; Lorimer, *Gazetteer of the Persian Gulf*, 2548.

71. FO 60/609 Dispatch (n. 91) from Sir Mortimer M. Durand, British Minister in Tehran, to the Marquess of Salisbury, September 16, 1899.

72. FO 60/609 Dispatch (n. 91) from Sir Mortimer M. Durand to the Marquess of Salisbury, September 16, 1899.

73. FO 60/609 Dispatch (n. 91) from Sir Mortimer M. Durand to the Marquess of Salisbury, September 16, 1899. [Enclosure n. 3] Reports of Meeting of Sanitary Council, August 24, 1899.

74. FO 60/609 Dispatch (n. 102) from Mr. Cecil Spring-Rice (in the absence of Sir Mortimer M. Durand, British Minister in Tehran) to the Marquess of Salisbury. Tehran, October 17, 1899. [Enclosure n. 1] Memorandum by Mr. Cecil Spring-Rice on Quarantine at Bushire.

75. Mary P. Sutphen, "Not Where, but When: Bubonic Plague and the Reception of Germ Theories in Hong Kong and Calcutta, 1894–1897," *Journal of the History of Medicine* 52 (1997): 98.

76. FO 60/609 Dispatch (n. 102) from Sir Mortimer M. Durand, British Minister in Tehran, to the Marquess of Salisbury, October 17, 1899. [Enclosure 3] Memorandum respecting Quarantine in Bushire.

77. FO 60/609 Dispatch (n. 102) from Sir Mortimer M. Durand to the Marquess of Salisbury, October 17, 1899. [Enclosure 3] Memorandum respecting Quarantine in Bushire.

78. FO 60/609 Dispatch (n. 91) from Sir Mortimer M. Durand, British Minister in Tehran, to the Marquess of Salisbury, September 16, 1899. [Enclosure n. 3] Dispatch (no. 93) from Sir Mortimer M. Durand, British Minister in Tehran, to the Marquess of Salisbury, September 17, 1899..

79. Marvin Entner, *Russo-Persian Commercial Relations, 1828–1914* (Gainesville: University of Florida Monographs, 1965), 17–51; Dutta, "Plague, Quarantine, and Empire," 79; FO 60/609 Dispatch (n. 111) from Mr. Cecil Spring-Rice (in the absence of Sir Mortimer M. Durand, British Minister in Tehran) to the Marquess of Salisbury. Tehran, November 14, 1899. [Enclosure n. 1] Memorandum by Mr. Cecil Spring-Rice on Quarantine in the Persian Gulf.

80. FO 60/732 Dispatch (n. 152) from Sir Arthur N. Hardinge, British Minister in Tehran, to the Marquess of Lansdowne, Secretary of State for Foreign Affairs, August 16, 1904.

81. FO 60/609 Dispatch (n. 103) from Sir Mortimer M. Durand, British Minister in Tehran, to the Marquess of Salisbury. Tehran, November 14, 1899. [Enclosure n. 1] Monthly summary from September 16 to October 16, 1899, by Lt.-Col. H. Picot.

82. Lorimer, *Gazetteer of the Persian Gulf*, 2548–2549; and FO 60/609 Persia and Arabia Confidential Dispatch (n. 91) from Sir Mortimer M. Durand, British Minister in Tehran, to the Marquess of Salisbury, September 16,1899. [Enclosure n. 3] Reports of Meeting of Sanitary Council, August 24, 1899.

83. Lorimer, *Gazetteer of the Persian Gulf*, 2548.

84. FO 60/732 Copy of Foreign Office Memorandum: "Russian Sanitary Cordon at Turbat-i Haidari," July 22, 1903.

85. FO 60/732 Dispatch (no. 1) from India Office to the Foreign Office, India Office, July 30, 1902. [Enclosure n. 3] Report by Captain H. Smyth on the Russian Quarantine Line, Meshed, May 2, 1902.

86. FO 60/732 Dispatch (no. 1) from India Office to the Foreign Office, India Office, July 30, 1902 [Enclosure n. 3] Report by Captain H. Smyth on the Russian Quarantine Line, Meshed, May 2, 1902.

87. L. P. Morris, "British Secret Service Activity in Khorassan, 1887–1908," *Historical Journal* 27, no. 3 (1984): 657–675.

88. William R. Roff, "Sanitation and Security: The Imperial Powers and the Nineteenth-Century Hajj," *Arabian Studies* 6 (1982): 143–160.

89. Alexander Ivanovitch Iyas was an officer of Finnish origin in the tsar's Lithuanian Regiment. After a short posting in Turkestan, he was appointed in 1901 as head of the Russian quarantine unit in Torbat-i Heydarieh. In 1912 he was appointed consul in Soujbulak, a Kurdish town south of Lake Urmiyeh, where he was murdered in 1914. See John Tchalenko, "Alexander Iyas and Vladimir Minorsky in Persian Kurdistan, 1912–1914," *Eastern Art Report* 55, no. 1 (2006): 44–48.

90. FO 60/732 Dispatch (n. 1) from India Office to the Foreign Office, India Office, July 30, 1902. [Enclosure n. 3] Report by Captain H. Smyth on the Russian Quarantine Line, Meshed, May 2, 1902.

91. FO 60/732 Copy of Foreign Office Memorandum: "Russian Sanitary Cordon at Turbat-i Haidari," July 22, 1903.

92. Memorandum, signed G.H. [Hamilton], March 1900, *FI* 338/00. Reference to Lee-Warner: Memorandum C 94, May 16, 1899, Political and Secret Library (IOL); cited in Briton Cooper Busch, *Britain and the Persian Gulf* (Berkeley: University of California Press, 1967), 235–268.

93. FO 60/732 Copy of letter by Sir Arthur N. Hardinge, British Minister in Tehran, to Atabeg i Azam, Iranian Prime Minister. Tehran, March 23, 1903.

94. FO 60/732 Copy of letter by Sir Arthur N. Hardingen to Atabeg i Azam. Tehran, March 23, 1903.

95. FO 60/732 Copy of letter by Sir Arthur N. Hardinge to Atabeg i Azam. Tehran. March 23, 1903.

96. FO 60/732 Dispatch (n. 1) from Sir Arthur N. Hardinge, British Minister in Tehran, to the Foreign Office. Dropmore, Maidenhead, undated [Received August 7,1902]. The Foreign Minister was preceded by Muhsin Khan Mushir al-Dawla.

97. FO 60/732 Dispatch (n. 1) from India Office to the Foreign Office, India Office, July 30, 1902. [Enclosure n. 1] Lieutenant-Colonel Chenevix-Trench, British Consul General in Khurasan and Sistan, to Sir Arthur N. Hardinge, British Minister in Tehran, Meshed, May 7, 1902; FO 60/732 Dispatch (n. 1) from India Office to the Foreign Office, India Office, July 30, 1902. [Enclosure n, 1] from Government of India [Curzon, A. P. Palmer, T. Raleigh, E. R. Elles, A. T. Arundel, Denzil Ibbetson, J. F. Finlay to Lord George F. Hamilton, British Secretary of State for India, Simla, July 3, 1902.

98. FO 60/732 Dispatch (n. 1) from Sir Arthur N. Hardinge, British Minister in Tehran, to the Foreign Office. Dropmore, Maidenhead, undated [Received August 7, 1902]; and FO 60/732 Government of India Foreign Department, Memorandum (n. 110) from Government of India [Curzon, A. P. Palmer, T. Raleigh, E. R. Elles, A. T. Arundel, Denzil Ibbetson, J. F. Finlay] to Lord George F. Hamilton, British Secretary of State for India. Simla, July 3, 1902.

99. FO 60/732 Dispatch (n. 1) from Sir Arthur N. Hardinge, British Minister in Tehran, to the Foreign Office. Dropmore. Maidenhead, undated [Received August 7,1902].

100. FO 60/732 Telegram from His Britannic Majesty's Consul General and Agent of the Government of India in Khurasan, Sistan, and Mashhad, to the Foreign Secretary. Simla. Mashhad, May 29, 1903.

101. FO 60/732 Telegram from His Britannic Majesty's Consul to the Foreign Secretary. Mashhad, May 29, 1903; and FO 60/732 Dispatch (n. 1) from Sir Arthur N. Hardinge, British Minister in Tehran, to the Marquess of Lansdowne, Secretary of State for Foreign Affairs, January 27, 1902.

102. FO 60/732 Dispatch (n. 1) from Sir Arthur N. Hardinge, British Minister in Tehran, to the Marquess of Lansdowne, Secretary of State for Foreign Affairs, January 27, 1902. [Enclosure 1] Consul General Trench, British Consul in Mashhad, to Sir Arthur N. Hardinge, British Minister in Tehran, Mashhad, December 3, 1901.

103. FO 60/732 Extract from the Diary of the Native Attaché, Mashhad Consulate-General, on Special Duty at Herat, for the Period April 5 to 18, 1903.

104. FO 60/732 Extract from the Diary of the Native Attaché, Mashhad Consulate-General, on Special Duty at Herat, for the Period April 5 to 18, 1903; and FO 60/732 Confidential Foreign Office Memorandum, "Russian Sanitary Cordon in Persia," July 27, 1903.

105. FO 60/732 Telegram (n. 63) from Lieutenant-Colonel J. F. Whyte, His Britannic Majesty's Consul General and Agent of the Government of India in Khurasan and Sistan, to the British Minister in Tehran, April 26, 1903; and Telegram (no. 64) from Lieutenant-Colonel J. F. Whyte, His Britannic Majesty's Consul General and Agent of the Government of India in Khurasan and Sistan, to the British Minister in Tehran, April 27, 1903.

106. FO 60/732 Telegram (n. 68) from Lieutenant-Colonel J. F. Whyte, His Britannic Majesty's Consul General and Agent of the Government of India in Khurasan and Sistan, to

the British Minister in Tehran, April 30, 1903; and FO 60/732 Telegram (n. 67) from His Britannic Majesty's Minister in Tehran to His Majesty's Secretary of State for Foreign Affairs, May 4, 1903.

107. FO 60/732 Telegram (n. 67) from Lieutenant-Colonel J. F. Whyte, His Britannic Majesty's Consul General and Agent of the Government of India in Khurasan and Sistan, to the British Minister in Tehran, April 29, 1903.

108. FO 60/732 Confidential Memorandum Printed for the use of the Cabinet: "Russian Sanitary Cordon in Persia," July 25, 1903.

109. FO 60/732 Confidential Memorandum Printed for the use of the Cabinet, July 25, 1903.

110. "Russia at the [Venice] conference clearly laid out that she reserved her liberty of action as regarded her Asiatic frontiers;" see FO 60/732 Confidential Memorandum Printed for the use of the Cabinet, July 25, 1903..

111. FO 60/732 Confidential Memorandum Printed for the use of the Cabinet, July 25, 1903.

112. FO 60/732 Confidential Memorandum Printed for the use of the Cabinet, July 25, 1903.

113. FO 881/8539 Memorandum by R. Graham, Persia Current Problems, December 11, 1905.

114. FO 881/8539 Memorandum by R. Graham, December 11, 1905. Amin al-Dawla helped establish the Belgian administration of Iran's Customs Department in 1898. He hoped that functionaries from a nonaligned country would be less vulnerable to external political pressures and more likely to bring order to Iran's main source of revenue collection. Their success in 1898 and 1899 motivated Amin al-Dawla to recruit more Belgians. Within several years, these functionaries expanded their operations to include the reorganization of the treasury and the postal service. See Annette Destrée, *Les fonctionnaires belges au service de La Perse* (Leiden: E. J. Brill, 1976), 33–34; Annette Destrée, "Belgian-Iranian Relations," *Encyclopedia Iranica* 4, fasc. 2 (1989): 124–126, updated online version accessed November 3, 2017, http://www.iranicaonline.org/articles/belgian-iranian-relations; Amin al-Dawla, *Khatirat-i siyasi-yi Mirza 'Ali Khan*, 256.

115. Gad G. Gilbar, "The Big Merchants and the Persian Constitutional Revolution of 1906," *Asian and African Studies* 2, no. 3 (1976): 290–295; Janet Afary, *The Iranian Constitutional Revolution, 1906–1911: Grassroots Democracy, Social Democracy, and the Origins of Feminism* (New York: Columbia University Press, 1996), 34–35. Also see Joseph Rabino, "Banking in Persia: Its Basis, History, and Prospects," *Journal of the Institute of Bankers* 13 (1892): 22.

116. Nikki R. Keddie, *Roots of Revolution: An Interpretive History of Modern Iran* (New Haven: Yale University Press, 1981), 70.

117. Destrée, *Les fonctionnaires belges*, 86–88.

118. FO 60/732 Confidential Dispatch (n. 121) from Sir Arthur Hardinge to the Marquess of Lansdowne K.G. Tehran, August 14, 1903.

119. The French medical officer Jean Jérôme Augustin Bussière (1872–1958) was formerly a military physician based in Senegal, where he called for the extension of the 1902 French metropolitan law for obligatory smallpox vaccination to the French colonies. See Jean Jérôme Augustin Bussière, *La vaccine et la variole au Sénégal, dans l'Inde et en Indo-Chine, rapport de M. le Médecin-Major Bussière* (Paris: Imprimerie de J. Gainche, 1903).

120. FO 60/732 Telegram (no. II.S.A.) from Lieutenant-Colonel C. A. Kemball, His Britannic Majesty's Resident at Bushehr, to Sir Arthur N. Hardinge, British Minister in Tehran, July 23, 1903.

121. FO 60/732 Confidential Dispatch (n. 121) from Sir Arthur Hardinge to the Marquess of Lansdowne K.G. Tehran, August 14, 1903.

122. FO 60/732 Confidential Dispatch (n. 121) from Sir Arthur Hardinge to the Marquess of Lansdowne K.G. Tehran, August 14, 1903.

123. FO 60/732 Extracts of Confidential Diary of Consul-General at Meshed, March 28, 1903.

124. FO 60/681 Confidential Dispatch (n. 17) from Sir Arthur Hardinge to the Marquess of Lansdowne K.G. Tehran, January 27, 1904. [Account of Journey from Baghdad to Tehran].

125. Amin al-Dawala, *Khatirat-i siyasi-yi Mirza 'Ali Khan*, 276.

126. FO 60/732 Telegram from Lieutenant-Colonel J. F. Whyte, His Britannic Majesty's Consul General and Agent of the Government of India in Khurasan and Sistan, to His Excellency the Viceroy of India, June 9, 1903.

127. FO 60/732 Telegram (n. 84) from Sir Arthur N. Hardinge, British Minister in Tehran, to the Marquess of Lansdowne, Secretary of State for Foreign Affairs, June 10, 1903.

128. FO 60/732 Telegram (n. 89) from Lieutenant-Colonel J. F. Whyte, His Britannic Majesty's Consul General and Agent of the Government of India in Khurasan and Sistan, to Sir Arthur N. Hardinge, British Minister in Tehran, May 24, 1903.

129. FO 60/732 Telegram (n. 729) from E. Naffilaert, Director General of the Iranian Customs at Muhammara, to Mr. Francis, British Quarantine Medical Officer, Muhammara, July 4, 1903.

130. FO 60/732 Telegram (n. II.S.A.) from Lieutenant-Colonel C. A. Kemball, His Britannic Majesty's Resident at Bushehr, to Sir Arthur N. Hardinge, British Minister in Tehran, July 23, 1903.

131. The conference met from October 10 to December 3, 1903. Nazar Aqa Yamin al-Saltana, the shah's ambassador to France, led the Iranian delegation at the conference. He was an Assyrian born in Iran and educated at the Lazarist Catholic School in Istanbul and the Polytechnic College (Dar al-funun) in Tehran. He served as a translator to the Iranian consulate in Tbilisi and the Iranian embassy in Paris before becoming the Iranian ambassador plenipotentiary to the French emperor in 1872, a post he held for several decades. See "Jinab-i Nazar Aqa [His Excellency Nazar Aqa]," *Sharaf*, no. 36, Safar 1303 AH [November 1885]; Howard-Jones, "The Scientific Background. 5," 463–468.

132. FO 251/76 Commercial Dispatch (n. 365) from Sir Edmund Monson to the Marquess of Lansdowne. Paris, December 23, 1903. [Enclosure 1] International Sanitary Convention, signed at Paris, December 3, 1903.

133. Stephanie Jones, *Two Centuries of Overseas Trading: The Origins and Growth of the Inchcape Group* (London: Palgrave MacMillan, 1989), 85.

134. FO 251/76 Commercial Dispatch (n. 365) from Sir Edmund Monson to the Marquess of Lansdowne. Paris, December 23, 1903. [Enclosure 1] International Sanitary Convention, signed at Paris, December 3, 1903.

135. Howard-Jones, "The Scientific Background. 5," 465; FO 251/76 Commercial Dispatch (n. 365) from Sir Edmund Monson to the Marquess of Lansdowne. Paris, December 23, 1903. [Enclosure 1] International Sanitary Convention, signed at Paris, December 3, 1903.

136. Ministère des Affaires étrangères, *Conférence Sanitaire Internationale de Paris, Procès-Verbaux [10 Octobre–3 Décembre 1903]*, 389–390.

137. The British unsuccessfully fought to have the international quarantine station in Basra rather than Hormuz, which had the potential of choking off shipping from India into the Persian Gulf. Ministère des Affaires étrangères, *Conférence Sanitaire Internationale de Paris, Procès-Verbaux [10 Octobre–3 Décembre 1903]*, 125–126; FO 251/76 Commercial Dispatch (n. 365) from Sir Edmund Monson to the Marquess of Lansdowne. Paris, December 23, 1903. [Enclosure n. 2] Conférence Sanitaire Internationale de Paris, Procès-verbal de signature. Paris, December 3, 1903.

138. Lorimer, *Gazetteer of the Persian Gulf*, 2542.

139. Ministère des Affaires étrangères, *Conférence Sanitaire Internationale de Paris, Procès-Verbaux [10 Octobre–3 Décembre 1903]*, 457.

140. Amir Khan Amir A'lam (1256–1340 AS [1877/78–1961]) was the son of Mirza 'Ali Mu'tamid al-Vuzara, Iran's envoy to Baghdad and Syria. He was among the first cohort of students sent to study medicine at the Military Medical School in Lyon after 1896 and became one of the most influential figures in Iranian public health in the twentieth century. During his decades of public service in Iran, Amir 'Alam held various positions, including professor at the Polytechnic College, deputy speaker in the National Assembly, dean of the Faculty of Medicine at the University of Tehran, member of the Imperial Senate, minister in various cabinets, and personal physician to the last Qajar monarch and Riza Shah Pahlavi. See Abbas Naficy, *La médecine en Perse des origines à nos jours: ses fondements theoriques d'apres l'Encyclopédie médicale de Gorgani* (Paris: Edition Véga, 1933), 61; Mahdi Bamdad, *Shahr-i hal-i rijal-i Iran dar qarn-i 12, 13, 14 hijri* [The biography of Iranian personalities during the 12, 13, 14 centuries], 4th ed. Tehran: Intisharat-i zavar, 1371 AS [1992], 5:32.

141. Ministère des Affaires étrangères, *Conférence Sanitaire Internationale de Paris, Procès-Verbaux [10 Octobre–3 Décembre 1903]*, 457–458.

142. 'Abd al-Majid Mirza 'Ayn al-Dawla (1845–1926) was a grandson of Fath 'Ali Shah and married to the eldest daughter of the reigning Muzaffar al-Din Shah. He became prime minister several times between 1904 and 1918 and held many other high government positions throughout his life. He opposed Muzaffar al-Din Shah's ratification of the Iranian Constitution and fiercely disliked his predecessor, 'Ali Asghar Khan Amin al-Sultan, in whose assassination he might have had a hand. See Bamdad, *Shahr-i hal-i rijal-i Iran*, 2:93–101.

143. FO 60/732 Confidential Dispatch (n. 170) from Sir Arthur N. Hardinge, British Minister in Tehran, to the Marquess of Lansdowne, Secretary of State for Foreign Affairs. Gulahek, September 9, 1904.

144. FO 416/17 Telegram (n. 134) from Sir Arthur N. Hardinge, British Minister in Tehran, to the Marquess of Lansdowne, Secretary of State for Foreign Affairs. [Enclosure n. 1] Dispatch (no. 18) from Colonel Kemball to Sir Arthur N. Hardinge. Bushehr February 3, 1904; Lorimer, *Gazetteer of the Persian Gulf*, 2549.

145. FO 248/818 Precis (n. 78) "Reports certain incidents in connection with quarantine and preventive measures in the Persian Gulf" from Major P. Z. Cox, British Residency and Consulate-General in Bushehr, to Sir Arthur N. Hardinge, British Minister in Tehran. Bushehr, May 27, 1904.

146. FO 248/818 Precis (n. 78) from Major P. Z. Cox to Sir Arthur N. Hardinge. Bushehr, May 27, 1904.

147. FO 248/818 Precis (n. 78) from Major P. Z. Cox to Sir Arthur N. Hardinge. Bushehr, May 27, 1904.

148. FO 248/818 Copy of Letter (n. 4528) from Monsieur Dambrain, Director General of Customs in Bushehr, to Major Cox, British Residency and Consulate-General in Bushehr. Bushehr, May 19, 1904; Copy of Letter (n. 517) from Major Cox, British Residency and Consulate-General in Bushehr, to Monsieur Dambrain, Director General of Customs in Bushehr. Bushehr, May 20, 1904; Precis (n. 85) "Reports what has transpired in connection with the question of sanitary control of the Gulf" from Major P. Z. Cox, British Residency and Consulate-General in Bushehr, to Sir Arthur N. Hardinge, British Minister in Tehran. Bushehr, June 25, 1904.

149. FO 248/818 Precis (n. 85) from Major P. Z. Cox to Sir Arthur N. Hardinge. Bushehr, June 25, 1904.

150. FO 248/818 Extracts from a Report by the Residency Surgeon, Bushehr, July 8, 1904, enclosed in Precis (n. 109) "In continuation of previous correspondence" from Major P. Z. Cox, British Residency and Consulate-General in Bushehr, to Sir Arthur N. Hardinge, British Minister in Tehran. Bushehr, July 23, 1904.

151. FO 248/818 Extracts from a Report by the Residency Surgeon, Bushehr, July 8, 1904, enclosed in Precis (n. 109) "In continuation of previous correspondence" from Major P. Z. Cox,

British Residency and Consulate-General in Bushehr, to Sir Arthur N. Hardinge, British Minister in Tehran. Bushehr, July 23, 1904; Lorimer, *Gazetteer of the Persian Gulf*, 2550.

152. FO 248/818 Letter (n. 842) from Major P. Z. Cox, British Residency and Consulate-General, to Monsieur E. Waffelaert, Director General of Customs, Bushehr, August 3, 1904, enclosed in Precis (n. 112) "In pursuance of H.E. telegram" from Major P. Z. Cox, British Residency and Consulate-General in Bushehr, to Sir Arthur N. Hardinge, British Minister in Tehran. Bushehr, August 4, 1904.

153. FO 60/732 Dispatch (n. 143) from Sir Arthur N. Hardinge, British Minister in Tehran, to the Marquess of Lansdowne, Secretary of State for Foreign Affairs, August 3, 1904.

154. Tehran paid approximately one hundred pounds per month for the operation of the quarantines. See IO L/P & S/10/123 (Register no. 3081, enclosure no. 226–229) Commercial Dispatch (n. 7) Sir C. Spring-Rice to Sir Edward Grey. Tehran, April 19, 1907.

155. FO 60/732 Dispatch (n. 143) from Sir Arthur N. Hardinge, British Minister in Tehran, to the Marquess of Lansdowne, Secretary of State for Foreign Affairs, August 3, 1904.

156. FO 60/732 Confidential Dispatch (n. 170) from Sir Arthur N. Hardinge, British Minister in Tehran, to the Marquess of Lansdowne, Secretary of State for Foreign Affairs. Gulahek, September 9, 1904.

157. FO 60/732 Confidential Dispatch (n. 170) from Sir Arthur N. Hardinge to the Marquess of Lansdowne. Gulahek, September 9, 1904.

158. FO 881/8539 Memorandum by R. Graham, Persia Current Problems, December 11, 1905.

159. Iranian historiography has had a dim view of Joseph Naus, who is portrayed as an instrument of Russia's imperialist policies. Annette Destrée, the most prominent biographer of the Belgian customs administrators in Iran, has a more charitable interpretation of Naus's downfall, seeing him as a victim of his character flaws and disorganized accounting. Naus had no choice but to be friendly with the Russians, who held sway in the most productive parts of the country in the North, and clash with the British, whose laissez-faire trade philosophy made them naturally inimical to Iranian customs tariffs. Destrée, *Les fonctionnaires belges*, 110–116, 146–147; Kazemzadeh, *Russia and Britain*, 496.

160. Browne, *The Persian Revolution of 1905–1909* (Cambridge: Cambridge University Press, 1910), 137–138.

161. Lorimer, *Gazetteer of the Persian Gulf*, 2542–2543.

162. FO 371/309 Dispatch (n. 114) from Sir C. Spring-Rice to Sir Edward Grey. Gulhek, May 23, 1907.

163. FO 60/732 Confidential Dispatch (n. 170) from Sir Arthur N. Hardinge, British Minister in Tehran, to the Marquess of Lansdowne, Secretary of State for Foreign Affairs. Gulahek, September 9, 1904.

164. FO 881/8539 Memorandum by R. Graham, Persia Current Problems, December 11, 1905.

165. Lorimer, *Gazetteer of the Persian Gulf*, 2251.

166. Anja Pistor-Hatam, "Progress and Civilization in the Nineteenth-Century Japan: The Far Eastern State as a Model for Modernization," *Iranian Studies* 29, nos. 1–2 (1996): 111; Roxan Haag-Higuchi, "A Topos and Its Dissolution: Japan in Some 20th Century Iranian Texts," *Iranian Studies* 29, nos. 1–2 (1996): 71–83.

167. "Raji'i 'imadih'i rus va angilis" [On the subject of the agreement between the Russians and the English], *Musavat*, no. 2, 19th Ramzan 1325 AH [October 27, 1907], 4; "Khalij-i Fars" [Persian Gulf], *Shams*, no. 24, 13 Jamadi al-Thani 1329 AH [July 24, 1907], 2.

168. IO L/P & S/10/123 (Register no. 3081, enclosure no. 226–229) Commercial Dispatch (no. 7) Sir C. Spring-Rice to Sir Edward Grey. Tehran, April 19, 1907.

169. FO 60/681 [Enclosure n. 3] "Quarantine in Persia in 1903." Confidential Dispatch (n. 41) from Sir Arthur Hardinge to the Marquess of Lansdowne K.G. Tehran, February 29, 1904.

170. Much has been written about the Anglo-Russian agreement. Authoritative works on this issue include Rogers Platt Churchill, *The Anglo-Russian Convention of 1907* (Cedar Rapids, IA: Torch Press, 1939); Busch, *Britain and the Persian Gulf*, 348–383; Firouz Kazemzadeh, "Anglo-Russian Convention of 1907," *Encyclopedia Iranica*, 1:68–70, updated online version accessed November 3, 2015, at http://www.iranicaonline.org/articles/anglo-russian-convention-of -1907-an-agreement-relating-to-persia-afghanistan-and-tibet; Kazemzadeh, *Russia and Britain*, 499–509.

171. IO L/PS/10/123 (Register no. 3384, enclosure no. 160–162) Copy of Foreign Office Memorandum to Russian Government, London, July 24, 1907; IO L/P & S/10/123 (Register no. 3384, enclosure no. 159) Deciphered copy of coded telegram, Sir C. Spring-Rice to Foreign Office, Tehran, July 27, 1907.

172. Conseil Sanitaire de L'Empire de Perse [Sanitary Council of the Persian Empire], 44ème Séance, December 2, 1907; enclosed in IO L/P & S/10/123 (Register no. 1987, enclosure no. 1) Under Secretary of State, Foreign Office to Under Secretary of State, India Office. London, January 20, 1908.

Chapter 4 · *Cholera, Germs, and the 1906 Constitutional Revolution*

1. Ministère des Affaires étrangères, *Conférence Sanitaire Internationale de Paris [10 Octobre–3 Décembre 1903]*, 457.

2. *Tarbiyat*, no. 322, 23 Rabi' al-Thani 1322 AH [July 7, 1904], 1485; *Tarbiyat* or Education was one of the foremost weeklies of its time and considerably influenced the strata of intellectuals in Iran. See Muhammad Sadr Hashemi, *Tarikh-i jarayid va majallat-i Iran* [The history of newspapers and magazines in Iran] (Isfahan: Intisharat-i kamal, 1332 AS [1953–1954]), 1:116–124; Edward Granville Browne, *The Press and Poetry of Modern Persia: Partly Based on the Manuscript Work of Mirza Muhammad 'Ali Khan "Tarbiyat" of Tabriz* (Cambridge: Cambridge University Press, 1914), 62; Abdolla Mostofi, *The Administrative and Social History of the Qajar Period*, trans. Nayer Mostofi Glenn (Costa Mesa: Mazda, 1997), 2:354.

3. *Tarbiyat*, no. 322, 23 Rabi' al-Thani 1322 AH [July 7, 1904], 1485.

4. Taj al-Saltana, *Crowning Anguish: Memoirs of a Persian Princess from the Harem to Modernity, 1884–1914*, ed. Abbas Amanat and trans. by Anna Vanzan and Amin Neshati (Washington, DC: Mage, 1993), 283.

5. *Tarbiyat*, no. 322, 23 Rabi' al-Thani 1322 AH [July 7, 1904], 1486.

6. *Tarbiyat*, no. 322, 23 Rabi' al-Thani 1322 AH [July 7, 1904], 1485.

7. *Tarbiyat*, no. 322, 23 Rabi' al-Thani 1322 AH [July 7, 1904], 1486.

8. Mahdi Bamdad, *Shahr-i hal-i rijal-i Iran dar qarn-i 12, 13, 14 hijri* [The biography of Iranian personalities during the 12, 13, 14 centuries], 4th ed. (Tehran: Intisharat-i zavar, 1371 AS [1992]), 3:275.

9. Hajj Mirza 'Abd-al Baqi Tabib Hakim Bashi I'tizad al-Attiba was the father of Khalil Khan A'lam al-Dawla Saqafi, Iran's junior delegate at the 1894 International Sanitary Conference. Bamdad, *Shahr-i hal-i rijal-i Iran*, 1:487.

10. Mahmud Nadjmabadi, "Tibb-i Dar al-funun va kutub-i darsi-yi an" [Medicine at the Polytechnic College and its medical books]," in *Amir Kabir va Dar al-funun* [Amir Kabir and the Polytechnic College], ed. Qudrat Allah Rushani Za'faranlu (Tehran: Majmu'a-yi kitab-khana-yi markazi va asnad-i Danishgah-i Tehran, 1354AS [1975]), 227.

11. Dieulafoy is remembered for his work in pathological medicine, including the 1884 discovery of critical submucosal arterial malformations in the gastrointestinal tract that bears his name (Dieulafoy lesions). Georges Dieulafoy, "Exulceratio simplex: Leçons 1–3," in *Clinique Médicale de L'Hôtel-Dieu de Paris*, ed. Georges Dieulafoy (Paris: Masson et Cie, 1898), 1–38.

12. Nadjmabadi, "Tibb-i Dar al-funun," 227.

13. Husayn Mahbubi Ardakani, *Tarikh-i mu'assasat-i tamadduni-yi jadid dar Iran* [The history of the institutions of modern civilization in Iran], 2nd ed. (Tehran: Anjuman-i Danishjuyan-i Danishgah-i Tehran, 1370 AS [1992]), 1:288.

14. Muhammad Hasan Khan I'timad al-Saltana, *Ruznama-yi khatirat-i I'timad al-Saltana* [The memoirs of I'timad al-Saltana], ed. Iraj Afshar (Tehran: Amir Kabir, 1345 AS [1976]), 396.

15. Mahdiquli Hidayat (Mukhbir al-Saltana), *Khatirat va khatarat* [The memoirs of Mahdiquli Hidayat Mukhbir al-Saltana] (Tehran: Intisharat-i zavar, 1375 AS [1996]), 68; Bamdad, *Shahr-i hal-i rejal-i Iran*, 3:276.

16. Mirza Muhammad Khan Kermanshahi caught cholera from a patient during the 1904 epidemic and recovered only to die in 1908 at the age of eighty-one. See Bamdad, *Shahr-i hal-i rijal-i Iran*, 3:277–278; Nadjmabadi, "Tibb-i Dar al-funun," 228.

17. Bamdad, *Sharh-i hal-i rijal-i Iran*, 3:278.

18. Mostofi, *The Administrative*, 1:291.

19. Kirmanshahi overestimated the length of the *Vibrio cholerae* to be 0.5 to 0.8 micrometer wide and 1.4 to 2.6 micrometers long in the form of slightly curved rods with polar flagella; see "Tahqiqat-i Jinab-i Jalalatmab Duktur Mirza Muhammad Kirmanshahi (maraz-i kulira)" [The research of His Excellency Doctor Mirza Muhammad Kirmanshahi (the disease of cholera)], *Tarbiyat*, no. 322, 23 Rabi' al-Thani 1322 AH [July 7, 1904], 1488.

20. *Tarbiyat*, no. 322, 23 Rabi' al-Thani 1322 AH [July 7, 1904], 1488.

21. *Tarbiyat*, no. 322, 23 Rabi' al-Thani 1322 AH [July 7, 1904], 1489.

22. *Tarbiyat*, no. 322, 23 Rabi' al-Thani 1322 AH [July 7, 1904], 1489.

23. *Tarbiyat*, no. 322, 23 Rabi' al-Thani 1322 AH [July 7, 1904], 1489.

24. *Tarbiyat*, no. 322, 23 Rabi' al-Thani 1322 AH [July 7, 1904], 1490.

25. This view, conveyed by the government newspaper *Ittila'*, persisted through the last decade of the nineteenth century; see "Jinn va vaba," *Ittila'*, no. 354, 11 Rabi' al-Awwal 1312 AH [September 11, 1894]; "Tasfiya-yi hava," *Ittila'*, no. 354, 11 Rabi' al-Awwal 1312 AH [September 11, 1894].

26. *Tarbiyat*, no. 322, 23 Rabi' al-Thani 1322 AH [July 7, 1904], 1490.

27. "Mujasama-yi Pastur" [The statue of Pasteur], *Tarbiyat*, no. 328, 6 Jamadi 1322 AH [August 25, 1904], 1535.

28. *Tarbiyat*, no. 322, 23 Rabi' al-Thani 1322 AH [July 7, 1904], 1488–1490.

29. "Surat-i Dastur-i Jinab-i Sa'id al-Attiba" [The prescription of His Excellency Sa'id al-Attiba], *Tarbiyat*, no. 322, 23 Rabi' al-Thani 1322 AH [July 7, 1904]. 1487.

30. The "cocktail of chloroform" was a dilution of two grams of chloroform for every one thousand grams of water. Caretakers had to give a teaspoon of the mixture to the patient every hour for the medicine to be efficacious; see *Tarbiyat*, no. 322, 23 Rabi' al-Thani 1322 AH [July 7, 1904]. 1487.

31. *Tarbiyat*, no. 322, 23 Rabi' al-Thani 1322 AH [July 7, 1904]. 1487. Laudanum is a tincture of opium containing approximately ten percent powdered opium by weight.

32. Mostofi, *The Administrative*, 1:291.

33. *Tarbiyat*, no. 322, 23 Rabi' al-Thani 1322 AH [July 7, 1904], 1487.

34. *Tarbiyat*, no. 322, 23 Rabi' al-Thani 1322 AH [July 7, 1904], 1487.

35. Iranians were aware of the periodic outbreaks of cholera in Mecca. See Qahraman Mirza 'Ain al-Saltana, *Ruznama-yi khatirat-i 'Ain al-Saltana* [The memoirs of 'Ain al-Saltana], ed. Massoud Salur and Iraj Afshar (Tehran: Intisharat-i asatir, 1376 AS [1997–1998]), 2:1511; United States Public Health and Marine Hospital Service, "Table Showing Progress of Cholera from Mecca in 1902 to Persia and the Transcaspian Provinces in 1904, Prepared by the Sanitary Council of Teheran," *Public Health Reports Issued by the Surgeon General* 20, no. 6 (1905) [Washington, DC: Government Printing Office]: 233.

36. The first cases in Karbala appeared on the third of Ramazan (November 17) 1903. See Charles-René Coppin, "Rapport sur l'épidémie de cholera qui sévit en Perse (1904) et sur l'état actuel de l'hygiène," *Annales d'Hygiène et de Médicine Coloniales* 8 (1905) : 272.

37. Coppin, "Rapport sur l'épidémie," 273.

38. Document (n. 13) telegraph from Mushir al-Dawla [Nasrallah Khan] to Farman Farma ['Abd al-Husayn Mirza]. Tehran, December 26, 1903, in *Vaba-yi alamgir. Asnad, madarik va mukatibat-i Abdul-Husayn Mirza Mirza Farman Farma dar hukumat-i Kermanshah 1321– 23 A.H.* [World conquering cholera. records, documents, and correspondences of Abdul-Husayn Mirza during his rule in Kermanshah, 1903/1904–1905/1906], ed. Mansoureh Ettehadieh, Esmail Shams, and Azam Ghafouri (Tehran: Nashr-i tarikh-i Iran, 2013/2014), 71–72.

39. Document (n. 15) telegraph from Mushir al-Dawla [Nasrallah Khan] to Farman Farma ['Abd al-Husayn Mirza]. Tehran, December 27, 1903; and Document (n. 16) telegraph from Mushir al-Dawla [Nasrallah Khan] to Farman Farma ['Abd al-Husayn Mirza]. Tehran, December 30, 1903, in Ettehadieh et al., *Vaba-yi alamgir*, 73–75; FO 60/681 Dispatch (no. 41) from Sir Arthur Hardinge to the Marquess of Lansdowne. Tehran, February 29, 1904. [Enclosure n. 2] Quarantine in Persia in 1903. For travel time between Tehran and the frontier, see Charles Issawi, ed., *The Economic History of Iran, 1800–1940* (Chicago: University of Chicago Press, 1971), 195.

40. Joseph Scott, "The Recent Cholera Epidemic in Persia," *British Medical Journal* 2 (1904): 621.

41. FO 60/732 Dispatch (n. 71) from Sir Arthur Hardinge to the Marquess of Lansdowne. Tehran, April 23, 1904.

42. "British Trade Abroad," *Board of Trade Journal* 41, no. 342 (1903): 540–541.

43. FO 60/732 Dispatch (n. 71) from Sir Arthur Hardinge to the Marquess of Lansdowne. Tehran, April 23, 1904. [Enclosure n. 1] Letter from Sir Arthur Hardinge to Dr. Scott. Tehran, April 9, 1904.

44. FO 60/681 Confidential Dispatch (n. 71) from Sir Arthur Hardinge, British Minister in Tehran, to the Marquess of Lansdowne. Tehran, April 23, 1904.

45. Scott, "The Recent Cholera Epidemic," 620.

46. FO 60/681 Confidential Dispatch (n. 71) from Sir Arthur Hardinge to the Marquess of Lansdowne. Tehran, April 23, 1904. [Enclosure n. 5] Extracts from the Kermanshah News Letter n. 10.

47. Scott, "The Recent Cholera Epidemic," 621.

48. Scott, "The Recent Cholera Epidemic," 621–622.

49. Document (n. 43) Telegraph from Mushir al-Dawla to Farman Farma. Tehran, April 6, 1904, in Ettehadieh et al., *Vaba-yi alamgir*, 102; FO 60/681 Confidential Dispatch (n. 71) from Sir Arthur Hardinge to the Marquess of Lansdowne. Tehran, April 23, 1904. [Enclosure n. 5] Extracts from the Kermanshah News Letter n. 10; Scott, "The Recent Cholera Epidemic," 621.

50. FO 60/682 Dispatch (n. 86) from Sir Arthur Hardinge to the Marquess of Lansdowne K.G. Tehran, May 19, 1904.

51. FO 60/681 Confidential Dispatch (n. 71) from Sir Arthur Hardinge to the Marquess of Lansdowne. Tehran, April 23, 1904. [Enclosure n. 5] Extracts from the Kermanshah News Letter n. 11.

52. Document (n. 49) and Document (n. 50) Telegraphs from Bihjat al-Mulk to Mushir al-Dawla. Kermanshah, April 9, 1904; Document (n. 57) Dispatch from the Foreign Ministry to the Ottoman Embassy. Tehran, April 14, 1904, in Ettehadieh et al., *Vaba-yi alamgir*, 107–110, 114–115.

53. Document (n. 24) Translation of order from Joseph Naus to Farman Farma. Kermanshah, January 27, 1904, in Ettehadieh et al., *Vaba-yi alamgir*, 82.

54. Document (n. 21) Translation of a summary attestation by consuls and other European officials. Kermanshah, January 10, 1904, in Ettehadieh et al., *Vaba-yi alamgir*, 77–78.

55. FO 60/681 Confidential Dispatch (n. 71) from Sir Arthur Hardinge to the Marquess of Lansdowne. Tehran, April 23, 1904. [Enclosure n. 5] Extracts from the Kermanshah News Letter n. 10.

56. FO 60/732 Extracts of the Kermanshah News Letter n. 9, April 9, 1904.

57. Scott, "The Recent Cholera Epidemic," 622; FO 60/681 Confidential Dispatch (n. 71) from Sir Arthur Hardinge, British Minister in Tehran, to the Marquess of Lansdowne. Tehran, April 23, 1904. Report of Ehtesham ul-Hokama, Sanitary Officer of Kermanshah.

58. Scott, "The Recent Cholera Epidemic," 621; FO 60/732 Extracts of the Kermanshah News Letter n. 9, April 9, 1904; FO 60/732 Extracts of the Kermanshah News Letter n. 13, April 13, 1904.

59. FO 60/732 Extracts of the Kermanshah News Letter n. 11, April 11, 1904; FO 60/732 Dispatch (n. 71) from Sir Arthur Hardinge to the Marquess of Lansdowne. Tehran, April 23, 1904.

60. FO 60/682 Dispatch (n. 86) from Sir Arthur Hardinge to the Marquess of Lansdowne K.G. Tehran, May 19, 1904.

61. FO 60/682 Dispatch (n. 86) From Sir Arthur Hardinge to the Marquess of Lansdowne K.G. Tehran, May 19, 1904.

62. FO 60/732 Dispatch (n. 71) from Sir Arthur Hardinge to the Marquess of Lansdowne. Tehran, April 23, 1904.

63. Nikki R. Keddie, "Iranian Politics, 1900–1905: Background to Revolution—II," *Journal of Middle Eastern Studies* 5, no. 2 (1969) p. 164.

64. FO 60/681 Confidential Dispatch (n. 71) from Sir Arthur Hardinge to the Marquess of Lansdowne. Tehran, April 23, 1904. [Enclosure n. 5] Extracts from the Kermanshah News Letter (n. 12). Telegraphic Correspondence between Aga Mamegani, the Farman Farmah, and the Ain-ed-Dowala.

65. FO 60/681 Confidential Dispatch (n. 71) from Sir Arthur Hardinge to the Marquess of Lansdowne. Tehran, April 23, 1904. [Enclosure n. 5] Extracts from the Kermanshah News Letter (n. 12). (Telegraphic) The Vazir-i-Azam, ['Ain al-Dowala] to Aga Fazel Mamegani, April 12, 1904.

66. FO 60/732 Extracts of the Kermanshah News Letter n. 12, April 12, 1904.

67. FO 60/732 Dispatch (n. 71) from Sir Arthur Hardinge to the Marquess of Lansdowne. Tehran, April 23, 1904.

68. FO 60/681 Telegraphic Correspondence between Aqa Mamaghani, the Farman Farma, and the Ain al-Dowala in Confidential Dispatch (n. 71) from Sir Arthur Hardinge, British Minister in Tehran, to the Marquess of Lansdowne. Tehran, April 23, 1904.

69. FO 60/732 Extracts of the Kermanshah News Letter n. 11, April 11, 1904.

70. FO 60/682 Dispatch from Joseph Scott to Sir Arthur Hardinge. Kermanshah, April 21, 1904.

71. FO 60/682 Dispatch (n. 86) from Sir Arthur Hardinge to the Marquess of Lansdowne K.G. Tehran, May 19, 1904; and FO 60/682 Dispatch from Joseph Scott to Sir Arthur Hardinge. Kermanshah, April 21, 1904.

72. FO 60/732 Extracts of the Kermanshah News Letter n. 13, April 13, 1904.

73. FO 60/682 Dispatch (n. 86) from Sir Arthur Hardinge to the Marquess of Lansdowne K.G. Tehran, May 19, 1904. [Enclosure n. 4] Telegrams from Kermanshah about Cholera. Extracts from reports by Mr. Rabino. Kermanshah, April 21, 1904.

74. FO 60/682 Extracts from reports by Mr. Rabino. Kermanshah, April 21, 1904; and FO 60/682 Dispatch from Joseph Scott to Sir Arthur Hardinge. Kermanshah, April 21, 1904.

75. FO 60/682 Extracts from reports by Mr. Rabino. Kermanshah, April 21, 1904.

76. FO 60/682 Dispatch (n. 86) from Sir Arthur Hardinge to the Marquess of Lansdowne K.G. Tehran, May 19, 1904.

77. FO 60/682 Dispatch (n. 86) from Sir Arthur Hardinge to the Marquess of Lansdowne K.G. Tehran, May 19, 1904.

78. Scott, "The Recent Cholera Epidemic," 622.

79. FO 60/682 Dispatch from Joseph Scott to Sir Arthur Hardinge. Kermanshah, April 21, 1904.

80. FO 60/682 Extracts from reports by Mr. Rabino. Kermanshah, April 21, 1904.

81. FO 60/682 Dispatch (n. 124) from Sir Arthur Hardinge to the Marquess of Lansdowne K.G. Tehran, July 4, 1904.

82. FO 60/682 Dispatch (n. 86) from Sir Arthur Hardinge to the Marquess of Lansdowne K.G. Tehran, May 19, 1904.

83. Mostofi, *The Administrative*, 2:349.

84. FO 60/682 Dispatch (n. 146) from Sir Arthur Hardinge to the Marquess of Lansdowne K.G. Gulahek, July 15, 1904.

85. FO 60/682 Dispatch (n. 124) from Sir Arthur Hardinge to the Marquess of Lansdowne K.G. Tehran, July 4, 1904.

86. FO 60/682 Dispatch (n. 126) from Sir Arthur Hardinge to the Marquess of Lansdowne K.G. Gulahek, July 17, 1904.

87. FO 60/682 Dispatch (n. 86) from Sir Arthur Hardinge to the Marquess of Lansdowne K.G. Tehran, May 19, 1904; Moojan Momen, *An Introduction to Shi'i Islam: The History and Doctrines of Twelver Shi'ism* (New Haven: Yale University Press, 1985), 246.

88. FO 60/682 Dispatch (n. 86) from Sir Arthur Hardinge to the Marquess of Lansdowne K.G. Tehran, May 19, 1904.

89. FO 60/682 Dispatch (n. 86) from Sir Arthur Hardinge to the Marquess of Lansdowne K.G. Tehran, May 19, 1904.

90. This included Muhammad Baqir Bihbahani, Muhammad Tabataba'i, Shaykh Fazlallah Nuri, and Ali Akbar Tafrashi.

91. FO 60/682 Dispatch from Joseph Scott to Sir Arthur Hardinge. Kermanshah. May 9, 1904.

92. Scott, "The Recent Cholera Epidemic," 622.

93. Scott, "The Recent Cholera Epidemic," 622.

94. FO 60/682 Dispatch (n. 126) from Sir Arthur Hardinge to the Marquess of Lansdowne K.G. Gulahek, July 17, 1904. [Enclosure n. 3] Letter from J. Rabino, Imperial Bank of Persia to Sir Arthur Hardinge.

95. For more on the Lynch Bakhtiyari road, see David McLean, *Britain and Her Buffer State: The Collapse of the Persian Empire, 1890–1914* (London: Royal Historical Society, 1979), 64–66; Henry Blosse Lynch, "Across Luristan to Ispahan," *Proceedings of the Royal Geographical Society* 12 (1890): 523–553.

96. FO 60/682 Dispatch (n. 126) from Sir Arthur Hardinge to the Marquess of Lansdowne K.G. Gulahek, July 17, 1904. [Enclosure n. 2] Paraphrase of telegraphic correspondence with H.M. Residency at Bushire respecting cholera.

97. FO 60/682 Dispatch (n. 146) from Sir Arthur Hardinge to the Marquess of Lansdowne K.G. Gulahek, July 15, 1904. [Enclosure n. 1] Dispatch (n. 33) from George Grahame to Sir Arthur Hardinge. Shiraz, July 27, 1904.

98. FO 60/682 Dispatch (n. 150) from Sir Arthur Hardinge to the Marquess of Lansdowne K.G. Gulahek, August 16, 1904. [Enclosure n. 1] Monthly Summary.

99. After the men were caught and found guilty, Shiraz's Court of Justice sentenced them to have their hands cut off. See FO 60/683 Dispatch (n. 189) from Mr. E. Grant Duff to the

Marquess of Lansdowne K.G. Tehran, October 12, 1904. [Enclosure n. 1] Monthly Summary of Events in Persia not reported in Separate Dispatches.

100. Church Mission Society Archives, University Birmingham Library, Birmingham (CMSA), Special Collections Department, Unofficial Papers, Diaries and Papers of Miss Mary Ellen Brighty of the Persia Mission, CMS F1-3 (Acc113) 1899–1937 Mary Ellen Brighty, Family Papers of Miss Brighty F3: Diaries, October 5, 1904.

101. CMSA, Family Papers of Miss Brighty F3: Diaries, October 5, 1904.

102. FO 60/683 Confidential Dispatch (n. 218) from Mr. E. Grant Duff to the Marquess of Lansdowne K.G. Tehran, December 9, 1904. [Enclosure n. 1] Monthly Summary of Events in Persia not reported in Separate Dispatches.

103. FO 60/685 Decyphered Commercial Telegram (n. 8) from Sir Arthur Hardinge to the Marquess of Lansdowne K.G. Tehran, June 19, 1904; FO 60/686 Commercial Dispatch (n. 11) from George, HBM Consul for Shiraz, to the Marquess of Lansdowne K.G. Shiraz, July 19, 1904; FO 60/683 Dispatch (n. 189) from Mr. E. Grant Duff to the Marquess of Lansdowne K.G. Tehran, October 12, 1904. [Enclosure n. 1] Monthly Summary of Events in Persia not reported in Separate Dispatches.

104. FO 60/683 Dispatch (n. 189) from Mr. E. Grant Duff to the Marquess of Lansdowne K.G. Tehran, October 12, 1904. [Enclosure n. 1] Monthly Summary of Events in Persia not reported in Separate Dispatches.

105. CMSA, Family Papers of Miss Brighty F3: Diaries, July 31, 1904. *Ruza khani* is a Shi'ite public mourning ritual recitation that commemorates the martyrdom of Husayn ibn 'Ali, the third Imam.

106. CMSA, Family Papers of Miss Brighty F3: Diaries, October 3, 1904.

107. CMSA, Family Papers of Miss Brighty F3: Diaries, September 21, 1904.

108. CMSA, Family Papers of Miss Brighty F3: Diaries, September 24, 1904.

109. FO 60/682 Dispatch (n. 126) from Sir Arthur Hardinge to the Marquess of Lansdowne K.G. Gulahek, July 17, 1904.

110. Taj al-Saltana, *Crowning Anguish*, 279.

111. FO 60/682 Confidential Dispatch (n. 120) from Sir Arthur Hardinge to the Marquess of Lansdowne K.G. Gulahek, June 22, 1904. [Enclosure n. 1] Summary of Events in Persia not reported in separate Dispatches; FO 60/685 Commercial Telegram (n. 87) from Mr. E. Grant-Duff to the Marquess of Lansdowne K.G. Tehran, July 8, 1904.

112. Taj al-Saltana, *Crowning Anguish*, 280.

113. FO 60/682 Confidential Dispatch (n. 126) from Sir Arthur Hardinge to the Marquess of Lansdowne K.G. Gulahek, July 17, 1904; AMAE, Correspondance Politique et Commerciale (Nouvelle Série), 1897–1918 (NS), Perse, vol. 1 (1896–1904). Dépêche (direction politique n. 47) Mr. Defrance, ministre de France à Téhéran à Son Excellence le ministre des affaires étrangères à Paris. Téhéran, 6 Août, 1904.

114. AMAE, Correspondance Politique et Commerciale (Nouvelle Série), 1897–1918 (NS), Perse, vol. 1 (1896–1904). Dépêche (direction politique n. 47) Mr. Defrance, ministre de France à Téhéran à Son Excellence le ministre des affaires étrangères à Paris. Téhéran, 6 Août, 1904.

115. AMAE, Correspondance Politique et Commerciale (Nouvelle Série), 1897–1918 (NS), Perse, vol. 1 (1896–1904), Dépêche (direction politique n. 47). 6 Août, 1904.

116. FO 60/682 Dispatch (n. 126) from Sir Arthur Hardinge to the Marquess of Lansdowne K.G. Gulahek, July 17, 1904. [Enclosure n. 3] Letter from J. Rabino, Imperial Bank of Persia, to Sir Arthur Hardinge; FO 60/682 Confidential Dispatch (n. 126) from Sir Arthur Hardinge to the Marquess of Lansdowne K.G. Gulahek, July 17, 1904; John G. Wishard, "Addresses and Original Communications," *Indiana Medical Journal* 23, no. 4 (1904): 130.

117. Wishard, "Addresses," 130.

118. FO 60/682 Dispatch (n. 126) from Sir Arthur Hardinge to the Marquess of Lansdowne K.G. Gulahek, July 17, 1904.

119. Wishard, "Addresses," 130; and Taj al-Saltana, *Crowning Anguish*, 280–281.

120. FO 60/682 Dispatch (n. 126) from Sir Arthur Hardinge to the Marquess of Lansdowne K.G. Gulahek, July 17, 1904. [Enclosure n. 4] Copy of Letter from J. Rabino, Imperial Bank of Persia to R. H New, Director of the Indo-European telegraph. Tehran, July 12, 1904.

121. "Dastur al-amal-i marizkhana-yi yingi dunya-yi bara-yi hifz-i sihat va ihtiyat" [The prescription of the American Hospital for hygiene and prevention], *Tarbiyat*, no. 324, 7 Shahr Jumadi Al-Awwal 1322 [July 21, 1904], 1503.

122. FO 60/682 Dispatch (n. 126) from Sir Arthur Hardinge to the Marquess of Lansdowne K.G. Gulahek, July 17, 1904. [Enclosure n. 4] Copy of Letter from J. Rabino, Imperial Bank of Persia to R. H New, Director of the Indo-European telegraph. Tehran, July 12, 1904. Hajji Siyah Mahalati and Mutamid al-Saltana were among the people from Tehran credited with contributing to the cholera relief fund. See *Tarbiyat*, no. 324, 7 Shahr Jumadi al-Awwal 1322 AH [July 21, 1904], 1503–1504.

123. *Tarbiyat*, no. 324, 7 Shahr Jumadi al-Awwal 1322 AH [July 21, 1904], 1504.

124. *Tarbiyat*, no. 324, 7 Shahr Jumadi al-Awwal 1322 AH [July 21, 1904], 1504; John G. Wishard, *Twenty Years in Persia: A Narrative of Life under the Last Three Shahs* (New York: Flemming H. Revell, 1908), 222–223.

125. Arthur C. Boyce, "Alborz College of Tehran and Dr. Samuel Martin Jordan, Founder and President," in *Cultural Ties between Iran and the United States*, ed. Ali Pasha Saleh (Tehran: Sherkat-e Chapkhaneh Bistopanj-e-Shahrivar, 1976), 169.

126. Wishard, "Addresses," 130.

127. Boyce, "Alborz College," 169; Wishard, *Twenty Years*, 220–221; Wishard, "Addresses," 130.

128. Wishard, "Addresses," 130–131.

129. Wishard, "Addresses," 131.

130. Wishard, "Addresses," 130.

131. Wishard, *Twenty Years*, 222.

132. Wishard, *Twenty Years*, 220; the booklet appears to have been similar to pamphlets on prevention distributed by charities and insurance companies throughout the United States. See G. S. Winston and W. R. Gillette, *Hints to Prevent Cholera and also its Treatment*, 3rd ed. (New York: Mutual Life Insurance Company of New York, 1885).

133. Wishard, "Addresses," 130.

134. *Tarbiyat*, no. 324, 7 Shahr Jumadi Al-Awwal 1322 AH [July 21, 1904], 1501.

135. One recommendation for those who absolutely had to drink was to "boil" their spirits. See *Tarbiyat*, no. 324, 7 Shahr Jumadi Al-Awwal 1322 AH [July 21, 1904], 1502.

136. FO 60/682 Dispatch (n. 146) from Sir Arthur Hardinge to the Marquess of Lansdowne K.G. Gulahek, July 15, 1904.

137. Coppin, "Rapport sur l'épidemie," 274–276.

138. Coppin, "Rapport sur l'épidémie," 274–276.

139. Coppin, "Rapport sur l'épidémie," 276.

140. Scott, "The Recent Cholera Epidemic," 622.

141. The American hospital was established during the 1892 cholera epidemic. A smaller religiously endowed hospital was founded in 1901 by a local cleric in Tehran. The Cossack Brigade's hospital and infirmary, under the direction of a Russian military physician, served only the needs of its own soldiers. A thirty-bed Russian Red Cross hospital, with the only trained (Russian) midwife in Tehran, began operating near the end of the 1904 cholera outbreak. The former minister of war also willed a considerable amount of money to build a vast hospital

three kilometers outside of Tehran. However, the government converted the facility to a sanatorium for people with mental illness and severe developmental disabilities because of the impossibility of transporting the medically ill over the long distance. See Scott, "The Recent Cholera Epidemic," 622; AMAE, Correspondance Politique et Commerciale (Nouvelle Série), 1897–1918 (CP), Perse, Français au Service de Perse: médecins et vétérinaires: dossier personnelles, 1908–1916, vol. 57, dossier 6A. "Rapport Sur L'Organisation de L'Enseignement Médical en Perse," contenu dans dépêche (direction politique n. 778) Dr. Schneider, Médecin Principal de 1ère Classe, au Ministère des Affaires étrangères, Téhéran, 19 Juin 1906.

142. "Zaban-i milli" [National language], *Nida-yi vatan*, no. 84, 11 Ramazan 1325 AH [October 19, 1907], 2.

143. Wishard, "Addresses," 130.

144. "Baghiya-yi tahghighat-i jinab-i jallalatmaab Duktur Mirza Muhammad Kermanshahi (maraz i kolira)" [The rest of the research of His Excellency Dr. Mirza Muhammad Kermanshahi (the disease of cholera)], *Tarbiyat*, no. 323, 30 Rabi' al-Thani 1322 AH [July 14, 1904], 1498.

145. Taj al-Saltana, *Crowning Anguish*, 282.

146. *Tarbiyat*, no. 324, 7 Shahr Jumadi al-Awwal 1322 AH [July 21, 1904], 1501–1503.

147. *Tarbiyat*, no. 326, 26 Jumadi al-Awwal 1322 AH [August 11, 1904] 1522–1523.

148. *Tarbiyat*, no. 326, 26 Jumadi al-Awwal 1322 AH [August 11, 1904] 1522–1523.

149. FO 60/732 Procès-Verbal, Conseil Sanitaire de L'Empire de Perse, August 6, 1904; *Tarbiyat*, no. 328, 6 Jamadi 1322 AH [August 25, 1904], 1536–1537.

150. FO 60/732 Copy of Confidential Dispatch (n. 143) from Sir Arthur N. Hardinge, British Minister in Tehran, to the Marquess of Lansdowne, Secretary of State for Foreign Affairs. Gulahek, August 3, 1904.

151. The rehabilitation of Schneider's patron, Amin al-Sultan, who recovered his premiership following the downfall of Amin al-Dawla in 1898, could explain the Iranian government's growing esteem for the French physician. Schneider's clinical acumen in treating the sickly monarch also increased his influence. In 1901 Schneider earned the shah's appreciation when he correctly diagnosed the cardiac cause of his palpitations; something his other physicians (Hugh Adcock and Hakim al-Mulk) had been unable to do. He also successfully treated one of the shah's favorite sons. Schneider's accomplishments put him at odds with Adcock, who accused him of committing "professional misconduct" when he evaluated the shah without being asked to do so by the monarch's chief physicians. This led to the French government's request for Schneider to be officially appointed as chief physician to the shah to stem any future conflicts of interests. See AMAE, Correspondance Politique et Commerciale (Nouvelle Série), 1897–1918 (CP), Perse, Français au Service de Perse: médecins dossier personnelles, 1896–1905, vol. 56, dossier 6A. Dépêche (direction politique n. 48) Dr. Schneider à Son Excellence le ministre des affaires étrangères à Paris. Paris, 27 Juillet, 1902; FO 60/681 Dispatch (n. 6) from E. Grant-Duff to the Marquess of Lansdowne K.G. Tehran, January 6, 1904. [Enclosure n. 1] Monthly summary of Events in Persia not reported in separate dispatches.

152. Godard, Docteurs appelés en consultation à Contrexéville , June 1900, Bibliothèque interuniversitaire de Santé (BIU Santé), Paris, Collection de portraits, cote: CIPB0443. The assembled international group of physicians determined that the shah had a "mild gouty diathesis" after examining the monarch for three and a half hours on the morning of June 18, 1900. It is interesting to note that most of them lacked an expertise in renal diseases. Eugen Hollander, a German surgeon, had been a pioneer of the facelift. Richard Friedrich Johannes Pfeiffer, also German, was a renowned bacteriologist. Albert-Emile Debout d'Estrées had been the chief physician at the Contrexéville Spa for more than thirty years. Hugh Adcock, the shah's English physician-in-chief, had treated the monarch as a generalist for twelve years before being replaced by Jean-Etienne Justin Schneider after this trip. Mahmud Khan Hakim al-Mulk served as both the shah's physician and court minister, and Khalil Khan A'lam al-Dawla Saqafi was the official

Iranian court doctor. Ibrahim Khan, another French-educated Iranian physician, was likely invited at the prompting of his uncle Mahmud Khan. Only Sigismond Jaccoud, a Swiss-born physician and well-regarded lecturer at several of Paris's hospitals, had written his dated thesis on the pathophysiology of kidney failure and albuminuria. See "La maladie du Shah de Perse," *La Chronique Medicate: Revue Bi-Mensuelle de Medecine Historique, litteraire et Anecdotique* 7, no. 1 (1900): 530–531.

153. AMAE, Correspondance Politique et Commerciale (Nouvelle Série), 1897–1918 (CP), Perse, Français au Service de Perse: médecins dossier personnelles, 1896–1905, vol. 56, dossier 6A. Dépêche (direction politique n. 48) Dr. Schneider à Son Excellence le ministre des affaires étrangères à Paris. Paris, 27 Juillet, 1902; FO 60/681 Dispatch (n. 6) from E. Grant-Duff to the Marquess of Lansdowne K.G. Tehran, January 6, 1904; FO 60/732 Telegram (n. 143) from Sir Arthur N. Hardinge, British Minister in Tehran, to the Marquess of Lansdowne, Secretary of State for Foreign Affairs, August 3, 1904.

154. "Medical News," *British Medical Journal* 1, no. 2316 (1905): 1104.

155. Besides Schneider, the participants included Muhtashim al-Saltana, from the Ministry of Foreign Affairs; Muhandis Bashi, from the Ministry of Interior; a delegate from the Ministry of Customs and Posts; Riza Quli Khan, from the Ministry of Public Instruction; Nazim al-Attiba, one of the shah's Iranian doctors; and the physicians to the British, American, German, and Russian legations in Tehran. The Cossack Brigade's physician also attended along with the shah's personal French pharmacist. Tehran's municipal administrators such as Sa'id al-Saltana, the minister of police, and I'temad al-Saltana, the municipalities minister, were invited but were unable to attend. See FO 60/732 Procès-Verbal, Conseil Sanitaire de L'Empire de Perse, August 6, 1904.

156. FO 60/732 Procès-Verbal, Conseil Sanitaire de L'Empire de Perse, August 6, 1904.

157. 60/732 Copy of Confidential Dispatch (n. 143) from Sir Arthur N. Hardinge, British Minister in Tehran, to the Marquess of Lansdowne, Secretary of State for Foreign Affairs. Gulahek, August 3, 1904.

158. 60/732 Copy of Confidential Dispatch (n. 143) from Sir Arthur N. Hardinge to the Marquess of Lansdowne. Gulahek, August 3, 1904.

159. "Akhbar-i dakhila" [Domestic news], *Tarbiyat*, no. 328, 6 Jamadi 1322 AH [August 25, 1904] 1536.

160. This included the absence of public laundry places, covered waterways, clean streets, abattoirs, and cemeteries outside the city walls. See "Dar al-khalafa va majlis-i hifz i sihhat" [The capital and the Sanitary Council], *Tarbiyat*, no. 353, 25 Zu'l Hijjah 1322 AH [March 2, 1905], 1742–1743; "Akhbar-i dakhila" [Domestic news], *Tarbiyat*, no. 328, 6 Jamadi 1322 AH [August 25, 1904] 1536–1537.

161. AMAE, Correspondance Politique et Commerciale (Nouvelle Série), 1897–1918 (NS), Perse, vol. 1 (1896–1904). Dépêche (direction politique n. 49) Mr. Defrance, ministre de France à Téhéran à Son Excellence le ministre des affaires étrangères à Paris. Téhéran, 28 Août, 1904.

162. FO 248/818 Residency Diary (n. 76) from Major Cox, HBM acting Consul General for Fars, Khusistan, etc., to Sir Arthur Hardinge. Bushire June 11, 1904. [Enclosure n. 1] Shiraz Newsletter June 9 to 29, 1904.

163. Document (n. 145) Kermanshah [n.d.], in Ettehadieh et al., *Vaba-yi alamgir*, 196.

164. "Persia Faces Famine. Minister Tells of Dire Want among His People. Many Facing Starvation," *Washington Post*, November 10, 1905.

165. Document (n. 148) Kermanshah [n.d.], in Ettehadieh et al., *Vaba-yi alamgir*, 199–200.

166. FO 60/682 Confidential Dispatch (n. 118) from Sir Arthur Hardinge to the Marquess of Lansdowne K.G. Gulahek, June 21, 1904.

167. FO 60/682 Dispatch (n. 150) from Sir Arthur Hardinge to the Marquess of Lansdowne K.G. Gulahek, August 16, 1904. [Enclosure n. 1] Monthly Summary.

168. FO 60/681 Confidential Dispatch (n. 58) from Sir Arthur Hardinge to the Marquess of Lansdowne K.G. Tehran, March 30, 1904. [Enclosure n. 1] Monthly Summary of Events in Persia not reported in separate dispatches; FO 60/683 Dispatch (n. 173) from Sir Arthur Hardinge to the Marquess of Lansdowne K.G. Gulahek, September 11, 1904. [Enclosure n. 1] Monthly Summary of Events in Persia; FO 60/686 Commercial Dispatch (n. 11) from George, HBM Consul for Shiraz, to the Marquess of Lansdowne K.G. Shiraz, July 19, 1904.

169. The fear of cholera had caused soldiers to abandon their roadway guard duty, leading to the prevalence of highway brigandage and insecurity throughout Iran. FO 248/820 Extracts from Kerman Diary for the 3rd week in November 1904 (n. 330) Major P. Molesworth Sykes, Consul at Kerman, to Evelyn Grant-Duff. Kerman, November 20, 1904.

170. FO 248/820 Extracts from Kerman Diary for the last week of October 1904 (n. 315) Major P. Molesworth Sykes, Consul at Kerman, to Evelyn Grant Duff. Kerman, November 1, 1904; FO 248/820 Extracts from Kerman Diary for the 1st week in November 1904 (n. 330) Major P. Molesworth Sykes, Consul at Kerman, to Evelyn Grant-Duff. Kerman, November 20, 1904; FO 60/688 Major P. Z Cox, Officiating Political Resident to Persian Gulf, to L. W. Dane, Secretary to the Government of India. Bushehr, September 18, 1904. [Enclosure n. 1] Confidential Ahwaz Diary (n. 9) August 30, 1904 (by Lt. Lorimer).

171. The Russo-Japanese War also contributed to the loss of trade revenue. Iran had lost more than £200,000 in customs dues in 1904; see FO 60/682 Confidential Dispatch (n. 118) from Sir Arthur Hardinge to the Marquess of Lansdowne K.G. Gulahek, June 21, 1904; Hooshang Amirahmadi, *The Political Economy of Iran under the Qajars* (London: I. B. Tauris, 2012), 152.

172. The moneychangers (*saraf*) of Tehran had seventy thousand tuman owed to them from the Treasury. See Vanessa Martin, *Islam and Modernism: The Iranian Revolution of 1906* (Syracuse: Syracuse University Press, 1989), 53.

173. Percy Sykes, *A History of Persia* (London: Macmillan, 1921), 2:400–401; Annette Destrée, *Les fonctionnaires belges au service de la Perse* (Leiden: E. J. Brill, 1976), 119.

174. AMAE, Correspondance Politique et Commerciale (Nouvelle Série), 1897–1918 (NS), Perse, vol. 1 (1896–1904). Dépêche (direction politique n. 49) Mr. Defrance, ministre de France à Téhéran à Son Excellence le ministre des affaires étrangères à Paris. Téhéran, 28 Août, 1904.

175. For more on secret societies in Iran, see Ann K. Lambton, "Secret Societies and the Persian Revolution of 1905–1906," in *Qajar Persia: Eleven Studies*, ed. Ann K. Lambton (Austin: University of Texas Press, 1988), 301–308; Janet Afary, *The Iranian Constitutional Revolution, 1906–1911: Grassroots Democracy, Social Democracy, and the Origins of Feminism* (New York: Columbia University Press, 1996), 37–50.

176. Members included Nazim al-Islam Kirmani, Mirza Ahmad Kirmani, Shaykh Muhammad Filsuf al-Dawala Shirazi, Sayid Burhan-i Khalkhali, and Shaykh Husayn 'Ali Bihbahani. See Edward Granville Browne, *The Persian Revolution of 1905–1909* (Cambridge: Cambridge University Press, 1910), 420.

177. Martin, *Islam and Modernism*, 60.

178. Nazim al-Islam Kirmani, *Tarikh-i bidari-yi Iranian* [The history of Iranian awakening], 4th ed. (Tehran: Intesharat-i agah, 1362 AS [1983–1984]), 2:246.

179. Kirmani, *Tarikh-i bidari*, 249–251.

180. This attack on Iran's poor educational system, particularly in the sciences, continued in the Constitutional period. See Kirmani, *Tarikh-i bidari*, 252–254; "Vizarat-i ulum" [The Ministry of Science], *Habl al-matin*, no. 2, 19 Rabi al-Thani 1325 AH [June 1, 1907], 1–2; "Tiligraf-i makhsus-i idarih" [Special office telegraph], *Habl al-matin*, no. 32, 22 Rabi al-Thani 1325 AH [June 4, 1907], 1.

181. Kirmani, *Tarikh-i bidari*, 260.

182. Kirmani, *Tarikh-i bidari*, 260–261.

183. The Secret Society went to great lengths to adopt a correct religious tone and rhetoric, even while addressing worldly concerns such as the lack of hygiene in the country. The society repeatedly emphasized that its views were not contrary to Islam. This, along with the fact that most of the members of the society were mid-ranking clerics, meant that Western notions of disease causation and sanitation began to overshadow traditional religious views of the subject. See Magnol Bayat, *Iran's First Revolution: Shi'ism Constitutional Revolution of 1905–1909* (Oxford: Oxford University Press, 1991), 73.

184. *Habl al-matin*, no. 2, March 23, 1905, cited in Ervand Abrahamian, "The Causes of the Constitutional Revolution in Iran," *International Journal of Middle Eastern Studies* 10 (1979): 404.

185. "Persia Faces Famine," *Washington Post*, November 10, 1905.

186. AMAE, Correspondance Politique et Commerciale (Nouvelle Série), 1897–1918 (NS), Perse, vol. 1 (1896–1904). Dépêche (direction politique n. 47) Mr. Defrance, ministre de France à Téhéran à Son Excellence le ministre des affaires étrangères à Paris. Téhéran, 6 Août, 1904.

187. Browne, *The Persian Revolution*, 111.

188. Bayat, *Iran's First Revolution*; Martin, *Islam and Modernism*, 110.

Chapter 5 · Wars, Plagues, and Institutional Developments in Health, 1906–1926

1. Ministère des Affaires étrangères, *Conférence Sanitaire Internationale de Paris, Procès-Verbaux [7 Novembre 1911–17 Janvier 1912]* (Paris: Imprimerie Nationale, 1912), 60.

2. "Persia. Annual Report, 1909," in *Iran Political Diaries, 1881–1965*, vol. 4: *1908–1909*, ed. Robert Michael Burrell and Robert L. Jarman (London: Archives Editions, 1997), 417–418.

3. Jean-Etienne Justin Schneider, *Médecine Persane: Les Médecins Français en Perse-Leur influence* (Paris: Société Anonyme des Imprimeries Wellhoff & Roche, 1911), 15.

4. Schneider, *Médecine Persane*. École du Service de santé militaire de Lyon was consolidated and renamed École de santé des armées in 2011 and is currently the French military's principal school of medicine.

5. Ministère des Affaires étrangères, *Conférence Sanitaire Internationale de Paris, Procès-Verbaux [10 Octobre–3 Décembre 1903]*, 456.

6. Homa Nategh, "Les Persans à Lyon," in *Pand-o Sokhan*, ed. Christoph Balay, Claire Kapper, and Ziva Vesel (Tehran: Institute Français de Recherche en Iran, 1995), 191.

7. AMAE, Correspondance Politique et Commerciale (Nouvelle Série) 1897–1918 (CP), Perse, Français au service de Perse: médecins et vétérinaires: dossier personnelles 1908–1916, vol. 57, Dossier 6A. Dépêche (direction politique n. 778) Dr. Schneider, Médecin Principal de 1ère Classe, au ministre des affaires étrangères, "Rapport Sur L'Organisation de L'Enseignement Médical en Perse," Téhéran, 19 Juin 1906.

8. Cyril Elgood, *A Medical History of Persia and the Eastern Caliphate from the Earliest Times until the Year A.D. 1932* (Cambridge: Cambridge University Press, 1951), 536.

9. The first-year curriculum at the Polytechnic College required the instruction of basic natural sciences, and the second year was dedicated to clinically oriented fields such as anatomy, physiology, and pathology. See Schneider, *Médecine Persane*, 5–11.

10. Ministère des Affaires étrangères, *Conférence Sanitaire Internationale de Paris, Procès-Verbaux [7 Novembre 1911–17 Janvier 1912]*, 791.

11. The decade preceding the First World War saw the ascent of French medical influence in Iran. Several French military and civilian physicians assumed influential positions as private clinicians to Iran's political notables including the prince-governor of Isfahan, the minister of war, the prince-governor of the province of Fars, and the prince-governor of the province of Gilan. In addition to their role in education, the French were hired to key positions in other

paramedical fields such as chief veterinarian of the royal stables, royal pharmacist, and the shah's dentist. See Schneider, *Médecine Persane*, 6–7, and Djalal Brimani, *Histoire des Relation Médicales Franco Iraniennes* (Tehran: Vahid, 1370 AS [1992]), 136.

12. National Library of Medicine, History of Medicine Collection, Bethesda (NLM), Procès-Verbal 169ème Séance Extraordinaire du Conseil Sanitaire de l'Empire de Perse, January 20, 1916, 418.

13. FO 60/683 Dispatch (n. 173) from Sir Arthur Hardinge to the Marquess of Lansdowne K.G. Gulahek, September 11, 1904. [Enclosure n. 1] Monthly Summery of Events in Persia; FO 60/683 Dispatch (n. 189) from Mr. E. Grant Duff to the Marquess of Lansdowne K.G. Tehran, October 12, 1904. [Enclosure n. 1] Monthly Summery of Events in Persia not reported in Separate.

14. Charles Marie Jules Viry, *Principes d'Hygiène Militaire* (Paris: L. Battaille Et Cie, 1896), 3–4. During the Franco-Prussian War, smallpox killed more than twenty thousand French soldiers but fewer than five hundred Germans due to the implementation of a vaccination program by the German military.

15. Moussa Khan, *Contribution a l'Etude hygiénique des Nouveaux Appareils de Chauffage Sans tuyau de dégagement* (Paris: A. Maloine, 1906), 5.

16. Nategh, "Les Persans à Lyon," 197–198.

17. NLM, Procès-Verbal 103ème Séance ordinaire du Conseil Sanitaire de l'Empire de Perse, 3 Juin 1912, 94.

18. NLM, Procès-Verbal 169ème Séance Extraordinaire du Conseil Sanitaire de l'Empire de Perse, 20 Janvier 1916, 416.

19. Archives Diplomatiques, Royaume de Belgique, Ministère des Affaires étrangères, Brussels (MAE), vol. 2890, Rapport de la Légation de Belgique à Tehran [Report of the Belgian Legation in Tehran], Decembre 27, 1898.

20. "Persia. Annual Report, 1926," in *Iran Political Diaries, 1881–1965*, vol. 7: *1924–1926*, ed. Robert Michael Burrell and Robert L. Jarman (London: Archives Editions, 1997), 604.

21. Amir Afkhami, "The Sick Men of Persia: The Importance of Illness as a Factor in the Interpretation of Modern Iranian Diplomatic History," *Journal of Iranian Studies* 36 (2003): 339–352.

22. During Muzaffar al-Din Shah's visit to Europe in 1900, Schneider and the French ambassador in Tehran convinced Paris to send a French physician to Tabriz to maintain France's influence over the crown prince. The French government selected Charles-René Coppin, a physician major in the French Colonial Troops, for this post. See Schneider, *Médecine Persane*, 4.

23. Afary, *The Iranian Constitutional Revolution*, xx.

24. "Persia. Annual Report, 1909," in Burrell and Jarman, *Iran Political Diaries*, 4:417–418.

25. "Persia. Annual Report, 1909," in Burrell and Jarman, *Iran Political Diaries*, 4:418.

26. L/P & S/10/123 (Register n. 3209, encl. n. 186–191) Commercial Dispatch (n. 101) Sir C. Spring-Rice to Sir Edward Grey. Tehran, May 19, 1907.

27. Government of India, *Administration Report of the Persian Gulf Political Residency for 1905–1906* (Calcutta: Office of the Superintendent of Government Printing, 1907), 6:14.

28. IO L/P & S/10/123 (Register n. 973-4, encl. n. 200) Telegram by Viceroy of India to Foreign Secretary. Calcutta, February 6, 1907; IO L/P & S/10/123 (Register n. 3081, encl. n. 226–229) Commercial Dispatch (n. 7) Sir C. Spring-Rice to Sir Edward Grey. Tehran, April 19, 1907; IO L/P & S/10/123 (Register n. 973, encl. n. 195) Telegram by Viceroy of India to Foreign Secretary (repeated to Tehran and Bushehr), Calcutta, June 5, 1907 and IO L/P & S/10/123, (Register n. 195, encl. n. 183) Reference Paper, Political and Secret Department, to Under Secretary of State (Foreign Office circular) London, June 24, 1907; Government of India, *Administration Report*, 416.

29. "Persia. Annual Report, 1911," in *Iran Political Diaries, 1881–1965*, vol. 5: *1910–1920*, ed. Robert Michael Burrell and Robert L. Jarman (London: Archives Editions, 1997), 395; Ministère des Affaires étrangères, *Conférence Sanitaire Internationale de Paris, Procès-Verbaux [7 Novembre 1911–17 Janvier 1912]*, 61.

30. IO L/P & S/10/184 (Register n. 2918, encl. n. 1785) Commercial Dispatch (n. 34) George Barclay to Sir Edward Grey. Tehran, December 28, 1909.

31. "Persia. Annual Report, 1910," in Burrell and Jarman, *Iran Political Diaries*, 5:128–129.

32. Ministère des Affaires étrangères, *Conférence Sanitaire Internationale de Paris, Procès-Verbaux [7 Novembre 1911–17 Janvier 1912]* (Paris: Imprimerie Nationale, 1912), 62; twenty thousand tuman was equivalent to thirty-five thousand dollars at the time, see Carroll Davidson Wright, ed., *The New Century Book of Facts: A Handbook of Ready Reference* (Springfield: The King Richardson Company, 1909), 553.

33. "Persia. Annual Report, 1910," in Burrell and Jarman, *Iran Political Diaries*, 5:128–129.

34. IO L/P & S/10/184 (Register n. 3883, encl. n. 31325) Commercial Dispatch (n. 29) A. R. Neligan to M. Marling Esq., His Britannic Majesty's Chargé d'Affaires. Gulhek, August 2, 1910.

35. "Persia. Annual Report, 1912," in Burrell and Jarman, *Iran Political Diaries*, 5:488.

36. Ministère des Affaires étrangères, *Conférence Sanitaire Internationale de Paris, Procès-Verbaux [7 Novembre 1911–17 Janvier 1912]*, 61 ; IO L/P & S/10/184 File 3004/1910 "Tehran Sanitary Council" (Register n. 13190) Procès-Verbal 87ème Séance du Conseil Sanitaire de l'Empire de Perse, 6 Février 1911, 8.

37. The tax was fixed at seven qiran per cadaver and increased to twelve qiran in 1912. See NLM, Procès-Verbal 103ème Séance ordinaire du Conseil Sanitaire de l'Empire de Perse, 3 Juin 1912, 96; "Persia. Annual Report, 1910," in Burrell and Jarman, *Iran Political Diaries*, 5:128–129.

38. NLM, Procès-Verbal 98ème Séance ordinaire du Conseil Sanitaire de l'Empire de Perse, 8 Janvier 1912, 76; "Persia. Annual Report, 1910," in Burrell and Jarman, *Iran Political Diaries*, 5:62; "Persia. Annual Report, 1911," in Burrell and Jarman, *Iran Political Diaries*, 5:397; "Persia. Annual Report, 1913," in Burrell and Jarman, *Iran Political Diaries*, 5:588.

39. Elgood, *A Medical History*, 532.

40. Elgood, *A Medical History*, 532. Dr. Von Regimentsarzt Feistmantel, "Oeffentliches Sanitätswesen. Zur Frage des sanitären Schutzes des Persichens Golfs," *Deutsche Medizinische Wochenschrift* 34, no. 1 (1908): 655–659; Dr. Von Regimentsarzt Feistmantel, "Die sanitären Verhältnisse in Persien," *Wiener Klinische Wochenschrift* 21 (1908): 323–327.

41. Elgood, *A Medical History*, 532; "Persia. Annual Report, 1909," in Burrell and Jarman, *Iran Political Diaries*, 4:416–417.

42. AMAE, Correspondance Politique et Commerciale (Nouvelle Série) 1897–1918 (CP), Perse, Français au Service de Perse: médecins dossier personnelles 1896–1905 volume 57, dossier 6A. "Compte-Rendu du médecin-major de 1ère classe Georges, en mission hors cadres sans solde auprès du gouvernement Persan, professeur à l'école de médecine de Téhéran," in Dépêche (direction politique n. 41) Le Chargé d'Affaires de France à Téhéran à Son Excellence le ministre des affaires étrangères à Paris. Téhéran, 25 Mai, 1906.

43. AMAE, Correspondance Politique et Commerciale (Nouvelle Série) 1897–1918 (CP), Perse, Français au Service de Perse: médecins dossier personnelles 1896–1905, volume 57, dossier 6A. Dépêche (direction politique n. 66) Le Chargé d'Affaires de France à Téhéran à Son Excellence le ministre des affaires étrangères à Paris. Téhéran, 9 Septembre, 1906.

44. AMAE, Correspondance Politique et Commerciale (Nouvelle Série) 1897–1918 (CP), Perse, Français au service de Perse: médecins et vétérinaires: dossier personnelles 1908–1916, volume 57, dossier 6A. "Rapport Sur L'Organisation de L'Enseignement Médical en Perse," contenu dans Dépêche (direction politique n. 778) Dr. Schneider, Médecin Principal de 1ère Classe, au ministre des affaires étrangères, Téhéran, 19 Juin 1906.

45. NLM, Procès-Verbal 111ème Séance ordinaire du Conseil Sanitaire de l'Empire de Perse, 2 Décembre 1912; NLM, Procès-Verbal 112ème Séance ordinaire du Conseil Sanitaire de l'Empire de Perse, 26 Décembre 1912; NLM, Procès-Verbal 113ème Séance ordinaire du Conseil Sanitaire de l'Empire de Perse, 4 Janvier 1913; NLM, Procès-Verbal 115ème Séance ordinaire du Conseil Sanitaire de l'Empire de Perse, 4 Mars 1913; NLM, Procès-Verbal 116ème Séance ordinaire du Conseil Sanitaire de l'Empire de Perse, 8 Avril 1913.

46. Afary, *The Iranian Constitutional Revolution*, 317–324; Edward Granville Browne, *The Persian Crisis of December, 1911; How It Arose and Whither It May Lead Us; Compiled for the Use of the Persia Committee* (Cambridge: Privately Printed at the Cambridge University Press, 1912); Charles Willis Thompson, "How Russia Came to Make War on W. Morgan Shuster," *New York Times*, November 26, 1911; William Morgan Shuster, *The Strangling of Persia: Story of the European Diplomacy and Oriental Intrigue That Resulted in the Denationalization of Twelve Million Mohammedans, a Personal Narrative* (New York: Century, 1912).

47. Jacques Philippe Gachet arrived in Tehran on April 1911 as Georges's replacement at the Polytechnic College. He was a French military physician (first-class physician in the navy) and professor at the French naval medical school (L'École de médecine navale de Rochefort). See NLM, Procès-Verbal 169ème Séance Extraordinaire du Conseil Sanitaire de l'Empire de Perse, 20 Janvier 1916, 418; "Persia. Annual Report, 1912," in Burrell and Jarman, *Iran Political Diaries*, 5:487.

48. "Persia. Annual Report, 1911," in Burrell and Jarman, *Iran Political Diaries*, 5:397; NLM, Procès-Verbal 104ème Séance ordinaire du Conseil Sanitaire de l'Empire de Perse, 1 Juillet 1912; NLM, Procès-Verbal 108ème Séance ordinaire du Conseil Sanitaire de l'Empire de Perse, 7 Octobre 1912, 107.

49. "Persia. Annual Report, 1913," in Burrell and Jarman, *Iran Political Diaries*, 5:588.

50. NLM, Procès-Verbal 100ème Séance ordinaire du Conseil Sanitaire de l'Empire de Perse, 28 Février 1912, 83.

51. NLM, Procès-Verbal 99ème Séance ordinaire du Conseil Sanitaire de l'Empire de Perse, 5 Février 1912, 80; IO L/P & S/10/184 File 3004/1910 "Tehran Sanitary Council" (Register n. 33759) Persian Sanitary Council, report of the 93rd meeting of. Commercial Dispatch (n. 17) Sir G. Barclay to Sir Edward Grey. Gulhek, August 15, 1911 [Enclosure n. 1] A. R. Neligan to G. Barclay. Gulhek, August 8, 1911.

52. "Persia. Annual Report, 1912," in Burrell and Jarman, *Iran Political Diaries*, 5:488; IO L/P & S/10/184 File 3004/1910 'Tehran Sanitary Council' (Register n. 41609) Commercial Dispatch (n. 20) Sir G. Barclay to Sir Edward Grey. Gulhek, September 9, 1911; IO L/P & S/10/184 File 3004/1910 'Tehran Sanitary Council' (Register n. 46109) Procès-Verbal 95ème Séance du Conseil Sanitaire de l'Empire de Perse, 2 Octobre 1911, 56–57; NLM, Procès-Verbal 160ème Séance ordinaire du Conseil Sanitaire de l'Empire de Perse, 2 Octobre 1915, 381–388.

53. Iran was a signatory to the 1906 Geneva Convention for the Amelioration of the Condition of the Wounded and Sick in Armies in the Field (second Geneva Convention) and a regular participant in the international Red Cross conferences, including its ninth gathering held in Washington D.C. in 1912. See American Red Cross, *Neuvième Conférence Internationale de la Croix Rouge tenue à Washington du 7 au 17 Mai 1912, Compte Rendu* (Washington, DC: Press of Judd & Detweiler, 1912).

54. United Kingdom. Parliament, Accession of Persia to the International Agreement of December 9, 1907, Respecting the Creation of an International Office of Public Health. Cd. 4901. 1909.

55. Josep L. Barona, *The Rockefeller Foundation, Public Health and International Diplomacy, 1920–1945* (London: Pickering & Chatto, 2015), 16–17.

56. IO L/P & S/10/184 File 3004/1910 'Tehran Sanitary Council' (Register n. 13190) Procès-Verbal 87ème Séance du Conseil Sanitaire de l'Empire de Perse, 6 Février 1911, 12.

57. Norman Howard-Jones, "The Scientific Background of the International Sanitary Conferences 1851–1938. 6." *WHO Chronicle* 28, no. 11 (1974): 495–496.

58. IO L/P & S/10/184 File 3004/1910 'Tehran Sanitary Council' (Register n. 41610) Persian Sanitary Council. Commercial Dispatch (n. 20) Sir G. Barclay to Sir Edward Grey. Gulhek, October 4, 1911 [Enclosure n. 1] A. R. Neligan to G. Barclay. Gulhek, October 3, 1911.

59. Ministère des Affaires étrangères, *Procès-Verbaux [10 Octobre–3 Décembre 1903]*, 457.

60. Ministère des Affaires étrangères, *Procès-Verbaux [7 Novembre 1911–17 Janvier 1912]*, 61.

61. Ministère des Affaires étrangères, *Procès-Verbaux [7 Novembre 1911–17 Janvier 1912]*, 793.

62. IO L/P & S/10/184 File 3004/1910 'Tehran Sanitary Council' (Register n. 33759) Persian Sanitary Council, report of the 93rd meeting of. Commercial Dispatch (n. 17) Sir G. Barclay to Sir Edward Grey. Gulhek, August 15, 1911 [Enclosure n. 1] A. R. Neligan to G. Barclay. Gulhek, August 8, 1911.

63. "Persia. Annual Report, 1910," in Burrell and Jarman, *Iran Political Diaries*, 5:128–129.

64. Howard-Jones, "The Scientific background. 6," 496–498; Ministère des Affaires étrangères, *Procès-Verbaux [7 Novembre 1911–17 Janvier 1912]*, 604–605.

65. Ministère des Affaires étrangères, *Procès-Verbaux [7 Novembre 1911–17 Janvier 1912]*, 210.

66. William John Ritchie Simpson, "The Croonian Lectures on Plague, Delivered Before the Royal College of Physicians on June 18, 20, 25 and 27. Lecture 3," *The Journal of Tropical Medicine and Hygiene* 10, no. 15 (August 1, 1907): 252; Ronald Ross, "The Inoculation Accident at Mulkowal," *Nature* 75, no. 1951 (1907): 486–487; W. E. van Heyningen and John R. Seal, *Cholera: The American Scientific Experience, 1947–1980* (Boulder, CO: Westview Press, 1983), 151–156.

67. FO 60/683 Dispatch (no. 189) From Mr. E. Grant Duff to the Marquess of Lansdowne K.G. Tehran, November 9, 1904. [Enclosure n. 1] Monthly Summery of Events in Persia not reported in Separate Dispatches.

68. Ministère des Affaires étrangères, *Procès-Verbaux [7 Novembre 1911–17 Janvier 1912]*, 62.

69. Schneider, Médecine Persane, 27–28.

70. Procès-Verbal 43ème Séance ordinaire du Conseil Sanitaire de L'Empire de Perse, 4 Novembre 1907 included in IO L/P & S/10/123 (Register no. 1987, encl. no. 1) Under Secretary of State, Foreign Office to Under Secretary of State, India Office. London, January 20, 1908.

71. Procès-Verbal 40ème Séance ordinaire du Conseil Sanitaire de L'Empire de Perse, 2 Septembre 1907 in IO L/P & S/10/123 (Register no. 1987, encl. no. 1) Under Secretary of State, Foreign Office to Under Secretary of State, India Office. London, January 20, 1908; Schneider, *Médecine Persane*, 28.

72. Ministère des Affaires étrangères, *Conférence Sanitaire Internationale de Paris*, 62.

73. Because of distance and transit time from Europe, the calf-lymph vaccines were frequently inert by the time they reached Tehran. See Elgood, *A Medical History*, 532.

74. Ministère des Affaires étrangères, *Conférence Sanitaire Internationale de Paris*, 62.

75. NLM, Procès-Verbal 99ème Séance ordinaire du Conseil Sanitaire de l'Empire de Perse, 5 Février 1912, 80.

76. NLM, Procès-Verbal 98ème Séance ordinaire du Conseil Sanitaire de l'Empire de Perse, 8 Janvier 1912, 76; NLM, Procès-Verbal 100ème Séance ordinaire du Conseil Sanitaire de l'Empire de Perse, 28 Février 1912, 84.

77. NLM, Procès-Verbal 111ème Séance ordinaire du Conseil Sanitaire de l'Empire de Perse, 2 Décembre 1912, 122.

78. The Sanitary Council received 2,694 tuman toward its vaccination activities in May 1912 alone. See NLM, Procès-Verbal 102ème Séance ordinaire du Conseil Sanitaire de l'Empire de Perse, 8 Janvier 1912, 91; NLM, Procès-Verbal 104ème Séance ordinaire du Conseil Sanitaire de l'Empire de Perse, 1 Juillet 1912; NLM, Procès-Verbal 108ème Séance ordinaire du Conseil Sanitaire de l'Empire de Perse, 7 Octobre 1912, 107 ; NLM, Procès-Verbal 133ème Séance ordinaire du

Conseil Sanitaire de l'Empire de Perse, 7 Avril 1914, 230; NLM, Procès-Verbal 127ème Séance ordinaire du Conseil Sanitaire de l'Empire de Perse, 4 Novembre 1913, 193.

79. NLM, Procès-Verbal 133ème Séance ordinaire du Conseil Sanitaire de l'Empire de Perse, 7 Avril 1914, 232.

80. NLM, Procès-Verbal 133ème Séance ordinaire du Conseil Sanitaire de l'Empire de Perse, 7 Avril 1914, 230; NLM, Procès-Verbal 127ème Séance ordinaire du Conseil Sanitaire de l'Empire de Perse, 4 Novembre 1913,187.

81. NLM, Procès-Verbal 126ème Séance ordinaire du Conseil Sanitaire de l'Empire de Perse, 7 Octobre 1913, 186.

82. NLM, Procès-Verbal 134ème Séance ordinaire du Conseil Sanitaire de l'Empire de Perse, 5 Mai 1914, 239–242.

83. NLM, Procès-Verbal 177ème Séance ordinaire du Conseil Sanitaire de l'Empire de Perse, 10 Juillet 1916, 141. The government, following this guideline, eventually built a public abattoir outside of Tehran's walls in the 1920s; see Gilmour, *Report on an Investigation*, 52.

84. NLM, Procès-Verbal 129ème Séance ordinaire du Conseil Sanitaire de l'Empire de Perse, 6 Janvier 1914, 208–209.

85. NLM, Procès-Verbal 129ème Séance ordinaire du Conseil Sanitaire de l'Empire de Perse, 6 Janvier 1914, 209–211.

86. Gilmour, *Report on an Investigation*, 29.

87. NLM, Procès-Verbal 147ème Séance ordinaire du Conseil Sanitaire de l'Empire de Perse, 6 Janvier 1915, 308.

88. NLM, Procès-Verbal 147ème Séance ordinaire du Conseil Sanitaire de l'Empire de Perse, 6 Janvier 1915, 308.

89. Peter Avery et al. ed., *The Cambridge History of Iran*, vol. 7: *From Nadir Shah to the Islamic Republic* (Cambridge: Cambridge University Press, 1991), 208; Michael Zirinsky, "American Presbyterian Missionaries at Urmia during the Great War," *Journal of Assyrian Academic Studies* 12, no. 1 (1998): 6–27.

90. Christopher Sykes, *Wassmuss: The German Lawrence* (New York: Longmans, Green, 1936); Leila Tarazi Fawaz, *A Land of Aching Hearts: The Middle East in the Great War* (Cambridge, MA: Harvard University Press, 2014), 58–59; Sykes, *A History of Persia*, 2:477.

91. Cosroe Chaqueri, *The Soviet Socialist Republic of Iran, 1920–1921* (Pittsburgh: University of Pittsburgh Press, 1995).

92. IO R/15/1/712 Administration Report of the Persian Gulf Political Residency for the Years 1915–1919, Annual Report of the Persian Gulf Political Residency for the Year 1917, 21.

93. Touraj Atabaki, "Persia/Iran," in *1914–1918-online. International Encyclopedia of the First World War*, ed. Ute Daniel et al. (Berlin: Freie Universität Berlin, 2016), accessed March 7, 2017, doi: 10.15463/ie1418.10899.

94. Mohammad Gholi Majd, *The Great Famine and Genocide in Persia, 1917–1919*, 2nd ed. (Lanham, MD: University Press of America, 2003).

95. The incidence of infectious diseases and mortality statistics in the two years preceding the First World War in Iran can be found in: NLM, Procès-Verbal 131ème Séance ordinaire du Conseil Sanitaire de l'Empire de Perse, 3 Février 1914, 217 and NLM, Procès-Verbal 150ème Séance ordinaire du Conseil Sanitaire de l'Empire de Perse, 6 Avril 1915, 325. Gonorrhea, the main cause of male sterility in Iran at this time, was widespread. Syphilis was also prevalent, although the disease strain in Iran appeared to be less aggressive than in Europe. Later syphilitic stages that caused aneurysms and paralysis (neurosyphilis) were uncommon and maternal-child "hereditary" transmissions rare. See Anthony R. Neligan, "Public Health in Persia. 1914–1924," *The Lancet* 207, no. 5352 (1926): 693.

96. Neligan, "Public Health," 690–692.

97. John F. Hutchinson, *Politics and Public Health in Revolutionary Russia 1890–1918* (Baltimore: Johns Hopkins University Press, 1990).

98. IO L/P & S/10/284 File 2612/1912 Pt 2 'Tehran Sanitary Council' (Register n. 140221) Procès-Verbal 155ème Séance du Conseil Sanitaire de l'Empire de Perse, 7 Août 1915, 359; IO L/P & S/10/284 File 2612/1912 Pt 2 'Tehran Sanitary Council' (Register n. 156023) Procès-Verbal 156ème Séance Extraordinaire du Conseil Sanitaire de l'Empire de Perse, 22 Août 1915, 364–367; IO L/P & S/10/284 (Register n. 156023) Procès-Verbal 157ème Séance Extraordinaire du Conseil Sanitaire de l'Empire de Perse, 25 Août 1915, 364–367; NLM, Procès-Verbal 166ème Séance ordinaire du Conseil Sanitaire de l'Empire de Perse, 20 Décembre 1915, 406 ; David R. Woodward, *World War I Almanac* (New York: Facts on File, Inc., 2009), 1917.

99. Majd, *The Great Famine*, 21–22.

100. IO L/P & S/10/284 File 2612/1912 Pt 2 'Tehran Sanitary Council' (Register n. 19853) Dispatch (n. 153) A. Clark Kerr (for the Minister) to Sir Edward Grey. Gulahek, December 4, 1915 [Enclosure n. 1] A. R. Neligan to Charles M. Marling. Tehran, December 1, 1915; Sykes, *A History of Persia*, 2:447.

101. IO L/P & S/10/284 File 2612/1912 Pt 2 'Tehran Sanitary Council' (Register n. 125719) Dispatch (n. 69) A. Clark Kerr (for the Minister) to Sir Edward Grey. Gulahek, June 2, 1916 [Enclosure n. 1] A. R. Neligan to Charles M. Marling. Tehran, May 2, 1916; [Enclosure n. 2] A. R. Neligan to Charles M. Marling. Tehran, May 10, 1916.

102. Edward J. Erickson, *Ordered to Die: A History of the Ottoman Army in the First World War* (Westport, CT: Greenwood Press, 2001), 152–153.

103. IO L/P & S/10/284 File 2612/1912 Pt 2 'Tehran Sanitary Council' (Register n. 33813) Dispatch (n. 7) Charles M. Marling to Sir Edward Grey. Gulahek, January 19, 1916 [Enclosure n. 1] A. R. Neligan to Charles M. Marling. Tehran, January 11, 1916.

104. IO L/P & S/10/284 File 2612/1912 Pt 2 'Tehran Sanitary Council' (Register n. 25719) Dispatch (n. 69) A. Clark Kerr (for the Minister) to Sir Edward Grey. Gulahek, June 2, 1916 [Enclosure n. 1] A. R. Neligan to C. M. Marling. Tehran, May 2, 1916.

105. Presbyterian Church in the U.S.A. Board of Foreign Missions, *The Eighty-Second Annual Report of the Board of Foreign Missions of the Presbyterian Church of the United States of America. Presented to General Assembly, May, 1919* (New York: Board of Foreign Missions of the Presbyterian Church in the U.S.A., 1919), 272.

106. Presbyterian Church in the U.S.A. Board of Foreign Missions, *Report on India and Persia of the Deputation: Sent by the Board of Foreign Missions of Presbyterian Church in the U.S.A. to visit these fields in 1921–22 Presented by Mr. Robert E. Speer and Mr. Russell Carter* (New York: Board of Foreign Missions of the Presbyterian Church in the U.S.A., 1922), 465.

107. "Consul's Death in Persia Told," *Los Angeles Times*, December 21, 1918.

108. Francis Arthur Cornelius Forbes-Leith, *Checkmate: Fighting Tradition in Central Persia* (New York: Robert M. McBride, 1927; repr., Arno Press 1973), 20–21.

109. Majd, *The Great Famine*, 45–48, 79.

110. Neligan, "Public Health" 690–691; Gilmour, *Report on an Investigation*, 41.

111. Amir Afkhami, "Compromised Constitutions: The Iranian Experience with the 1918 Influenza Pandemic," *Bulletin of the History of Medicine*, 77, no. 2 (2003): 373–377; FO 371/3892 Percy Cox to George N. Curzon, Tehran, March 8, 1920. Insert (n. 1) Anthony R. Neligan to Percy Cox.

112. F. Hale, *From Persian Uplands* (New York: E. P. Dutton, 1920), 235–237.

113. FO 371/3892 Percy Cox to George N. Curzon, Tehran, March 8, 1920. Insert (n. 1) Anthony R. Neligan to Percy Cox.

114. Sykes, *A History of Persia*, 2:515.

115. Sykes, *A History of Persia*, 2:515.

116. Iran lost anywhere from 902,400 to 2,431,000 of its population to the 1918 influenza pandemic. See Afkhami, "Compromised Constitutions," 391–392.

117. The vice presidents who directed the Sanitary Council's affairs during the war years included Luqman al-Mamalik (the shah's physician-in-chief), Von Feistmantel (Austro-Hungarian delegate), Hydar Mirza, 'Ali Khan (professor of medicine at the Polytechnic College and minister of public instruction), Anthony Richard Neligan (British delegate), Qazala Bey (Ottoman delegate). See NLM, Procès-Verbal 169ème Séance Extraordinaire du Conseil Sanitaire de l'Empire de Perse, 20 Janvier 1916, 418.

118. IO L/P & S/10/284 File 2612/1912 Pt 2 'Tehran Sanitary Council' (Register n. 33818) Dispatch (n. 12) Charles M. Marling to Sir Edward Grey. Tehran, January 27, 1916 [Enclosure n. 1] A. R. Neligan to Charles M. Marling. Tehran, January 20, 1916.

119. Rapport de la Commission de défense sanitaire (Séance de 9 Septembre 1916), in NLM, Procès-Verbal 180ème Séance ordinaire du Conseil Sanitaire de l'Empire de Perse, 30 Octobre 1916, 494–496.

120. Rapport de la Commission de défense sanitaire (Séance de 9 Septembre 1916), in NLM, 494–496.

121. Rapport de la Commission de défense sanitaire (Séance de 9 Septembre 1916), in NLM, 494–496.

122. Rapport de la Commission de défense sanitaire (Séance de 9 Septembre 1916), in NLM, 494–496. By 1924, Tehran had 13 public (unpaid) vaccinators who were doctors, while the suburbs of the capital had 9. The provinces had 180 vaccinators, with 163 having fixed posts and 17 traveling in the districts. Vaccinations occurred during the spring and autumn months because of the impassability of northern roads during the winter. See Gilmour, *Report on an Investigation*, 29.

123. NLM, Procès-Verbal 177ème Séance ordinaire du Conseil Sanitaire de l'Empire de Perse, 10 Juillet 1916, 461–468.

124. NLM, Procès-Verbal 177ème Séance ordinaire du Conseil Sanitaire de l'Empire de Perse, 10 Juillet 1916, 461–468.

125. NLM, Procès-Verbal 177ème Séance ordinaire du Conseil Sanitaire de l'Empire de Perse, 10 Juillet 1916, 461–468.

126. NLM, Procès-Verbal 158ème Séance ordinaire du Conseil Sanitaire de l'Empire de Perse, 31 Août 1915, 369–372; NLM, Procès-Verbal 159ème Séance ordinaire du Conseil Sanitaire de l'Empire de Perse, 15 Septembre 1915, 375–379; NLM, Procès-Verbal 160ème Séance ordinaire du Conseil Sanitaire de l'Empire de Perse, 2 Octobre 1915, 381–387; NLM, Procès-Verbal 163ème Séance ordinaire du Conseil Sanitaire de l'Empire de Perse, 3 Novembre 1915, 389–395; NLM, Procès-Verbal 164ème Séance ordinaire du Conseil Sanitaire de l'Empire de Perse, 9 Novembre 1915, 396–400.

127. M. J. Sheikh-ol-Islami, "Ahmad Shah Qajar," *Encyclopedia Iranica* 1, fasc. 6 (1984): 657–660, updated online version accessed November 18, 2015, http://www.iranicaonline.org/articles/ahmad-shah-qajar-1909-1925-the-seventh-and-last-ruler-of-the-qajar-dynasty.

128. "Persia. Annual Report, 1922," in *Iran Political Diaries, 1881–1965*, vol. 6: *1921–1923*, ed. Robert Michael Burrell and Robert L. Jarman (London: Archives Editions, 1997), 413–414.

129. Elgood, *A Medical History*, 560.

130. "Persia. Annual Report, 1922," in Burrell and Jarman, *Iran Political Diaries*, 6:415.

131. Schneider, *Médecine Persane*, 28.

132. Jean-Pierre Dedet, *Les Instituts Pasteur d'Outre-Mer: Cent vingt ans de microbiologie française dans le monde* (Paris: Harmattan, 2000).

133. Afkhami, "Compromised Constitutions," 380.

134. League of Nations Archives, Private Archives (1884–1986), United Nations Office, Geneva, Papers of Prince Firouz Nosratdoleh, Letter #173 Letter de Maurice Paléologue, Direction

Politique, Ministère des Affaires étrangères, Republique Francais à Son Excellence Samad khan Momtazos Saltaneh, ministre de Perse à Paris, Paris, 7 Avril 1920.

135. The delegation to the Pasteur Institute included Muhammad 'Ali Forughi (Zulka al-Mulk), Iran's representative to the Peace Conference; Luqman al-Dawla, dean of the Faculty of Medicine at the Polytechnic College; his physician-brother Hassan Adham (Hakim al-Dawla); and Mahmud Khan Mutamid, physician to Firuz Farman Farma (Nusrat al-Dawla), Iran's minister of foreign Affairs; see Mehdi Ghodssi, "L'Institut Pasteur dans le Monde," *Bulletin de l'association des Anciens Elèves de l'Institut Pasteur* 22 (1964): 34.

136. Gilmour, *Report on an Investigation*, 26.

137. Ghodssi, "L'Institut Pasteur," 34.

138. Marcel Baltazard, "L'Institut Pasteur de L'Iran," *Archives de L'Institut Pasteur de l'Iran* 1, no. 1 (1948): 2.

139. Institut Pasteur, Service des Archives, Paris, Fonds privés des chercheurs de l'Institut Pasteur (AIP), Fonds Félix Mesnil (Ref: FR AIP MES), Lettres Jean Kérandel à Felix Mensil (1906–1933), Kérandel à Mensil, Tehran, 12 Juin 1927; Brimani, 146.

140. Gilmour, *Report on an Investigation*, 26–27.

141. In addition to the vaccination service, the smallpox vaccines were distributed to the Army Medical Services and civilian practitioners. See Gilmour, *Report on an Investigation*, 27.

142. Gilmour, *Report on an Investigation*, 27.

143. Gilmour, *Report on an Investigation*, 27.

144. Gilmour, *Report on an Investigation*, 41.

145. AIP, Fonds Félix Mesnil (Ref: FR AIP MES), Lettres Jean Kérandel à Felix Mensil (1906–1933), Kérandel à Mensil, Tehran, 29 Octobre 1927.

146. Cyrus Ghani, *Iran and the Rise of Reza Shah From Qajar Collapse to Pahlavi Power* (London: I. B. Tauris, 1998), 202.

147. Hassan Adham was the son of Mirza Zayn al-'Abidin Khan Adham Luqman al-Mamalik who led the Sanitary Council on an interim basis for much of 1908. 'Ali Asghar Nafisi Mu'adab al-Dawla (1870–1961) was the eldest son of Nizam al-Attiba, a court physician to Muzaffar al-Din Shah. Nafisi, who studied medicine in Lyon, was a trusted member of Riza Pahlavi's political circle, and was later assigned to supervise his son, Crown Prince Muhammad Riza Pahlavi, during his schooling in Switzerland. See Ghani, *Iran and the Rise of Reza Shah*, 202; Mohammad Ali Faghih, "Behdari" *Encyclopedia Iranica* 4, fasc. 1 (1989): 100–104, updated online version accessed November 19, 2015, http://www.iranicaonline.org/articles/behdari; Elgood, *A Medical History*, 562; "Persia. Annual Report, 1922," in Burrell and Jarman, *Iran Political Diaries*, 6:414.

148. Elgood, *A Medical History*, 562; Gilmour, *Report on an Investigation*, 18.

149. Homayoun Katouzian, "Nationalist Trends in Iran, 1921–1926," *International Journal of Middle East Studies* 10, no. 4 (1979): 533–551.

150. Elgood, *A Medical* History, 562–564.

151. "Persia. Annual Report, 1926," in Burrell and Jarman, *Iran Political Diaries*, 7:604.

152. "Persia. Annual Report, 1926," in Burrell and Jarman, *Iran Political Diaries*, 7:604.

153. "Persia. Annual Report, 1926," in Burrell and Jarman, *Iran Political Diaries*, 7:604; "Persia. Annual Report, 1923," in Burrell and Jarman, *Iran Political Diaries*, 6:726–728.

154. "Persia. Annual Report, 1926," in Burrell and Jarman, *Iran Political Diaries*, 7:604.

155. Elgood, *A Medical* History, 560–561.

156. Iran's delegation was led by 'Ali Partow (Hakim-i 'Azam), an internist and former undersecretary at the Iranian Ministry of Public Instruction, who had been instrumental, along with Amir Khan Amir A'lam, in the passage of the medical licensing law by the National Assembly in 1911; see Ministère des Affaires étrangères, *Conférence Sanitaire Internationale de Paris, Procès-Verbaux [10 Mai–21 Janvier 1926]* (Paris: Imprimerie Nationale, 1927), 32; Cyrus

Schayegh, *Who Is Knowledgeable Is Strong: Science, Class, and the Formation of Modern Iranian Society, 1900–1950* (Berkeley: University of California Press, 2009), 55.

157. Elgood, *A Medical* History, 568–569.

158. Husayn Bahrami (Ahya al-Saltana) came from a large and prominent family that embraced Riza Shah's nationalism. His brother, Faraj Allah Bahrami (Dabir A'zam) was Riza Shah's private secretary in 1922 and his cousin 'Abdullah Bahrami was the undersecretary of education from 1923 to 1925. His daughter would eventually wed one of shah's younger sons. His combative personality led to blows during the Fourth National Assembly (1923–1925) when he slapped Sayyid Hasan Mudaris, a representative of the "religious block" who opposed Riza Khan's centralizing policies. See Mostofi, *The Administrative*, 3:1114; Ghani, *Iran and the Rise*, 313; Mehrdad Amanat, "Bahrami, Faraj-Allah" *Encyclopedia Iranica* 3, fasc. 5 (1988): 525–526, updated online version accessed November 19, 2015, http://www.iranicaonline.org/articles/bahrami -faraj-allah-dabir-azam-1878-79.

159. Elgood, *A Medical* History, 570.

160. Elgood, *A Medical* History, 571–572.

161. Elgood, *A Medical* History, 575.

162. Elgood, *A Medical* History, 575–576.

163. David Greasley and Les Oxley "Discontinuities in Competitiveness: The Impact of the First World War on British Industry," *Economic History Review* 49 (1996): 83–88.

164. Arthur C. Millspaugh, *The American Task in Persia* (New York: Century, 1925), 54, 310; Esfandiar Bahram Yaganegi, "Recent Financial and Monetary History of Persia" (PhD diss., Columbia University, 1934), 42.

165. Arthur C. Millspaugh, *The Financial and Economic Situation of Persia* (Tehran: Imperial Persian Government, 1926), 28.

166. Amin Banani, *The Modernization of Iran, 1925–1941* (Stanford: Stanford University Press, 1961), 67.

167. "Fly with Cholera Vaccine to End Epidemic in Persia," *Chicago Sunday Tribune*, August 7, 1927, 7.

168. Elgood, *A Medical* History, 572–574.

169. "Documents Divers L'Institut D'Hessarek de 1931 à 1945," *Archives of Razi Institute* 3, no. 1 (1946): 95–104; Frank B. Berry, "The Razi State Serum and Vaccine Institute," *Bulletin of the New York Academy of Medicine* 44, no. 6 (1968): 755–758.

170. Dhiman Barua, "Chapter 1. History of Cholera," in *Cholera*, ed. Dhiman Barua and William B. Greenough III (New York: Plenum, 1992), 15; John J. Heagerty, "World Health during 1939," *Canadian Public Health Journal* 31, no. 10 (1940): 475; Mohammad-Hossein Azizi and F. Azizi, "History of Cholera Outbreaks in Iran during the 19th and 20th Centuries," *Middle East Journal of Digestive Diseases* 2, no. 1 (2010): 51–55.

Epilogue

1. Iran's last pandemic outbreak was in 1923, but it faced regional epidemics in 1927, 1931, and 1939; World Health Organization, *Minutes of the First Meeting of the Fifteenth Session of the Regional Committee for the Eastern Mediterranean Held at the Africa Hall, Addis-Ababa on Wednesday 22 September 1965*, EM/RC15A/Prog.min.1 (November 1965), 20, accessed December 7, 2015, http://apps.who.int/iris/bitstream/10665/124087/1/em_rc15_a_prog_min_1_en.pdf.

2. Mehdi Ghodssi, "Cholera El-Tor en Iran," *Acta Medica Iranica* 12, nos. 1-2 (1969): 1.

3. B. Cvjetanovic and D. Barua, "The Seventh Pandemic of Cholera," *Nature* 239, no. 5368 (1972): 137–138.

4. The Iranian response to the 1965 outbreak is covered in detail in World Health Organization, *Minutes of the First Meeting of the Fifteenth Session of the Regional Committee for the Eastern Mediterranean*, and Ghodssi, "Cholera El-Tor en Iran."

5. World Health Organization, *Minutes of the First Meeting of the Fifteenth Session of the Regional Committee for the Eastern Mediterranean*, 29; John Krauskopf, *Iran: Stories from the Peace Corps* (San Francisco: John Krauskopf, 2013), 163.

6. Krauskopf, *Iran*, 158–159.

7. Earl Ubell, "Cholera: Now It Turns Up in Crimea," *New York Times*, August 16, 1970; "Travel to Iran Barred," *Washington Post*, August 3, 1965.

8. World Health Organization, *Minutes of the First Meeting of the Fifteenth Session of the Regional Committee for the Eastern Mediterranean*, 28; Krauskopf, *Iran*, 159.

9. "Drug Sale Continued but F.D.A. Puts Warnings on Antibiotic Label," *New York Times*, January 27, 1961.

10. Krauskopf, *Iran*, 160.

11. E. J. Gangarosa, H. Saghari, J. Emile, A. Santi, H. Siadat, and Y. Watanabi, "Search for a Mass Chemotherapeutic Drug for Cholera Control. A Study of Vibrio Excretion following Single and Multiple Dose Treatment," *Bulletin of the World Health Organization* 35, no. 5 (1966): 669, 672.

12. "Iran Takes Steps to End Epidemic," *Baltimore Sun*, August 1, 1965; "Cholera Kills 89 in Iran Epidemic," *Baltimore Sun*, August 2, 1965.

13. Ghodssi, "Cholera El-Tor en Iran," 4–5.

14. Ghodssi, "Cholera El-Tor en Iran," 4–5; World Health Organization, *Minutes of the First Meeting of the Fifteenth Session of the Regional Committee for the Eastern Mediterranean*, 28; "Iran Fights Cholera Peril," *New York Times*, August 28, 1966.

15. Krauskopf, *Iran*, 162–163.

16. H. Mirchamsy, "Technical Considerations of Cholera Vaccines and Some Results of Cholera El Tor Mass Immunization," *Archives of Razi Institute* 18, no. 1 (1966): 92–93.

17. Ghodssi, "Cholera El-Tor en Iran," 9; World Health Organization, "Quarantinable Diseases Infected Areas as on 30 September 1965," *Weekly Epidemiological Record* 40, no. 39 (1965): 487, http://apps.who.int/iris/bitstream/10665/215055/1/WER4039.PDF; World Health Organization, *Minutes of the Seventeenth Session of the Regional Committee for the Western Pacific, Manila, 21–27 September 1966, Annex 1: Global Incidence of Cholera Cases and Deaths*, WPR/RC17/6 (July 1, 1966), 7–8, http://iris.wpro.who.int/bitstream/handle/10665.1/9014/WPR_RC017_06_Cholera_1966_en.pdf.

18. "Cholera Reported in Soviet," *New York Times*, August 24, 1965; "U.N. Told of Cholera in Soviet," *New York Times*, August 22, 1965.

19. Vladimir Shkolnikov, France Mesle, and Jacques Vallin, "Health Crisis in Russia I. Recent Trends in Life Expectancy and Causes of Death from 1970 to 1993," *Population: An English Selection* 8 (1996): 133.

20. M. I. Narkevich, G. G. Onischenko, J. M. Lomov, E. A. Moskvitina, L. S. Podosinnikova, and G. M. Medinsky, "The Seventh Pandemic of Cholera in the USSR, 1961–89," *Bulletin of the World Health Organization* 71, no. 2 (1993): 190–192; Ubell, "Cholera: Now It Turns Up in Crimea."

21. The Ministry of Public Health grew out of the Public Health Department, which had absorbed the Sanitary Council in the twilight of the Qajar dynasty.

22. Wellcome Institute for the History of Medicine (London), Contemporary Medical Archives Center (GC/46/ 14/1-11). Amir Khan Amir Alam, "Concise History of Medicine and Hygiene of Iran in These Four Decades: Summery of Conferences Made during the Medical Congress of Iran 1945," 3–4.

23. World Health Organization, *Minutes of the First Meeting of the Fifteenth Session of the Regional Committee for the Eastern Mediterranean*, 29; Krauskopf, *Iran*, 164–165.

24. William E. Warne, *Mission for Peace Point 4 in Iran* (Indianapolis: Bobbs-Merrill, 1956), 151–169; Julian Bharier, *Economic Development in Iran 1900–1970* (London: Oxford University

Press, 1971), 226–229; "Tehran Gets Pure Piped Water but Mourns Quaint Old Customs," *New York Times*, November 9, 1955; Firuz Tawfiq, "Census i. In Iran," *Encyclopedia Iranica* 5, fasc. 2: 142–152, updated online version accessed November 18, 2016, http://www.iranicaonline.org /articles/census-i.

25. Frederick F. Aldridge, Eugene Baird, and Vigen Gevorkian, "Modern Sanitation in an Iranian Village," *Public Health Reports* 69, no. 3 (1954): 305–308.

26. "Tehran's Biggest Wastewater Treatment Plant Coming Up" *Financial Tribune* (Tehran), February 8, 2015, https://financialtribune.com/articles/people/10707/tehran-s-biggest-wastewater -treatment-plant-coming-up; Eric Pace, "Tehran Project Face Challenges," *New York Times*, June 6, 1976.

27. Warne, *Mission for Peace Point 4 in Iran*, 151–169.

28. These investments began in the last phase of Iran's Third Development Plan (1962–1968) and expanded in the 1970s during a period of massive urbanization in the country. World Health Organization, *Minutes of the First Meeting of the Fifteenth Session of the Regional Committee for the Eastern Mediterranean*, 29; Harald Mehner, "Development and Planning in Iran After World War II," in *Iran under the Pahlavis*, ed. George Lenczowski (Stanford: Hoover Institution Press, 1975), 170–171.

29. Ministère des Affaires étrangères, *Conférence Sanitaire Internationale de Paris, Procès-Verbaux [10 Mai–21 Janvier 1926]*, 347–348.

30. This is also reflected in Iran's enthusiastic adoption of the Bacillus Calmette-Guerin (BCG) vaccine against tuberculosis, which remains a controversial method of preventing the disease because of its limited immunogenicity. See Mohammad Hossein Azizi and Moslem Bahadori, "A Brief History of Tuberculosis in Iran during the 19th and 20th Centuries," *Archives of Iranian Medicine* 14, no. 3 (2011): 215–219.

31. H. Mirchamsy, "Technical Considerations of Cholera Vaccines and Some Results of Cholera El Tor Mass Immunization." *Archives of Razi Institute* 18, no. 1 (1966): 85–94; Mehdi Ghodssi, "Cholera El-Tor en Iran." *Acta Medica Iranica* 12, nos. 1–2 (1969): 4–7.

32. World Health Organization, *Minutes of the First Meeting of the Seventeenth Session of the Regional Committee for the Eastern Mediterranean Held at the Royal Teheran Hilton Hotel, Teheran on Monday, 25 September 1967*, 15, http://apps.who.int/iris/bitstream/10665/124233/1/em _rc17_a_min_1_en.pdf.

33. World Health Organization, "Cholera 2007," *Weekly Epidemiological Record* 83, no. 31 (2007): 276, http://www.who.int/wer/2008/wer8331.pdf?ua=1.

34. "Iran Battles Cholera Outbreak," *New York Times*, September 7, 1979; "18 Dead and 557 Ailing in Iran Cholera Upsurge," *New York Times*, September 22, 1979.

35. Iran's history of drug addiction and its evolving treatment modalities in the twentieth century are covered in Amir Afkhami, "From Punishment to Harm Reduction: Resecularization of Addiction in Contemporary Iran," in *Contemporary Iran: Economy, Society, Politics*, ed. Ali Gheissari (Oxford: Oxford University Press, 2009), 194–210.

36. Afkhami, "From Punishment to Harm Reduction," 201–202.

37. Afkhami, "From Punishment to Harm Reduction," 203–206.

38. Bill Samii, "Iran: Government Reverses Course in War on Drugs," Radio Free Europe/ Radio Liberty, September 29, 2005, https://www.rferl.org/a/1061753.html.

39. Ali Akbar Dareini, "Drug Abuse in Iran on Rise despite Executions, Police Raids," Associated Press, February 12, 2015, http://www.timesofisrael.com/drug-abuse-in-iran-on-rise -despite-executions-police-raids/.

40. Mehrnaz Samimi, "Iran Has a Growing Drug Problem, and Young, Well-Educated Women Are among Its Biggest Victims," *Quartz*, May 29, 2016, https://qz.com/693031/iran-has -a-growing-drug-problem-and-young-well-educated-women-are-among-its-biggest-victims/.

41. "UN: Freeze Funding of Iran Counter-Narcotics Efforts" *Human Rights Watch News*, December 17, 2014, https://www.hrw.org/news/2014/12/17/un-freeze-funding-iran-counter-narcotics-efforts.

42. Paul Richter, "Iran Has Interest in a Stable Afghanistan, Clinton Says," *Los Angeles Times*, March 31, 2009; United States Library of Congress, Congressional Research Service, *Afghanistan: Narcotics and US Policy*, by Christopher M. Blanchard, RL32686 (2009), 36–37.

43. Bill Samii, "Analysis: Iranian Counternarcotics Agencies In Bureaucratic Struggle," *Radio Free Europe/Radio Liberty*, September 29, 2005, https://www.rferl.org/a/1056914.html.

Archival Sources

Belgium

Archives Diplomatiques, Royaume de Belgique, Ministère des Affaires étrangères, Brussels (MAE)

Rapport de la Légation de Belgique à Tehran, vol. 2890 (1898)

France

Archives du Ministère des Affaires étrangères, Paris Quai d'Orsay (AMAE)

Affaires Divers Politiques, 1815–1896 IV (ADP), Perse, marge 4, numéro 93

Correspondance Politique, 1871–1896, Perse, vols. 44, 45

Correspondance Politique et Commerciale (Nouvelle Série), 1897–1918 (NS). Politique Intérieur, NS 1, Dossier Général (1896–1904)

Correspondance Politique et Commerciale (Nouvelle Série), 1897–1918 (CP), Perse, Français au Service de Perse: médecins dossier personnelles, 1896–1905, vols. 56, 57

Bibliothèque de l'Académie National de Médicine, Paris

Archives et manuscrits de la bibliothèque, cote 50.426, Joseph Désiré Tholozan, *Rapport à Sa Majesté le Chah sur l'état actuel de l'hygiène en Perse; progrès à réaliser; moyens de les effectuer; résultats obtenus depuis un an* [Report to His Majesty the Shah on the current sanitary state of Persia, progress to be realized, modes of bringing about progress, and results obtained since a year ago], 1869

Bibliothèque interuniversitaire de Santé, Paris (BIU Sante)

Collection de portraits

Bibliothèque Nationale de France, Richelieu, Paris

Département des Manuscrits

Institut Pasteur, Service des Archives, Paris (AIP)

Fonds privés des chercheurs de l'Institut Pasteur, Fonds Félix Mesnil (Ref: FR AIP MES)

Lettres Jean Kérandel à Felix Mensil (1906–1933)

Musée du Service de Santé des Armées au Val-de-Grâce, Paris

Section des Archives, cote C/1082, dossier 1541

Great Britain

British Library, London. Oriental and India Office Collections, India Office (IO)

L/P & S Political and Secret Correspondences Series: L/P & S/20/C248B (1906); L/P & S/10/123, n. 195, 973, 3209, 3081, 3384 (1907); 1987 (1908); L/P & S/10/184 n. 2918 (1909); n. 3883 (1910); IO L/P & S/10/184 file 3004/1910 "Tehran Sanitary Council" (1909–1911); IO L/P & S/10/284 file 2612/1912 pt. 2 "Tehran Sanitary Council" (1914–1916, 1920)

Persian Gulf Residencies: R/15/1-6 Records of the Bushire, Bahrain, Kuweit, Muscat, and the
    Trucial Coast Agencies, 1763–1951; R/15/1 Series, vols. 182 (1871) Records of the Persian
    Gulf Resident at Bushire; R/15/1/712 Administration Report of the Persian Gulf Political
    Residency for the Years 1915–1919
Church Mission Society Archives, Birmingham (CMSA)
    University Birmingham Library, Special Collections Department, Unofficial Papers, Diaries
    and Papers of Miss Mary Ellen Brighty of the Persia Mission, CMS F1-3 (Acc113) 1899–
    1937 Mary Ellen Brighty
Middle East Centre Archive, St Antony's College, Oxford
    Joseph Tholozan, photographic collection, album 3
    Herbert Rushton Sykes, photographic collection, GB165-0272
Public Record Office, Kew. Foreign Office (FO), General Correspondence
    FO 60 Series, vols. 20 (1821); 123, 124, 126, 127 (1846); 141 (1848); 174, 176 (1852); 179, 181 (1853);
        259 (1861); 286 (1863); 300 (1866); 305, 306 (1867); 313 (1868); 320, 323 (1869); 334, 336
        (1871); 342, 343, 345 (1872); 382 (1876); 505, 508 (1889); 510 (1890); 532, 533, 534, 535 (1892);
        542, 545, 546 (1893); 586 (1897); 608, 609 (1899); 681, 682, 683, 685, 686, 688, 732
        (1902–1904)
    FO 248 Series, vols. 232 (1866); 248 (1871); 271 (1871); 297 (1874); 545, 547, 548 (1892); 818, 820
        (1904)
    FO 251 Series, vol. 76 (1903)
    FO 371 Series, vols. 309 (1907); 3892 (1920)
    FO 416 Series, vol. 17 (1904)
    FO 539 Series, vol. 6411
    FO 881 Series, vols. 3186, 3332, 3600 (1877); 8539 (1905); 8780 (1906)
Royal Geographical Society, London
    Map Room Collection, mr Iran S.28, Najm al- Mulk, 'Abd al-Ghaffar. *Map of Tehran*. Scale
        ca. 1:4000. Tehran: Polytechnic College [Dar al-funun], April 1892
Wellcome Institute for the History of Medicine, London
    Contemporary Medical Archives Center (GC/46/ 14/1-11). Amir Khan Amir Alam, "Concise
        History of Medicine and Hygiene of Iran in These Four Decades: Summery of Confer-
        ences Made during the Medical Congress of Iran 1945"

Iran

Kitabkhana-yi Markazi Danishgah-i Tehran [Central Library, University of Tehran] (Danishgah)
Kitabkhana-yi Melli [National Library of Iran], Tehran (Melli Library)
Ministry of Foreign Affairs, The Center for Documents and the History, Tehran (IMFA)
    Documents Relating to Iran/Italy Relations

Switzerland

League of Nations Archives, United Nations Office, Geneva
    Private Archives (1884–1986), Papers of Prince Firuz Nosratdoleh

United States

National Archives, Washington, DC (NA)
    United States, State Department, Diplomatic Series (Dip. Ser.) M223, enclosure 1, no. 383, 389,
        390, 394, 397, 400, 408, 418, 433, 452. 1889
National Archives of the Presbyterian Church in the USA, Presbyterian Historical Society, Phil-
    adelphia (PHS)
    Board of Foreign Missions Correspondence and Reports, 1833–1911, Persia, Letters 1889–1890,
        vol. 6, no. 45

National Library of Medicine, History of Medicine Collection, Bethesda (NLM)
  Procès-Verbal Séance Extraordinaire/ordinaire du Conseil Sanitaire de l'Empire de Perse
    (1912–1917), nos. 98, 99, 100, 101, 102, 103, 104, 108, 111, 112, 113, 114, 115, 116, 124, 126, 127,
    129, 131, 133, 134, 137, 141, 147, 150, 156, 157, 158, 159, 160, 163, 164, 166, 169, 177, 180, 183
Yale University Library, Manuscripts and Archives, New Haven
  Ghassem Ghani Collection (MS 235), ser. 5, box 2, folder 18

Persian Manuscripts

Fakhr al-Hukama, Hajji Mirza Musa. *Dastur al-attiba' fi daf' al-ta'un va ilaj al-vaba*. Melli Library, Tehran, MS no. 675F, 1269 AH [1852–1853].

Kazulani, Hakim [Casolani, Fortunato]. *'Arizi-yi majlis-i mamuran bi padishah-i Inglis* [Petition of the assembled commission to the King of England]. Melli Library, Tehran, MS no. 313F, n.d.

Mar'ashi, Sayyid 'Ali 'ibn-i Muhammad Tabrizi. *Qanun al-'alaj* [The cannon of treatments]. Danishgah, Tehran, MS no. 4115, 1269 AH [1852–1853].

Polak, Jakob Eduard. *Bimari-yi vaba* [The illness of cholera]. Melli Library, Tehran, MS no. 349F, 1269 AH [1852–1853].

Qajar, Muhammad Husayn va Mons. Jibril. *Risala dar vaba*. Melli Library, Teheran, MS no. 735F, 1262 AH [1845–1846].

Tabib, 'Abd al-Rasul Husayni. *Mémoire sur l'épidémie de choléra qui éclata à Téhéran en 1892* [An account of the cholera epidemic that broke out in Tehran in 1892] (handwritten Persian manuscript). Bibliothèque Nationale de France, Richelieu, Paris, Département des Manuscrits, cote: Supplément Persan 1290.

Published Primary and Secondary Sources in Persian and Arabic

'Ain al-Saltana, Qahraman Mirza. *Ruznama-yi khatirat-i 'Ain al-Saltana* [The memoirs of 'Ain al-Saltana], edited by Massoud Salur and Iraj Afshar. 3 vols. Tehran: Intisharat-i asatir, 1376 AS [1997–1998].

Amin al-Dawala, Mirza 'Ali Khan. *Khatirat-i siyasi-yi Mirza 'Ali Khan Amin al-Dawla* [The political memoirs of Mirza 'Ali Khan Amin al-Dawla], edited by Hafez Farmayan. 3rd ed. Tehran: Amir Kabir, 1370 AS [1992].

———. *Tarikh-i mu'assasat-i tamadduni-yi jadid dar Iran* [The history of the institutions of modern civilization in Iran]. 3 vols. 2nd ed. Tehran: Anjuman-i Danishjuyan-i Danishgah-i Tehran, 1370 AS [1992].

Bamdad, Mahdi. *Shahr-i hal-i rijal-i Iran dar qarn-i 12, 13, 14 hijri* [The biography of Iranian personalities during the 12, 13, 14 AH centuries]. 6 vols. 4th ed. Tehran: Intisharat-i zavar, 1371 AS [1992].

"*Bayan-i kiyfiyat-i tap-i vaba-yi va ta'un ki dar tibb-i akbar muzu' ast nivisht-i mishavad*; Or History of the Pestilential Fever and Plague from the Tib-e-Akhbar." In *A Practical Treatise on Epidemic Cholera, Ague, and Dysentery; Illustrating The Principals of Treatment By Their Anatomical Physiology Pointing Out Their Contagion and Westering Inclination; to Which Is Added A Persian Treatise on Plague and Cholera*, by W. G. Maxwell, 1–23. Calcutta: T. Ostell and Co. British Library, 1838.

Beygi, M. Hassan. *Tehran-i qadim* [Old Tehran]. Tehran: Qafnus, 1366 AH [1946–1947].

Ebrahimnejad, Hormoz. "Un traité d'epidémiologie de la médecine traditionnelle persane: *Mofarraq ol-Heyze Va'l-Vaba de Mirza Mohammad-Taqi Shirazi* (ca. 1800–1873)." *Studia Iranica* 27 (1998): 83–107.

Ettehadieh, Mansoureh, Esmail Shams, and Azam Ghafouri, eds. *Vaba-yi alamgir. Asnad, madarik va mukatibat-i Abdul-Husayn Mirza Mirza Farman Farma dar hukumat-i Kermanshah 1321–23 A.H.* [World conquering cholera. Records, documents, and correspondences of Abdul-Husayn Mirza during his rule in Kermanshah, 1903/04–1905/06]. Tehran: Nashr-i tarikh-i Iran, 2013/2014.

Hashemi, Muhammad Sadr. *Tarikh-i jarayid va majallat-i Iran* [The history of newspapers and magazines in Iran]. 4 vols. Isfahan: Intisharat-i kamal, 1332 AS [1953–1954].

Hidayat, Mahdiquli (Mukhbir al-Saltana). *Khatirat va khatarat* [The memoirs of Mahdiquli Hidayat Mukhbir al-Saltana]. Tehran: Intisharat-i zavar, 1375 AS [1996].

Iqbal, Abbas. "Abila kubi" [Smallpox inoculation]. *Yadigar* 4, no. 3 (1326 AS [1947]).

Isfahani, Sayyid Jamal al-Din Va'iz. *Libas al-tiqwa* [Virtuous garb]. Edited by Huma Rizvani. Tehran: Nashir-i tarikh-i Iran, 1363 AS [1984].

Isfahani, Sayyid Jamal al-Din Va'iz, and Malik al-Mutakallimin. *Ru'ya-yi sadiqah* [True dream]. Edited by Sadiq Sajjadi. Tehran: Nashir-i Tarikh-i Iran, 1363 AS [1984].

I'timad al-Saltana, Muhammad Hasan Khan. *Chehel sal tarikh-i Iran dar durih-yi padishahi-i Nasir al-Din Shah (al-ma'sir va al-athar)* [Forty year history of Iran during Nasir al-Din Shah's Reign]. 3 vols. Edited by Iraj Afshar. Tehran: Intishart-i asatir, 1368 AS [1989–1990].

———. *Ruznama-yi khatirat-i I'timad al-Saltana* [The memoirs of I'timad al-Saltana]. Edited by Iraj Afshar. Tehran: Amir Kabir, 1345 AS [1976].

———. *Tarikh-i muntazam-i Nasiri* [History of Nasir al-Din Shah's reign]. Edited by Muhammad Isma'il Rizvani, 3 vols. Tehran: Dunya-yi kitab, 1367 AS [1988].

Kirmani, Nazim al-Islam. *Tarikh-i bidari-yi Iranian* [The history of Iranian awakening]. 2 vols. 4th ed. Tehran: Entesharat-i agah, 1362 AS [1983–1984].

Maragha'i, Zayn al-'Abidin. *Siyahatnama-yi Ibrahim Big*. Tehran: Kitabha-yi sadaf, 1344 AS [1965].

Mir, Muhammad Taqi. *Pizishkan-i nami Pars* [Persian physicians]. 2nd ed. Tehran, 1363 AS [1984–1985].

Modarressi Tabataba'i, Hossein. *Ashna-yi ba chand nuskhaha-yi khatti* [introduction to several written manuscripts]. Qom: Chapkhana-yi mihr, 1355 AS [1976].

Nadjmabadi, Mahmud. "Duktur Khalil Khan Saqafi (A'lam al-Dawla)" [Doctor Khalil Khan Saqafi (A'lam al-Dawla)]. *Jahan-i pizishki* 11, no. 12 (1958): 447–456.

———. "Tarikh-i tibb va danishkada-yi pizishki-yi Iran" [History of medicine and the Medical College of Iran]. *Jahan-i pizishki* 1, no. 3 (1326 AS [1947]): 43–46.

———. "Tibb-i Dar al-Funun va kutub-i darsi-yi an" [Medicine at the Polytechnic College and its medical books]. In *Amir Kabir va Dar al-Funun* [Amir Kabir and the Dar al-Funun], edited by Qudrat Allah Rushani Za'faranlu, 202–237. Tehran: Majmu'a-yi kitabkhana-yi markazi va asnad-i Danishgah-i Tehran, 1354 AS [1975].

Nategh, Homa. *Musibat-i vaba va bala-yi hukumat* [The calamity of cholera and the pain of government]. Tehran: Nashr-i gustarish, 1358 AS [1979].

Polak, Jakob Eduard. *Safarnama-yi Polak: Iran va Iranian*. Translated by KeyKavoos Jahandari. Tehran: Khawrazmi, 1361 AH [1982]. Originally published as *Persien, das Land und seine Bewohner. Ethnographische Schilderungen*. Leipzig: Broukhaus, 1865.

Qasemi, Sayyid Farid. *Sarguzasht-i matbu'at-i Iran* [The history of the press in Iran]. 2 vols. Tehran: Sazman-i Chap va Intisharat-i Vizarat-i Farhang va Irshad-i Islami, 1380 AS [2001–2002].

Sadvandian, Sirus, and Mansoureh Ettehadieh, eds. *Amar-i dar al-khalafih-yi Tehran: asnadi az tarikh-i ijtimai-yi Tehran dar asr-i Qajar* [Statistics of Tehran: Manuscript pertaining to the social history of Tehran in the Qajar era]. Tehran: Nashr-i tarikh-i Iran, 1368 AS [1989].

Safa'i, Ibrahim. *Rahbaran-i mashruta* [The leaders of the constitutional revolution], 2nd ed. Tehran: Intishart-i javidan, 1364 AS [1985].

Shaykh Rezaei, Ensiya, and Shahla Azari, eds. *Guzarishha-yi nazmiya az mahalat-i Tehran: rapurt-i vaqa'i mukhtalif-i mahalat-i dar al-khalafih (1303 A.H.–1305 A.H.)* [Police reports on the various happenings in the various districts of Tehran, 1886–1888]. 2 vols. Tehran: Intisharat-i sazman-i asnad-i milli-yi Iran, 1377 AS [1999].

Newspapers and Other News Sources

Akhtar
Associated Press
Baltimore Sun
Chicago Daily Tribune
Chicago Sunday Tribune
Christian Observer
Financial Tribune (Tehran)
Habl al-matin
Human Rights Watch News
Illustrated London News
Ittila'
London Illustrated News
Los Angeles Times

Musavat
National New York Times
Nida-yi vatan
Quartz
Radio Free Europe / Radio Liberty
Ruznama-yi dawlat-i 'alliya-yi Iran
Ruznama-yi vaqayi'-i ittifaqiya
Shams
Sharaf
Tarbiyat
Times (London)
Washington Post

Official Sources

American Red Cross. *Neuvième Conférence Internationale de la Croix Rouge tenue à Washington du 7 au 17 Mai 1912, Compte Rendu.* Washington, DC: Press of Judd & Detweiler, 1912.

Baker, James E. (M.R.C.S., English Medical Superintendent, Her Britannic Majesty's Telegraph Staff in Persia). *A few Remarks on the most prevalent Diseases and the Climate of the North of Persia, appendix to Herbert, Report on the present State of Persia and her Mineral Resources.* House of Commons, Parliamentary Papers, Accounts and Papers 67 (1886), 323–326.

Burrell, Robert Michael, and Robert L. Jarman, eds. *Iran Political Diaries, 1881–1965*, vol. 4: *1908–1909.* London: Archives Editions, 1997.

———. *Iran Political Diaries, 1881–1965*, vol. 5: *1910–1920.* London: Archives Editions, 1997.

———. *Iran Political Diaries, 1881–1965*, vol. 6: *1921–1923.* London: Archives Editions, 1997.

———. *Iran Political Diaries, 1881–1965*, vol. 7: *1924–1926.* London: Archives Editions, 1997.

Conférence Sanitaire International. *Rapport sur Les Mesures à Prendre en Orient Pour Prévenir de Nouvelles Invasions du Cholera en Europe* [Out 1866]. Constantinople: Imprimerie du Levant Herald, 1866.

Gilmour, John. *Report on an Investigation into the Sanitary Conditions in Persia: Undertaken on behalf of the Health Committee of the League of Nations at the Request of the Persian Government.* Geneva: League of Nations, 1925.

Great Britain General Board of Health. *Report of the General Board of Health on the Epidemic Cholera of 1848 & 1849.* London: Her Majesty's Stationery Office, W. Clowes, 1850.

Government of India. *Administration Report of the Persian Gulf Political Residency for 1905–1906.* Vol. 6. Calcutta: Office of the Superintendent of Government Printing, 1907.

Lorimer, John Gordon. *Gazetteer of the Persian Gulf, Oman and Central Arabia.* 2 vols. Calcutta: India Superintendent Government Printing, 1915.

Millspaugh, Arthur C. *The Financial and Economic Situation of Persia.* Tehran: Imperial Persian Government, 1926.

Ministère des Affaires étrangères. *Conferences Sanitaire Internationale de Paris, Procès-Verbaux [7 Février–3 Avril 1894].* Paris: Imprimerie National, 1894.

———. *Conference Sanitaire Internationale de Paris, Procès-Verbaux [10 Octobre–3 Décembre 1903].* Paris: Imprimerie National, 1904.

———. *Conférence Sanitaire Internationale de Paris, Procès-Verbaux [7 Novembre 1911–17 Janvier 1912].* Paris: Imprimerie Nationale, 1912.

———. *Conférence Sanitaire Internationale de Paris, Procès-Verbaux [10 Mai–21 Janvier 1926].* Paris: Imprimerie Nationale, 1927.

Ministero degli affari esteri. *Conférence Sanitaire International de Venise, Procès-Verbaux [16 Février–19 Mars 1897]*. Rome: Forzani et Cie. Imprimeurs du Senat, 1897.

Moreau de Jonnès, Alexandre. *Rapport au Conseil supérieur de santé sur le choléra-morbus pestilentiel: les caractères pathologiques de cette maladie, les moyens curatifs et hygiéniques qu'on lui oppose, sa mortalité, son mode de propagation et ses irruptions dans l'Indoustan, l'Asie orientale, l'archipel indien, l'Arabie, la Syrie, la Perse, l'empire russe et la Pologne.* Paris: Imprimerie de Cosson, 1831.

Presbyterian Church in the U.S.A. Board of Foreign Missions. *The Eighty-Second Annual Report of the Board of Foreign Missions of the Presbyterian Church of the United States of America. Presented to General Assembly, May, 1919.* New York: Board of Foreign Missions of the Presbyterian Church in the U.S.A., 1919.

———. *Report on India and Persia of the Deputation: Sent by the Board of Foreign Missions of Presbyterian Church in the U.S.A. to Visit These Fields in 1921–22 Presented by Mr. Robert E. Speer and Mr. Russell Carter.* New York: Board of Foreign Missions of the Presbyterian Church in the U.S.A., 1922.

Radcliffe, John Netten. *Recent diffusion of cholera in Europe.* London: Local Government Board [printed by T. Harrison], 1872.

Sutherland, John. *Appendix (A) to the Report of the General Board of Health on the Epidemic Cholera of 1848 & 1849.* London: Her Majesty's Stationery Office, W. Clowes, 1850.

United Kingdom. Parliament. *Accession of Persia to the International Agreement of December 9, 1907, Respecting the Creation of an International Office of Public Health.* Cd. 4901. 1909.

United States Library of Congress. Congressional Research Service. *Afghanistan: Narcotics and US Policy*, by Christopher M. Blanchard. RL32686. 2009.

United States Public Health and Marine Hospital Service. "Table Showing Progress of Cholera from Mecca in 1902 to Persia and the Transcaspian Provinces in 1904, Prepared by the Sanitary Council of Teheran." *Public Health Reports Issued by the Surgeon General* 20, no. 6 (1905): 233. [Washington, DC: Government Printing Office.]

Woodworth, John Maynard. *The Cholera Epidemic of 1873 in the United States.* Washington, DC: Government Printing Office, 1875.

World Health Organization. "Cholera 2007." *Weekly Epidemiological Record* 83, no. 31 (2007): 269–284. Available online at http://www.who.int/wer/2008/wer8331.pdf?ua=1.

———. *Minutes of the First Meeting of the Fifteenth Session of the Regional Committee for the Eastern Mediterranean Held at the Africa Hall, Addis-Ababa on Wednesday 22 September 1965.* EM/RC15A/Prog.min.1 (November 1965). Available online at http://apps.who.int/iris/bitstream/10665/124087/1/em_rc15_a_prog_min_1_en.pdf.

———. *Minutes of the First Meeting of the Seventeenth Session of the Regional Committee for the Eastern Mediterranean Held at the Royal Teheran Hilton Hotel, Teheran on Monday, 25 September 1967.* EM/RC17A/Min.1 (September 25, 1967). Available online at http://apps.who.int/iris/bitstream/10665/124233/1/em_rc17_a_min_1_en.pdf.

———. *Minutes of the Seventeenth Session of the Regional Committee for the Western Pacific, Manila, 21–27 September 1966, Annex 1: Global Incidence of Cholera Cases and Deaths.* WPR/RC17/6 (July 1, 1966). Available online at http://iris.wpro.who.int/bitstream/handle/10665.1/9014/WPR_RC017_06_Cholera_1966_en.pdf.

———. "Quarantinable Diseases: Infected Areas as on 30 September 1965." *Weekly Epidemiological Record* 40, no. 39 (October 1, 1965): 481–492. Available online at http://apps.who.int/iris/bitstream/10665/215055/1/WER4039.PDF.

Websites

Encyclopedia Iranica (http://www.iranica.com)

International Encyclopedia of the First World War (http://encyclopedia.1914-1918-online.net)

Mahnama-yi iliktruniki-yi baharistan [Baharistan online monthly from the Institute for Contemporary Iranian Studies] (http://www.iichs.ir/p/Publication.aspx?PName=baharestan)

Tavoos Art Magazine (http://www.tavoosonline.com/)

Published Works in European Languages

Abrahamian, Ervand. "The Causes of the Constitutional Revolution in Iran." *International Journal of Middle Eastern Studies* 10 (1979): 381–414.

Afary, Janet. *The Iranian Constitutional Revolution, 1906–1911: Grassroots Democracy, Social Democracy, and the Origins of Feminism.* New York: Columbia University Press, 1996.

Afkhami, Amir A. "Compromised Constitutions: The Iranian Experience with the 1918 Influenza Pandemic." *Bulletin of the History of Medicine* 77, no. 2 (2003): 367–392.

———. "Disease and Water Supply: The Case of Cholera in Nineteenth-Century Iran." In *Transformations of Middle Eastern Natural Environments,* edited by Jeff Albert, Magnus Bernhardsson, and Roger Kenna, 206–220. New Haven: Yale University Press, 1998.

———. "From Punishment to Harm Reduction: Resecularization of Addiction in Contemporary Iran." In *Contemporary Iran: Economy, Society, Politics,* edited by Ali Gheissari, 194–210. Oxford: Oxford University Press, 2009.

———. "The Sick Men of Persia: The Importance of Illness as a Factor in the Interpretation of Modern Iranian Diplomatic History." *Journal of Iranian Studies* 36 (2003): 339–352.

Ahmadi, Kourosh. *Islands and International Politics in the Persian Gulf: The Abu Musa and the Tunbs in Strategic Perspective.* London: Routledge, 2008.

Aldridge, Frederick F., Eugene Baird, and Vigen Gevorkian. "Modern Sanitation in an Iranian Village." *Public Health Reports* 69, no. 3 (1954): 305–308.

Algar, Hamid. *Mirza Malkum Khan.* Berkeley: University of California Press, 1973.

Amanat, Abbas. "Between the *Madrasa* and the Marketplace: The Designation of Clerical Leadership in Modern Shi'ism." In *Authority and Political Culture in Shi'ism,* edited by Said Amir Arjomand, 98–132. Albany: State University of New York Press, 1988.

———. *Pivot of the Universe: Nasir al-Din Shah and the Iranian Monarchy, 1831–1896.* Berkeley: University of California Press, 1997.

———. "The Study of History in Post-Revolutionary Iran: Nostalgia, or Historical Awareness?" *Iranian Studies* 22, no. 4 (1989): 3–18.

———. *Resurrection and Renewal: The Making of the Babi Movement in Iran, 1844–1850.* Ithaca: Cornell University Press, 1989.

Amirahmadi, Hooshang. *The Political Economy of Iran under the Qajars.* London: I. B. Tauris, 2012.

Ashraf, Ahmad, and H. Hekmat. "Merchants and Artisans and the Development Processes of Nineteenth-Century Iran." In *The Islamic Middle East, 700–1900,* edited by A. L. Udovitch, 731–733. Princeton: Darwin Press, 1981.

Avery, Peter, Gavin Hambly, and C. P. Melville, eds. *The Cambridge History of Iran.* Vol. 7: *From Nadir Shah to the Islamic Republic.* Cambridge: Cambridge University Press, 1991.

Axworthy, Michael. *History of Iran: Empire of the Mind.* New York: Basic Books, 2008.

Azizi, Mohammad-Hossein, and F. Azizi. "History of Cholera Outbreaks in Iran during the 19th and 20th Centuries." *Middle East Journal of Digestive Diseases* 2, no. 1 (2010): 51–55.

Azizi, Mohammad-Hossein, and Moslem Bahadori. "A Brief History of Tuberculosis in Iran during the 19th and 20th Centuries." *Archives of Iranian Medicine* 14, no. 3 (2011): 215–219.

Bakhash, Shaul. *Iran: Monarchy, Bureaucracy and Reform under the Qajars, 1858–1896*. London: Ithaca Press, 1978.

Balay, Christoph, Claire Kapper, and Ziva Vesel, eds. *Pand-o Sokhan*. Tehran: Institute Français de Recherche en Iran, 1995.

Baldwin, Peter. *Contagion and the State in Europe, 1830–1831*. Cambridge: Cambridge University Press, 1999.

Baltazard, Marcel. "L'Institut Pasteur de l'Iran." *Archives de l'Institut Pasteur de l'Iran* 1, no. 1 (1948): 1–23.

Banani, Amin. *The Modernization of Iran, 1925–1941*. Stanford: Stanford University Press, 1961.

Barona, Josep L. *The Rockefeller Foundation, Public Health and International Diplomacy, 1920–1945*. London: Pickering & Chatto, 2015.

Barua, Dhiman. "History of Cholera." In *Cholera*, edited by Dhiman Barua and William B. Greenough III, 1–35. New York: Plenum, 1992.

Bayat, Magnol. *Iran's First Revolution: Shi'ism Constitutional Revolution of 1905–1909*. Oxford: Oxford University Press, 1991.

Bell, Charles W. "Report on the Epidemic of Ague or 'Fainting Fever' of Persia, A Species of Cholera, occurring in Tehran in the Autumn of the year 1842." *British and Foreign Medical Review, or Quarterly Journal of Practical Medicine and Surgery* 16 (1843): 558–566.

Bell, Gertrude. *Persian Pictures*. 1894. New York: Boni and Liveright, 1926.

Bendiner, Elmer. "Alexandre Yersin: Pursuer of Plague." *Hospital Practice* 24 (1989): 121–138.

Berry, Frank B. "The Razi State Serum and Vaccine Institute." *Bulletin of the New York Academy of Medicine* 44, no. 6 (1968): 755–758.

Bharier, Julian. *Economic Development in Iran, 1900–1970*. London: Oxford University Press, 1971.

Boyce, Arthur C. "Alborz College of Tehran and Dr. Samuel Martin Jordan, Founder and President." In *Cultural Ties between Iran and the United States*, edited by Ali Pasha Saleh, 155–234. Tehran: Sherkat-e Chapkhaneh Bistopanj-e-Shahrivar, 1976.

Brimani, Djalal. *Histoire des relation médicales franco iraniennes*. Tehran: Vahid, 1370 AS [1992].

"British Trade Abroad." *Board of Trade Journal* 41, no. 342 (1903): 540–541.

Browne, Edward Granville. *The Persian Crisis of December, 1911; How It Arose and Whither It May Lead Us; Compiled for the Use of the Persia Committee*. Cambridge: Privately Printed at the Cambridge University Press, 1912.

———. *The Persian Revolution of 1905–1909*. Cambridge: Cambridge University Press, 1910.

———. *The Press and Poetry of Modern Persia: Partly Based on the Manuscript Work of Mirza Muhammad 'Ali Khan "Tarbiyat" of Tabriz*. Cambridge: Cambridge University Press, 1914.

———. *A Year amongst the Persians: Impressions to the Life, Character, and Thought of the People of Persia Received during Twelve Months' Residence in That Country in the Year 1887–1888*. Cambridge: Cambridge University Press, 1927.

Busch, Briton Cooper. *Britain and the Persian Gulf, 1894–1914*. Berkeley: University of California Press, 1967.

Bussière, Jean Jérôme Augustin. *La vaccine et la variole au Sénégal, dans l'Inde et en Indo-Chine, rapport de M. le Médecin-Major Bussière*. Paris: Imprimerie de J. Gainche, 1903.

Camposampiero, Dr. "On The Recent Outbreak of Cholera in Persia. Report by Dr. Camposampiero, Ottoman Sanitary Delegate at Tehran to the Constantinople Board of Health. Summarized and reported to the Epidemiological Society by Dr. E. D. Dickson, Physician to the British Embassy at Constantinople." *Transactions of the Epidemiological Society of London* 13 (1893–1894): 154–159.

Celestin, Louis-Cyril. *Charles-Edouard Brown-Séquard: The Biography of a Tormented Genius*. London: Springer, 2014.

Chaqueri, Cosroe. *The Soviet Socialist Republic of Iran, 1920–1921*. Pittsburgh: University of Pittsburgh Press, 1995.

Churchill, Roger Platt. *The Anglo-Russian Convention of 1907*. Cedar Rapids, IA: Torch Press, 1939.

Clawson, Patrick, and Michael Rubin. *Eternal Iran: Continuity and Chaos*. New York: Palgrave Macmillan, 2005.

Clemow, Frank G. *Cholera epidemic of 1892 in the Russian Empire: with notes upon treatment and methods of disinfection in cholera, and a short account of the conference on cholera held in St. Petersburg in December 1892*. London: Longmans, Green, 1893.

———. "The Constantinople Board of Health." *Lancet* 204 (1923): 1074–1076, 1126–1127, 1170–1171, 1180–1181.

———. "A Contribution to the Epidemiology of Cholera in Russia." *Transactions of the Epidemiological Society of London* 13 (1893–1894): 60–82.

Cliff, Andrew David, Matthew Smallman-Raynor, Peter Haggett, Donna F. Stroup, and Stephen B. Thacker. *Infectious Diseases: Emergence and Re-emergence; A Geographical Analysis*. Oxford: Oxford University Press, 2009.

Cloquet, Louis André Ernest. "Sur le Choléra en Perse." *Bulletin de l'Académie Nationale de Médecine* 18 (1852–1853): 1190–1192.

Colwell, Rita R. "Global Climate and Infectious Disease: The Cholera Paradigm." *Science* 274, no. 5295 (1996): 2025–2031.

———. "Infectious Disease and Environment: Cholera as a Paradigm for Waterborne Disease." *International Microbiology* 7 (2004): 285–289.

Coppin, Charles-René. "Rapport sur l'epidémie de cholera qui sévit en Perse (1904) et sur l'état actuel de l'hygiène." *Annales d'Hygiène et de Médecine Coloniales* 8 (1905): 271–282.

Curzon, George Nathaniel. "The Karun River and the Commercial Geography of Southwest Persia." *Proceedings of the Royal Geographical Society* 9 (1890): 509–532.

———. *Persia and The Persian Question*. 2 vols. New York: Longmans, Green, 1892.

Cvjetanovic, B., and D. Barua. "The Seventh Pandemic of Cholera." *Nature* 239, no. 5368 (1972): 137–138.

Dedet, Jean-Pierre. *Les Instituts Pasteur d'Outre-Mer: Cent vingt ans de microbiologie française dans le monde*. Paris: Harmattan, 2000.

Dequevauviller, Jean François. *Notices sur le Docteur Ernest Cloquet, Médecin et Conseiller du Schah de Perse, Membre Correspondant de l'Académie Impériale de Médecine, Officier de la Légion D'honneur, Etc*. Paris: Imprimerie de L. Martinet, 1856.

Destrée, Annette. *Les fonctionnaires belges au service de la Perse*. Leiden: E. J. Brill, 1976.

Dickson, E. D. "On Cholera in Persia, 1866–1868." *Transactions of the Epidemiological Society of London* 3 (1866–1876): 257–264.

———. "The Outbreak of Cholera in Mesopotamia and Syria in 1889, 1890, and 1891." *Transactions of the Epidemiological Society of London* 13 (1893–1894): 127–153.

Dieulafoy, Georges, ed. *Clinique Médicale de l'Hôtel Dieu de Paris*. Paris: Masson et Cie, 1898.

"Documents Divers l'Institut d'Hessarek de 1931 à 1945." *Archives of Razi Institute* 3, no. 1 (1946): 95–104.

Duffy, John. *The Sanitarians: A History of American Public Health*. Urbana: University of Illinois Press, 1990.

Dutta, Sanchari. "Plague, Quarantine, and Empire: British-Indian Sanitary Strategies in Central Asia, 1897–1907." In *The Social History of Health and Medicine in Colonial India*, edited by Biswamoy Pati and Mark Harrison, 74–92. London: Routledge, 2009.

Ebrahimnejad, Hormoz. *Medicine in Iran: Profession, Practice, and Politics, 1800–1925*. New York: Palgrave Macmillan, 2014.

———. *Medicine, Public Health, and the Qajar State: Patterns of Medical Modernization in Nineteenth-Century Iran*. Leiden: E. J. Brill, 2004.

Echenberg, Myron. *Africa in the Time of Cholera: A History of Pandemics from 1817 to the Present*. Cambridge: Cambridge University Press, 2011.

———. "Pestis Redux: The Initial Years of the Third Bubonic Plague Pandemic, 1894–1901." *Journal of World History* 13, no. 2 (2002): 429–449.

———. *Plague Ports. The Global Urban Impact of Bubonic Plague, 1894–1901*. New York: New York University Press, 2007.

Ekhtiar, Maryam Dorreh. "The Dar al-Funun: Educational Reform and Cultural Development in Iran." PhD diss., New York University, 1994.

———. "Nasir al-Din Shah and the Dar al-Funun: The Evolution of an Institution." *Iranian Studies* 34 (2001): 153–163.

Elgood, Cyril. *A Medical History of Persia and the Eastern Caliphate from the Earliest Times until the Year A.D. 1932*. Cambridge: Cambridge University Press, 1951.

———. *Safavid Medical Practice; or, The Practice of Medicine, Surgery and Gynaecology in Persia between 1500 A.D. and 1750 A.D.* London: Lucaz, 1970.

———. *Safavid Surgery*. Oxford: Pergamon Press, 1966.

Entner, Marvin. *Russo-Persian Commercial Relations, 1828–1914*. Gainesville: University of Florida Monographs, 1965.

Erickson, Edward J. *Ordered to Die: A History of the Ottoman Army in the First World War*. Westport, CT: Greenwood Press, 2001.

Evans, Richard J. "Epidemics and Revolutions: Cholera in Nineteenth Century Europe." *Past and Present* 120 (1988): 123–146.

Eyler, John M. *Victorian Social Medicine: The Ideas and Methods of William Farr*. Baltimore: Johns Hopkins Press, 1979.

———. "William Farr on the Cholera: The Sanitarian's Disease Theory and the Statistician's Method." *Journal of the History of Medicine* 28 (1973): 79–100.

Farahani, Mirza Mohammad Hosayn. *A Shi'ite Pilgrimage to Mecca, 1885–1886: The Safarnameh of Mirza Mohammad Husayn Farahani*. Edited and translated by Hafez Farmayan and Elton L. Daniel. Austin: University of Texas Press, 1990.

Farmanfarmaian, Roxane. *War and Peace in Qajar Persia: Implications Past and Present*. London: Routledge, 2008.

Farmayan, Hafez. "Portrait of a Nineteenth-Century Iranian Statesman: The Life and Times of Grand Vizier Amin ud-Dawlah, 1844–1904." *International Journal of Middle East Studies* 15, no. 3 (1983): 337–351.

Farr, William. *Report on the Mortality of Cholera in England, 1848–49*. London: W. Clowes, 1852.

Farzaneh, Mateo Mohammad. *The Iranian Constitutional Revolution and the Clerical Leadership of Khurasani*. Syracuse: Syracuse University Press, 2015.

Fauvel, Antoine Sulpice. *Le Choléra: Étiologie et Prophylaxie*. Paris: J.-B. Ballière et Fils, 1868.

Fawaz, Leila Tarazi. *A Land of Aching Hearts: The Middle East in the Great War*. Cambridge, MA: Harvard University Press, 2014.

Feistmantel, Dr. Von Regimentsarzt. "Die sanitären Verhältnisse in Persien." *Wiener Klinische Wochenschrift* 21 (1908): 323–327.

———. "Oeffentliches Sanitätswesen. Zur Frage des sanitären Schutzes des Persichens Golfs." *Deutsche Medizinische Wochenschrift* 34, no. 1 (1908): 655–659.

Feuvrier, Joannes. *Trois Ans à la Cour de Perse* [Three years at the Persian court]. Paris: F. Juven, 1894.

Floor, William. "The Economic Role of the Ulama in Qajar Persia." In *The Most Learned of the Shi'a*, edited by Linda S. Walbridge, 53–81. Oxford: Oxford University Press, 2001.

———. *Public Health in Qajar Iran*. Washington, DC: Mage, 2004.

Forbes-Leith, Francis Arthur Cornelius. *Checkmate: Fighting Tradition in Central Persia*. New York: Robert M. McBride, 1927; repr., Arno Press 1973.

Fraser, James Baillie. *Narrative of a Journey into Khorasan, in the Years 1821 and 1822: Including Some Account of the Countries to the North-East of Persia; with remarks upon the national character, government, and resources of that kingdom*. London: Longman, Hurst, Rees, Orme, Brown, and Green, 1825.

Gangarosa, E. J., H. Saghari, J. Emile, A. Santi, H. Siadat, and Y. Watanabi. "Search for a Mass Chemotherapeutic Drug for Cholera Control: A Study of Vibrio Excretion Following Single and Multiple Dose Treatment." *Bulletin of the World Health Organization* 35, no. 5 (1966): 669–674.

Gardane, Claude Mathieu de. *Mission du général Gardane en Perse sous le premier empire*. Documents historiques publiés par son fils, le comte Alfred de Gardane. Paris: Librairie de AD. Lainé, 1865.

Ghani, Cyrus. *Iran and the Rise of Reza Shah from Qajar Collapse to Pahlavi Power*. London: I. B. Tauris, 1998.

Ghodssi, Mehdi. "Cholera El-Tor en Iran." *Acta Medica Iranica* 12, nos. 1–2 (1969): 1–13.

———. "L'Institut Pasteur dans le monde." *Bulletin de l'association des Anciens Elèves de l'Institut Pasteur* 22 (1964): 34–41.

Gilbar, Gad G. "The Big Merchants and the Persian Constitutional Revolution of 1906." *Asian and African Studies* 2, no. 3 (1976): 290–295.

Gordon, Thomas Edward. *Persia Revisited*. London: E. Arnold, 1896.

Greasley, David, and Les Oxley. "Discontinuities in Competitiveness: The Impact of the First World War on British Industry." *Economic History Review* 49 (1996): 82–100.

Gunergun, Feza. "Science in the Ottoman World." In *Imperialism and Science: Social Impact and Interaction*, edited by George N. Vlahakis, Isabel Maria Malaquias, Nathan M. Brooks, François Regourd, Feza Gunergun, and David Wright, 71–118. Santa Barbara: ABC-CLIO, 2006.

Haag-Higuchi, Roxan. "A Topos and Its Dissolution: Japan in Some 20th Century Iranian Texts." *Iranian Studies* 29, nos. 1–2 (1996): 71–83.

Hakimian, Hassan. "Wage Labor and Migration: Persian Workers in Southern Russia, 1880–1914." *International Journal of Middle East Studies* 17 (1985): 443–462.

Hale, F. *From Persian Uplands*. New York: E. P. Dutton, 1920.

Hardy, Arthur S. "Persia. Sanitary Report from Tehran." *Public Health Reports* 14, no. 10 (March 10, 1899): 330–332.

Harrison, Mark. *Public Health in British India: Anglo-Indian Preventive Medicine, 1859–1914*. Cambridge: Cambridge University Press, 1994.

Hayter, Henry Heylyn. *Victorian Yearbook, 1892*. 2 vols. Melbourne: Printed for the Government Printer by Sands & McDougall Limited, 1893.

Heagerty, John J. "World Health during 1939." *Canadian Public Health Journal* 31, no. 10 (1940): 473–476.

Howard-Jones, Norman. "The Scientific Background of the International Sanitary Conferences, 1851–1938. 2." *WHO Chronicle* 28, no. 5 (1974): 229–247.

———. "The Scientific Background of the International Sanitary Conferences, 1851–1938. 5. The Ninth Conference: Paris, 1894." *WHO Chronicle* 28, no. 10 (1974): 455–470.

———. "The Scientific Background of the International Sanitary Conferences, 1851–1938. 6." *WHO Chronicle* 28, no. 11 (1974): 495–508.

Hutchinson, John F. *Politics and Public Health in Revolutionary Russia, 1890–1918*. Baltimore: Johns Hopkins University Press, 1990.

Ibn Sina, Abu 'Ali al-Husayn Ibn 'Abdullah. *The Canon of Medicine of Avicenna [al-Qanun fi al-tibb]*. Edited and translated by Oskar Cameron Gruner. New York: AMS Press, 1973.

Issawi, Charles, ed. *The Economic History of Iran, 1800–1940*. Chicago: University of Chicago Press, 1971.

———. "The Tabriz-Trabzon Trade, 1830–1900: Rise and Decline of a Route." *International Journal of Middle East Studies* 1 (1970): 18–27.

Jones, Stephanie. *Two Centuries of Overseas Trading: The Origins and Growth of the Inchcape Group*. London: Palgrave MacMillan, 1989.

Kashani-Sabet, Firoozeh. *Conceiving Citizens: Women and the Politics of Motherhood in Iran*. New York: Oxford University Press, 2011.

Katouzian, Homayoun. "Nationalist Trends in Iran, 1921–1926." *International Journal of Middle East Studies* 10, no. 4 (1979): 533–551.

Kazemzadeh, Firuz. *Russia and Britain in Persia, 1864–1914*. New Haven: Yale University Press, 1968.

Keddie, Nikki R. "Iranian Politics, 1900–1905: Background to Revolution—II." *Journal of Middle Eastern Studies* 5, no. 2 (1969): 151–167.

———. *Religion and Rebellion in Iran: The Tobacco Protest of 1891–1892*. London: Frank Cass, 1966.

———. *Roots of Revolution: An Interpretive History of Modern Iran*. New Haven: Yale University Press, 1981.

Khan, Moussa. *Contribution à l'étude hygiénique des nouveaux appareils de chauffage sans tuyau de dégagement*. Paris: A. Maloine, 1906.

Kohn, George C., ed. *Encyclopedia of Plague and Pestilence*. New York: Facts on File, 1995.

Kondo, Nobuaki. *Islamic Law and Society in Iran: A Social History of Qajar Tehran*. New York: Routledge, 2017.

Krauskopf, John. *Iran: Stories from the Peace Corps*. San Francisco: John Krauskopf, 2013.

"La maladie du Shah de Perse." *La Chronique Médicale: Revue Bi-Mensuelle de Médecine Historique, littéraire et Anecdotique* 7, no. 1 (1900): 530–531.

Lambton, Ann K. "Secret Societies and the Persian Revolution of 1905–1906." In *Qajar Persia: Eleven Studies*, edited by Ann K. Lambton, 301–318. Austin: University of Texas Press, 1988.

Lankarani, Kamran B., and Seyed Moayed Alavian. "Lessons Learned from Past Cholera Epidemics, Interventions Which Are Needed Today." *Journal of Research in Medical Sciences: The Official Journal of Isfahan University of Medical Sciences* 18 (2013): 630–631.

Levy, Habib. *Comprehensive History of the Jews of Iran: The Outset of the Diaspora*. Costa Mesa: Mazda, 1999.

Lynch, Henry Blosse. "Across Luristan to Ispahan." *Proceedings of the Royal Geographical Society* 12 (1890): 523–553.

Mahalanabis, Dilip, A. M. Molla, and David Sack. "Clinical Management of Cholera." In *Cholera*, edited by Dhiman Barua and William B. Greenough III, 253–283. New York: Plenum, 1992.

Majd, Mohammad Gholi. *The Great Famine and Genocide in Iran: 1917–1919*. 2nd ed. Lanham, MD: University Press of America, 2013.

Malcolm, John. *The History of Persia from the Most Early Period to the Present Times*. Vol. 2. London: John Murray, 1815.

Maloney, Suzanne. *Iran's Political Economy since the Revolution*. New York: Cambridge University Press, 2015.

Mandell, Gerald L. *Principles and Practice of Infectious Diseases*. 5th ed. London: Churchill Livingstone, 2000.

Marashi, Afshin. *Nationalizing Iran: Culture, Power, and the State, 1870–1940*. Seattle: University of Washington Press, 2008.

Martin, Vanessa. *Islam and Modernism: The Iranian Revolution of 1906*. Syracuse: Syracuse University Press, 1989.

Massé, Henri. *Croyances et coutumes persanes: Suivi de contes et chansons populaires* [Persian beliefs and customs]. 2 vols. Paris: Librairie Orientale et Américaine, 1938.

Maulana, Muhammad Ali. *A Manual of Hadith*. Lahore: Ahmadiyya Anjuman, 1945.

Maxwell, W. G. *A Practical Treatise on Epidemic Cholera, Ague, and Dysentery; Illustrating The Principals of Treatment By Their Anatomical Physiology Pointing Out Their Contagion and Westering Inclination; to Which Is Added A Persian Treatise on Plague and Cholera*. Calcutta: T. Ostell and Co. British Library, 1838.

McDonald, J. C. "The History of Quarantine in Britain during the 19th Century." *Bulletin of the History of Medicine* 25, no. 1 (1951): 22–44.

McLean, David. *Britain and Her Buffer State: The Collapse of the Persian Empire, 1890–1914*. London: Royal Historical Society, 1979.

"Médecine et Chirurgie: Prix Bréant." *Comptes rendus hebdomadaires des séances de l'Académie des sciences* 75 (1872): 1362–1369.

"Medical News." *British Medical Journal* 1, no. 2316 (1905): 1104.

Mehner, Harald. "Development and Planning in Iran after World War II." In *Iran under the Pahlavis*, edited by George Lenczowski, 167–197. Stanford: Hoover Institution Press, 1975.

Melville, Charles. "The Persian Famine of 1870–1872: Prices and Politics." *Disasters* 12, no. 4 (1988): 309–325.

Menashri, David. *Education and the Making of Modern Iran*. Ithaca: Cornell University Press, 1992.

Millspaugh, Arthur C. *The American Task in Persia*. New York: Century, 1925.

Mirchamsy, H. "Technical Considerations of Cholera Vaccines and Some Results of Cholera El Tor Mass Immunization." *Archives of Razi Institute* 18, no. 1 (1966): 85–94.

Mollaret, Henri H., and Jacqueline Brossollet. *Yersin, un pasteurien en Indochine*. Paris: Belin, 1993.

Momen, Moojan. *An Introduction to Shi'i Islam: The History and Doctrines of Twelver Shi'ism*. New Haven: Yale University Press, 1985.

Mordtmann, Andreas David. "Statistique de L'Épidémie Cholérique en Perse en 1889 et 1892." *Gazette Médicale D'Orient* 35 (1892–1893): 297–300.

Morier, James. *A Second Journey Through Persia, Armenia, and Asia Minor, to Constantinople, Between the Years 1810 and 1816*. London: Longman 1818.

Morris, L. P. "British Secret Service Activity in Khorassan, 1887–1908." *Historical Journal* 27, no. 3 (1984): 657–675.

Morrison, Alexander. *Russian Rule in Samarkand, 1868–1910: A Comparison with British India*. Oxford: Oxford University Press, 2008.

Mostofi, Abdollah. *The Administrative and Social History of the Qajar Period*. 4 vols. Translated by Nayer Mostofi Glenn. Costa Mesa: Mazda, 1997.

Nadjmabadi, Mahmud. "Les relations médicales entre la Grande Bretagne et l'Iran et les médecins anglais serviteurs de la médecine contemporaine de l'Iran," 704–708. In *Proceedings of the XXIII International Congress of the History of Medicine*. London: Wellcome Institute of the History of Medicine, 1974.

Naficy, Abbas. *La médecine en Perse des origines à nos jours: Ses fondements theoriques d'apres l'Encyclopédie médicale de Gorgani*. Paris: Edition Véga, 1933.

Narkevich, M. I., G. G. Onischenko, J. M. Lomov, E. A. Moskvitina, L. S. Podosinnikova, and G. M. Medinsky. "The Seventh Pandemic of Cholera in the USSR, 1961–89." *Bulletin of the World Health Organization* 71, no. 2 (1993): 189–196.

Nashat, Guity. *The Origins of Modern Reform in Iran, 1870–80*. Urbana: University of Illinois Press, 1982.

Nasr, Seyyed Hossein. *Science and Civilization in Islam*. 2nd ed. Cambridge: Islamic Texts Society, 1987.

Nategh. Homa. "Les Persans à Lyon." In *Pand-o Sokhan*, edited by Christoph Balay, Claire Kapper, and Ziva Vesel, 191–199. Tehran: Institute Français de Recherche en Iran, 1995.

Neligan, Anthony R. "Public Health in Persia, 1914–1924." *Lancet* 207, no. 5352 (1926): 690–694.

Okazaki, Shoko. "The Great Persian Famine of 1870–71. I." *Bulletin of the School of Oriental and African Studies* 49 (1986): 183–192.

Olsen, Roger T. "Persian Gulf Trade and the Agricultural Economy of Southern Iran in the Nineteenth Century." In *Modern Iran: The Dialectics of Continuity and Change*, edited by Michael E. Bonine and Nikki R. Keddie, 173–189. Albany: State University of New York Press, 1981.

Onley, James. *The Arabian Frontier of the British Raj: Merchants, Rulers, and the British in the Nineteenth Century Gulf*. Oxford: Oxford University Press, 2007.

Pacini, Filippo. "Osservazioni microscopiche e deduzioni patologiche sul cholera asiatico." *Gazzetta Medica Italiana: Federativa Toscana* 2, ser. 4 (1854): 397–401, 405–412.

Peckham, Robert. *Epidemics in Modern Asia*. Cambridge: Cambridge University Press, 2016.

Pettenkofer, Max Joseph von. "Cholera." *Lancet* 2 (1884): 769–771, 816–819, 861–864, 904–905, 1042–1043, 1086–1088.

———. *Relations of The Air to The Cloths we Wear, the House we Live in, and The Soil We Dwell On: Three Popular Lectures Delivered Before The Albert Society at Dresden*. Translated by August Hess. London: N. Trubner, 1873.

Polak, Jakob Eduard. "La médicine militaire en Perse.—Par le docteur J.-E. Polak, ancien médecin particulier du schah de Perse." *Revue scientifique et administrative des médecins des armées de terre et de mer* 7 (1865): 649–651.

Pistor-Hatam, Anja. "Progress and Civilization in the Nineteenth-Century Japan: The Far Eastern State as a Model for Modernization." *Iranian Studies* 29, nos. 1–2 (1996): 111–126.

Proust, Adrien Achille. *Exposé Des Titres Et Travaux Scientifiques du Docteur A. Proust*. Paris: Imprimerie Emile Martinet, 1877.———. *La défense de l'Europe contre le choléra*. Paris: G. Masson, 1892.

———. "Le choléra de Mésopotamie, de Perse et de Syrie, en 1889 et 1890," *Bulletin de l'Académie de Médecine* 26 (1891): 136–154.

Rabino (di Borgomale), H. L. "La presse persane depuis ses origines jusqu'à nos jours." *Revue du monde musulman* 22 (1913): 287–315.

Rabino, Joseph. "Banking in Persia: Its Basis, History, and Prospects." *Journal of the Institute of Bankers* 13 (1892): 1–56.

Rawlinson, Henry Creswicke. *England and Russia in the East: A Series of Papers on the Political and Geographical Condition of Central Asia*. London: John Murray, 1875.

Richmond, Elsa, ed. *The Earlier Letters of Gertrude Bell*. London: Ernest Benn, 1937.

Roff, William R. "Sanitation and Security: The Imperial Powers and the Nineteenth-Century Hajj." *Arabian Studies* 6 (1982): 143–160.

Rosenberg, Charles E. *The Cholera Years: The United States in 1832, 1849, and 1866*. Chicago: University of Chicago Press, 1962.

Ross, Ronald. "The Inoculation Accident at Mulkowal." *Nature* 75, no. 1951 (1907): 486–487.

Saghaphi, Mirza Mahmoud Khan. *In the Imperial Shadow: Page to the Shah*. New York: Doubleday, Doran, 1928.

Shedd, J. H. "Persia: The Cholera. Persecution. Mar Shimon." *The Independent. Devoted to the consideration of politics, of social and economic tendencies, of history, literature, and the arts* 44, no. 2294 (November 17, 1892): 17.

Schayegh, Cyrus. *Who Is Knowledgeable Is Strong: Science, Class, and the Formation of Modern Iranian Society, 1900–1950*. Berkeley: University of California Press, 2009.

Schlimmer, Johannes L. *Du Présage et de L'Avortement de L'Imminence Cholérique*. Rotterdam: Nijgh & Van Ditmar, 1874.

——. *Terminologie médico-pharmaceutique et anthropologique français-persane, avec des traductions anglais et allemand des termes français*. Tehran: Lithographie d'Ali Gouli Khan, 1874.

Schneider, Jean-Etienne Justin. *Médecine persane: Les médecins français en Perse—Leur influence*. Paris: Société Anonyme des Imprimeries Wellhoff & Roche, 1911.

Scott, Joseph. "The Recent Cholera Epidemic in Persia." *British Medical Journal* 2 (1904): 620–622.

Seaton, Edward Cator. "A brief account of the proceedings of the international sanitary conference held at Vienna in 1874." *Transaction of the Epidemiological Society of London* 3 (1874): 556–570.

Seyf, Ahmad. "Iran and Cholera in the Nineteenth Century." *Middle Eastern Studies* 38, no. 1 (2002): 169–178.

——. "Iran and the Great Plague." *Studia Islamica* 69 (1989): 151–166.

——. "The Plague of 1877 and the Economy of Gilan." *Iran* 27 (1989): 81–86.

Shahnavaz, Shabaz. *Britain and South-West Persia, 1880–1914: A Study in Imperialism and Economic Dependence*. New York: Routledge Curzon, 2005.

Sheil, Mary Leonora Woulfe. *Glimpses of life and manners in Persia: With Notes on Russia, Koords, Toorkomans, Nestorians, Khiva, and Persia*. London: John Murray, 1856.

Shuster, William Morgan. *The Strangling of Persia: Story of the European Diplomacy and Oriental Intrigue That Resulted in the Denationalization of Twelve Million Mohammedans, a Personal Narrative*. New York: Century, 1912.

Shkolnikov, Vladimir, France Mesle, and Jacques Vallin. "Health Crisis in Russia I. Recent Trends in Life Expectancy and Causes of Death from 1970 to 1993." *Population: An English Selection* 8 (1996): 123–154.

Simpson, William John Ritchie. "The Croonian Lectures on Plague, Delivered before the Royal College of Physicians on June 18, 20, 25 and 27. Lecture 3." *Journal of Tropical Medicine and Hygiene* 10, no. 15 (August 1, 1907): 252.

Snow, John. "On the Pathology and Mode of Communication of the Cholera." *London Medical Gazette* 44 (1849): 730–732, 745–752, 923–929.

Snowden, Frank M. "Cholera in Barletta, 1910." *Past and Present* 132 (1991): 67–103.

——. *Naples in the Time of Cholera, 1884–1911*. Cambridge: Cambridge University Press, 1995.

Storey, Charles Ambrose. *Persian Literature: A Bio-bibliographical Survey*. Vol. 2, pt. 2: E. Medicine. London: Luzac [for] the Royal Asiatic Society of Great Britain and Ireland, 1971.

Sutphen, Mary P. "Not Where, but When: Bubonic Plague and the Reception of Germ Theories in Hong Kong and Calcutta, 1894–1897." *Journal of the History of Medicine* 52 (1997): 81–113.

Sykes, Christopher. *Wassmuss: The German Lawrence*. New York: Longmans, Green, 1936.

Sykes, Percy. *A History of Persia*. 2 vols. London: Macmillan, 1921.

Taj al-Saltana. *Crowning Anguish: Memoirs of a Persian Princess from the Harem to Modernity, 1884–1914*. Edited by Abbas Amanat. Translated by Anna Vanzan and Amin Neshati. Washington, DC: Mage, 1993.

Tchalenko, John. "Alexander Iyas and Vladimir Minorsky in Persian Kurdistan, 1912–1914." *Eastern Art Report* 55, no. 1 (2006): 44–48.

Theodorides, Jean. "Tholozan et la Perse [Tholozan and Persia]." *Histoire des Sciences Médicales* 32, no. 3 (1998): 287–296.

———. "Un grand épidémiologiste franco-mauricien: Joseph Désiré Tholozan (1820–1897) [A great Franco-Mauritian epidemiologist: Joseph Désiré Tholozan (1820–1897)]." *Bulletin de la Société de Pathologie Exotique* 91, no. 1 (1998): 104–108.

Tholozan, Joseph Désiré. *Des Métastases Thése Présentée et Soutenue à la Faculté de Médicine* [Metastases thesis presented and defended at the Faculty of Medicine]. Paris: Librairie de A. Delahaye et E. Chatel, 1857.

———. *Histoire de La Peste Bubonique en Perse ou Determination de Son Origine, de sa Marche, du Cycle de Ses Apparitions et de la cause de sa Prompte Extinction.* Paris: G. Masson, 1874.

———. "La grippe en Perse en 1889–1890." *Bulletin de l'Académie de Médecine* 26 (1891): 250–262.

———. *La Peste en Turquie Dans Les Temps Modernes: Sa Prophylaxie Défectueuse, Sa Limitation Spontanée.* Paris: G. Masson, 1880.

———. *Prophylaxie du choléra en Orient: l'hygiène et la réforme sanitaire en Perse.* Paris: Victor Masson et Fils, 1869.

———. "Rapport à S.M. le Shah sur l'état actuel de l'hygiène en Perse, progrès à réaliser, moyens de les effectuer, résultats obtenus [Texte présenté à la séance du 11 octobre par H. Larrey]." *Comptes rendus hebdomadaires des séances de l'Académie des sciences* 69 (1869): 838–840.

———. "Recherche sur quelques points d'anatomie et de physiologie et de pathologiques du cholera." *Gazette Médical de Paris* 4, no. 3 (1849): 557–558.

van Heyningen, W. E., and John R. Seal. *Cholera: The American Scientific Experience, 1947–1980.* Boulder, CO: Westview Press, 1983.

Viry, Charles Marie Jules. *Principes d'Hygiène Militaire.* Paris: L. Battaille Et Cie, 1896.

Warne, William E. *Mission For Peace: Point 4 in Iran.* Indianapolis: Bobbs-Merrill, 1956.

Weber, Hermann. "On Professor Pettenkofer's Theory of the Mode of Propagation Cholera." *Transactions of the Epidemiological Society of London* 2 (1867): 404–413.

Werner, Christoph. *An Iranian Town in Transition: A Social and Economic History of the Elites of Tabriz, 1747–1848.* Wiesbaden: Harrassowitz Verlag, 2000.

Williamson, Graham. "The Turko-Persian War of 1821–1823: Winning the War but Losing the Peace." In *War and Peace in Qajar Persia: Implications Past and Present,* edited by Roxane Farmanfarmaian, 88–109. London: Routledge, 2008.

Winston, G. S., and W. R. Gillette. *Hints to Prevent Cholera and also its Treatment.* 3rd ed. New York: Mutual Life Insurance Company of New York, 1885.

Wishard, John G. "Addresses and Original Communications." *Indiana Medical Journal* 23, no. 4 (1904): 129–131.

———. *Twenty Years in Persia: A Narrative of Life under the Last Three Shahs.* New York: Flemming H. Revell, 1908.

Wills, Charles James. *In the Land of the Lion and Sun, Or Modern Persia: Being Experiences of Life in Persia from 1866 to 1881.* London: Macmillan, 1883.

Wills, Christopher. *Yellow Fever, Black Goddess.* Reading: Helix Books, 1996.

Wilson, Herbert Wrigley, and John Alexander Hammerton, eds. *The Great War: The Standard History of the All-Europe Conflict.* 13 vols. London: Amalgamated Press, 1914–1919.

Wilson, Samuel Graham. *Persian Life and Customs: with scenes and incidents of residence and travel in the land of the lion and the sun.* New York: F. H. Revell, 1895.

Woodward, David R. *World War I Almanac.* New York: Facts on File, 2009.

Wright, Carroll Davidson, ed. *The New Century Book of Facts: A Handbook of Ready Reference.* Springfield: King Richardson, 1909.

Wright, Denis. *The English amongst the Persians.* London: Heinemann, 1977.

Yaganegi, Esfandiar Bahram. "Recent Financial and Monetary History of Persia." PhD diss., Columbia University, 1934.

Yate, C. E. *Khurasan and Sistan.* Edinburgh: William Blackwood and Sons, 1900.

Zandjani, Habibola. "Téhéran et sa population: Deux siècles d'histoire." In *Téhéran: Capitale bi-centenaire*, edited by Chahryar Adle and Bernard Hourcade, 251–66. Paris: Institut Français de Recherche en Iran, 1992.

Zirinsky, Michael. "American Presbyterian Missionaries at Urmia during the Great War." *Journal of Assyrian Academic Studies* 12, no. 1 (1998): 6–27.

Page numbers in *italics* refer to figures and tables.